AF361308

COLONIAL GEOGRAPHY

GERMAN AND EUROPEAN STUDIES

General Editor: Jennifer L. Jenkins

Colonial Geography

Race and Space in German East Africa, 1884–1905

MATTHEW UNANGST

UNIVERSITY OF TORONTO PRESS
Toronto Buffalo London

ISBN 978-1-4875-4340-2 (cloth)   ISBN 978-1-4875-4341-9 (EPUB)
                                  ISBN 978-1-4875-4342-6 (PDF)

Library and Archives Canada Cataloguing in Publication

Title: Colonial geography : race and space in German East Africa,
    1884–1905 / Matthew Unangst.
Names: Unangst, Matthew, author.
Series: German and European studies ; 47.
Description: Series statement: German and European studies ; 47 |
    Includes bibliographical references and index.
Identifiers: Canadiana (print) 20220181713 | Canadiana (ebook) 20220181802 |
    ISBN 9781487543402 (cloth) | ISBN 9781487543419 (EPUB) |
    ISBN 9781487543426 (PDF)
Subjects: LCSH: German East Africa – Historical geography. | LCSH:
    German East Africa – Race relations. | LCSH: German East Africa –
    Colonization. | LCSH: German East Africa – History. | LCSH: Germany –
    Colonies – Africa – History.
Classification: LCC DT438.7.U53 2022 | DDC 967.6/03–dc23

The German and European Studies series is funded by the DAAD with funds
from the German Federal Foreign Office.

DAAD Deutscher Akademischer Austauschdienst
German Academic Exchange Service

We wish to acknowledge the land on which the University of Toronto Press
operates. This land is the traditional territory of the Wendat, the Anishnaabeg,
the Haudenosaunee, the Métis, and the Mississaugas of the Credit First Nation.

University of Toronto Press acknowledges the financial support of the
Government of Canada, the Canada Council for the Arts, and the Ontario Arts
Council, an agency of the Government of Ontario, for its publishing activities.

To MKD and DTU

Contents

Illustrations

Acknowledgments

The majority of the sources I use in this book bear the imprint of multiple people. As documents passed through different offices, readers left comments in the margins or even covering the original text. Readers of those documents or print sources in archives and libraries left their own marks on text and image, which shaped my interpretation of the sources as not the creation of one person alone, but as the work of many. This book, in contrast, has until now existed in a form that hides the contributions that are not mine, both everywhere and nowhere as a seemingly infinite collection of digital files. Though their imprints have been rendered invisible, a great number of people have made important contributions to this project as I have crisscrossed the United States and done research in Germany and Tanzania.

This book began at Temple University over a decade ago. Both my fellow graduate students and faculty in the Department of History guided me through the winding path of research and writing. Ben Talton and Kathy Biddick served on the committee that steered me to completion. Rita Krueger, Harvey Neptune, and participants in the Spring 2014 history research seminar and the Spring 2015 Center for the Humanities seminar, both led by Petra Goedde, provided feedback at important stages of the process. And Jay Lockenour oversaw the often halting progress through the whole process to completion.

From Philadelphia, I moved to eastern Washington and taught at Washington State University. While there, I did the majority of the work of reworking my chapters. My colleagues in the Roots of Contemporary Issues program, and the history department more widely, were invaluable in the project's transformation. Sean Wempe provided key feedback for the brief time where WSU was probably the only history department in the country with two German colonial specialists. Jesse Spohnholz's feedback helped especially with reframing chapters.

He and Ashley Wright assisted with the transition to writing while teaching full time. Finally, Jared Secord was the first person to read the entire manuscript. Our biweekly writing groups kept me on a writing schedule and ensured that I would finish at some point.

From there, I moved to almost the farthest point in the continental United States – Jacksonville, Florida. In Florida, my writing group with Ashley Parcells and Chris Rominger kept me moving along. Jesse Hingson helped with securing the time needed to complete the project.

Staff at the archives used in this project – the Tanzania National Archives, the Bundesarchiv Berlin-Lichterfelde, Columbia University Libraries, the Institut für Länderkunde, and the Politisches Archiv of the Auswärtigen Amt – dealt with requests from the strange American researcher with his mediocre German and atrocious Swahili. Matthew Greene at the George L. Mosse Program in History assisted with images. Funding from the Central European History Society, the American Historical Association, Temple University, Washington State University, SUNY Oneonta, and Columbia University Libraries contributed to the completion of the research for this project.

My research in Tanzania would not have been possible without the assistance of Oswald Masebo, Kate Raum, Blandina and Jim Giblin, and Sean Bloch, who all helped me navigate the Tanzanian bureaucracy and find my way around Dar es Salaam. Many others beyond my institutional homes also helped me organize ideas and continue writing. Participants in the Washington German Historical Institute's 2014 Transatlantic Doctoral Seminar assisted with the early stages of writing, and comments and questions from a number of other conference attendees improved things along the way. The three anonymous readers and the manuscript review committee at the University of Toronto Press strengthened the book immeasurably. Stephen Shapiro has been the best editor I could hope for on this project, steering it through the long publication process and improving it significantly from what I sent him some time ago. Michelle Moyd has been especially helpful over many years. She has provided essential specialist feedback and guidance for many years.

A great number of people have also provided essential support that I cannot tie directly to sections of the book. In addition to many of the people already mentioned, Sarah Robey, Matt Shannon, Meredith Hohe, Patrick Grossi, Manna Duah, Melanie Newport, Albert Merkin, Steven Greenstein, Ruby Goodall, Skye Doney, Clif Stratton, David Jamison, Matt Groe, Sean Vohra, and Jason Wolfe helped me keep my sanity and my enthusiasm through many years. My family – Felicia, Nathan, and Rosaura – have done the same for yet longer. Rosaura's painting on the cover of the book will hopefully convince a few people to judge it rather than what is inside. Jennifer Hayes has let me drag her across the country three times now, putting up with my silly career choice and finding ways to thrive in new places.

It is to my parents that I dedicate this book. Though they may have been dubious of their son's decision to become a historian and float from job to job between the corners of the United States, they have been supportive through it all.

COLONIAL GEOGRAPHY

1 Introduction

In the second half of the 1880s, Europeans fixed their attention on eastern Africa. What held so many people rapt was the story of one man, Emin Pasha. Emin was born in 1840 in Silesia, which became part of the German Empire in 1871, but with Germany lacking a colonial empire, he joined the Ottoman service in search of adventure. He found it, eventually making his way up the Nile to become the Egyptian governor of the province of Equatoria. Emin simulated a German landscape in Equatoria, cultivating German crops and sending regular reports to scientists back home that promoted the potential for German agriculture in tropical Africa. In 1884, war in the Sudan cut off Emin's contact with Europe. Emin's story captured the essence of the period's civilizing missions: multiple men undertook expeditions to rescue him and the province from African primitivism and Arab barbarity for European civilization. For Germans, Emin was a figure onto whom they could project the hopes and concerns of their new nation. His work in Equatoria was proof that Germans could be successful colonizers. But he was also something more – a figure who connected Germans to an imagined pre-unification world free from the class, confessional, or political fault lines that were tearing at the social fabric of Imperial Germany. Emin had brought German *Kultur* to Equatoria, recreating a fantastical pre-modern agrarian Germany in the heart of Africa.

In 1884, the mainland part of today's Tanzania, together with Rwanda and Burundi, became the colony of German East Africa, *Ostafrika* in German, and would remain a German colony until the end of the First World War.[1] Over the next two decades, Germans in East Africa imagined that they could transform space to solve social problems. They would use a variety means, applying concepts from cultural geography to create an ideal world. German colonialists developed world-historical narratives based in a competition for space and the cultural diffusion that justified elaborate projects to cultivate the supposed racial characteristics of Ostafrika's population to refashion landscapes. East Africans became permanently paired with landscapes in the ubiquitous phrase "*Land und Leute*" (land and people), an idea from the works of cultural geographers.[2] Germans justified colonial rule with claims that Europeans had advanced through time to a higher level of economic and cultural activity than had the region's previous inhabitants. The ability of societies to create profitable economies and perform intellectual labour, claimed Germans, was apparent in the appearance of the landscapes they inhabited, especially in their agriculture. From a survey of agricultural techniques, Germans believed they could quickly but accurately judge the societies inhabiting a region and their place on a hierarchy of civilization. Agricultural techniques were of course malleable, but in this vision of the world agricultural changes required generations to solidify as part of a society's lifeways. African agriculture was cast as belonging to a distant, static past.[3] Though not explicitly biological, German thinking about *Kultur* applied the heritability of other European racial conceptions to relationships with the soil. In this vision of things, world-historical narratives broke down into judgments of spaces.

This book charts changes in how Germans conceptualized race and space from Ostafrika's 1884 founding to the early twentieth century. I argue that German colonizers in East Africa shifted from an assumption that possessing and cultivating the land would suffice to make it German to a belief that it was necessary to transform the land by reshaping its people. As a result of a series of crises, Germanizing East Africa increasingly came to mean manipulating the racial characteristics of the colony using methods that continued to change from the 1890s through the early 1900s. In the 1880s, the colony's private founders imagined that simply taking territory for Germany could make it German. German territory would in turn Germanize its inhabitants, who would practise agriculture following German models. Following challenges to private colonial rule, the German government assumed control of the colony. The 1890s were characterized by tensions between the prior belief that successful colonialism depended primarily on making territory German and growing calls for making use of the perceived racial characteristics of the colony's population. After the turn of the century, development plans increasingly depended on the latter idea as German administrators attempted to accelerate processes of African social and economic evolution. In the new model, African agriculturalists who had become more like Germans would make landscapes more German. This series of failed projects constituted the first German state project to manipulate race to manage territory, but they did not unfold without contestation. African actors continued to make space on terms not controlled by Germans and reshaped German projects to other ends.

Kultur as Colonial Ideology

At the core of discussions of space was the idea of *Kultur*. Like the English word "culture," *Kultur* contains a wide range of meanings, including religion, language, and ways of living. The most important facet of *Kultur* in Ostafrika was another meaning of the term, most directly translated as "cultivation," meaning the cultivation of the soil, or settled agriculture. Some Germans imagined settled agriculture as the antidote to the perceived social problems that accompanied Germany's rapid industrialization and urbanization in the second half of the nineteenth century. Agrarianism became both the primary goal of German administrators and the main criterion by which Germans measured East African societies. *Kultur* thus became an ideology for German colonialism, believed to be the only way that successful economic development could happen without attendant social problems.[4] Colonialists debated the relationship between people and space, whether environments determined racial characteristics or whether racial characteristics shaped environments, and whether state intervention could move the process along.

A focus on *Kultur* as ideology sheds light on several aspects of German history, East African history, and the history of geography. First, in the adoption of *Kultur* as the primary means of judging societies, we can more clearly see changes in how Germans articulated their own national identity in the late nineteenth century. Germany did not unify as a country until 1871. In the wake of unification, inhabitants of the new nation struggled over how to define what made them "German."[5] National identity was fractured, as different classes, confessions, and genders proposed different meanings for Germanness.[6] Some attempted to differentiate what made them German from the idea of *civilisation* at the core of French identity, given the central role Napoleon's invasion of German states had been in the formation of nationalism.[7] Leading up to, and for some time after, Germany's unification in 1871, many Protestants identified with Arminius, the Germanic chief who defeated the Roman army at the Battle of the Teutoburg Forest in 9 CE.[8] Later in the nineteenth century and into the twentieth, German nationalists abandoned the idea of Germany as a nation of barbarians resisting imperial invasion and defined their separation from France as one of *Kultur* versus *civilisation*. This new idea claimed German superiority through the organic growth of a community of shared *Kultur*, compared to a France unified only by politics.[9] Germans involved in colonizing Ostafrika worked to create an ideal space that would unify Germans across confessional and class lines.

Concerns about finding a deeper German identity intersected with worries about changes in modern society. Conservatives in Imperial Germany, the *Kaiserreich*, worried about the spread of socialism among the German working classes, which they believed posed the primary political and social threat of industrial society.[10] Agrarianism developed as a political ideology as a means of countering both the spread of socialism and the political power of industrial interests, with a focus on the less industrialized German-controlled Poland. In the eyes of agrarians, cities and socialism had destroyed authentic communal life, as socialists urged Germans to think of themselves as members of classes rather than communities. Agrarians posited family farms as the thread holding the fabric of German society together, promoting proper respect for the paternal order and economic self-sufficiency. *Kultur* was thus the frame on which German society was built. While the Agrarian League's politics were patrimonial, many members of the agrarian movement were radicals and called for major changes in German society, such as land redistribution. Conservatives in eastern Germany agitated in favour of the interests of the *Junkers*, the holders of large agricultural estates who employed Polish peasants. National Liberals, in contrast, demanded that some of these large estates be broken up and the land redistributed to landless Germans as a means of reinvigorating German citizenship by restoring connections between Germans and the soil. Agrarians claimed that a society of German farmers on German land would ensure a powerful German state and a cohesive nation. Agriculture was the best means to escape the alienation and artificiality of modern urban life.[11] As in Poland, attempts to renew German society through agriculture reform in East Africa failed. The situation on the ground did not match German imaginaries.

Studies of other European empires have shown that Europeans claimed that they had progressed further than the rest of the world and had entered a new, modern period while the rest of the world remained stuck in the past. Those claims depended on presenting other parts of the world as behind in time.[12] In his study of anthropology and its role in colonialism, Johannes Fabian described a "persistent and systematic tendency to place the referent … in a Time other than the present of the producer." Europeans imagined colonized peoples as being in a different stage of human history from which they themselves had advanced.[13] By examining the ways in which European invented these historical narratives, we can better understand how they have functioned politically.[14] For German administrators in Ostafrika, the criterion on which to judge African societies' progress was settled agriculture. East African landscapes not under cultivation, in German eyes, were untouched wilderness with no signs of human advancement. People who did not practise agriculture recognizable to Europeans were remnants of the human past. On that basis, administrators believed that Germans had a mission to improve settled agriculture in the region as the most important form of progress. They thought the creation of settled agriculture following European models would introduce civilization to the East African wilderness.

A central aspect of racialized colonialism across the globe was the belief that a society's evolution could best be read in how it used land. The notion that settled agriculture was an indicator of civilization was not unique to Germans in East Africa. Indeed, from the eighteenth century on, British colonists promoted ideas of progress through the introduction of European models of land ownership and agriculture. British officials split up land into private property to be redistributed for farming in settler colonies in Africa, Asia, Oceania, and the Americas.[15] The best known case is probably in North America, where white settlers moving west justified the conquest of territory from Indigenous peoples with claims that Indigenous societies had failed to improve the land, demonstrating their lack of evolutionary progress in contrast to "energetic" and "practical" Anglo-Saxon settlers.[16] Agrarianism was central to these processes, as settlers claimed that Indigenous peoples were incapable of farming without centuries of social evolution, denying evidence to the contrary.[17] Similar ideologies were also present in other parts of the globe. French settlers in North Africa, for instance, asserted French authority to remake landscapes with justifications that North Africans were not using the land properly and that the region needed to progress to agriculture following French models.[18]

The history of these processes in Ostafrika demonstrates both similarities with other cases and how the particular forms they took were deeply embedded in national culture and identity. *Kultur* was a philosophy for German colonialism comparable to the "civilizing missions" of other colonial powers in the period, a doctrine that both created differences between colonizer and colonized and underlay attempts to improve colonized societies.[19] As with what John Weaver describes as the "English will to possess and manipulate land" or Frederick Jackson Turner's American "frontier process," Germans developed their own explanations and ideologies of settled agriculture. What unfolded in Ostafrika is notable, first of all, for how consciously German colonialists drew on models developed in other empires. The colony's founders began their venture avowedly following British models for taking control of and transforming land. Carl Peters, the dominant figure in Ostafrika's early history, was a rabid nationalist who saw colonialism primarily as a means of national aggrandizement. Peters had never travelled out of Europe, so his understanding of colonialism developed entirely through studies of British colonial documents during a period living in the United Kingdom and reading popular publications about Africa. At points over the following two decades, German colonialists adopted models of racialized colonialism from the western United States, British South Africa and India, and the Dutch East Indies. But Ostafrika took a different path. The biggest difference was the lack of a large European settler community that was the central human element in most comparable contexts. Colonialists had to quickly abandon their hopes for waves of German settlers who would transform the landscape with German agricultural models, as most Germans who moved abroad preferred the security of relocating to places with a longer history of European settlement. The lack of a large settler community meant also that demands for the assimilation of African communities to German society were nearly absent. Finally, the absence of a settler community meant the transition to a land system based in private property was less pressing. German officials did not insist on the division of communal property or institute the cadastral mapping that was so important in other colonial contexts.[20] On the last point, the focus of German geography in Ostafrika was different in significant ways from that practised in other contemporary empires. The idea that cartography was important to the creation of modern European empires has become common. Historians of cartography, beginning with Brian Harley in the 1980s, have closely interrogated the role of maps in "anticipating" empire through the promotion of colonialism and creation of an imperial consciousness in European nations.[21] Through maps, "the distant land was made domestic and the unknowable conquerable."[22] Cartographers asserted European control over territory and shaped public consciousness about empire, providing a physical representation of imperial power. While cartography was important to the German colonial empire, cultural geography was even more so. Its importance is clear in procedures for taking and remaking territory. Whereas maps erased Indigenous people from landscapes, a process that functioned to "take the past out of space" and make them appear as people without history,[23] cultural geographies more clearly grappled with the pre-European past. German geographers developed historical narratives of spatial change that incorporated long-term histories of East Africa predating German arrival to claim and remake territory in the 1880s. When those failed to produce the desired results, geographers developed new explanations of social change rooted in biology that stretched even deeper into the past.

Spatial History

Colonial Geography is a spatial history, by which I mean it is concerned with ways of thinking about and making space.[24] It looks at the German project to "endow [space in East Africa] with value," the process by which space becomes place.[25] My guide here is work in critical geography, particularly the writings of Henri Lefebvre, David Harvey, and Edward Soja. Their work highlights the connection between conceptions of space and economic changes. Harvey introduces the concept of a "spatial consciousness," which people use to relate to the spaces around them. According to Harvey, spaces have meaning only in terms of their relationships with other spaces.[26] Soja argues for the existence of a "socio-spatial dialectic," a "mutually influential and formative relation between the social and the spatial dimensions of human life." He sees space, society, and history as mutually constitutive.[27] Lefebvre argued that space is created through the interaction of three elements: representational space, representations of space, and practice, what he defines as the "spatial triad." He defines representational space as theories and ideas of space. Representations of space are the physical depictions of space, most clearly found in maps. Finally, practice is the actions that people take in the spaces around them that make space reality.[28]

While critical geography in this line provides a framework for thinking about German discourses about East African space, I also offer a critique of the model of stages of space-making matching economic systems. Many recent theories of spatial evolution repeat colonial era claims about human difference and societal evolution. The idea that space goes through stages of development was central to German colonialism in East Africa; in the German colonial imaginary, space and history intersected. German colonialists justified their interventions with claims that they were assisting the evolution of East African spaces. The idea that the progress of societies can be read in the spaces they occupy has long outlived the theories of nineteenth-century geographers and pervades more recent critiques of historical processes of space-making. While Lefebvre's argument has led historians to ask new questions about the making of space, it replicates older conceptions of the non-European world as behind in time. He explicated a four-stage process through which premodern spaces evolved to modern ones. For Lefebvre, the "primitive nomad" lived in spite of the spaces around him, struggling against natural sources for survival. Primitive societies existed in a state of nature and did not change wilderness landscapes. From there, human states territorialized space to match their economic systems, transforming nature as they did so. This process culminated in the capitalist space of the modern period, a space totally transformed by human action.[29] Similarly, Frederic Jameson argues that each stage of capitalism has "generated a type of space unique to it." Over recent centuries, spaces have evolved with changing economic relations, a process led by the United States and Western Europe.[30] The ideas at work in Ostafrika, then, continue to shape how scholars conceptualize historical spaces. Human history can be read in space, a project that posits the West as more modern than the rest of the world.

To explain German representational space, I draw on Edward Said's work on Orientalism for the concept of a "spatial imaginary." Representational space encompasses the discourses about space in which Germans participated. Germans imagined a representational space that only in part matched reality on the ground. Colonialists based their plans for development on these imagined spaces even as East African communities continued to practise completely different spaces. Said argued that European discourses about the Middle East and South Asia produced representations of those places that only in part reflected their reality and were largely made up of imagined conceptions of them and the people who lived there. In this line of thinking, Europe became the norm, the site of modernity where societies had escaped the constraints of physical geography. These ideas provided rhetorical backing for European exploitation of colonies in the name of modernization.[31] At the core of colonialism was a hierarchy of European and Oriental spaces, with clear differences in location and a definite hierarchy with the former at the top. Europeans imagined spaces elsewhere as fundamentally different, thus justifying the creation of measures to reorganize and remake those spaces to bring them into modernity.[32]

But Said's claims about discourse's power to make space, like the materialist focus discussed above, are not fully convincing. First, the process is more complicated than what Said describes. What unfolded in Ostafrika bears many of the characteristics Said outlines, but the situation was more complicated than a straightforward comparison between Europe and a singular Other. Colonialists struggled to reconcile their ideas of the interior and coast as different kinds of spaces, one fully African, the other Oriental. The coast, with its use of Swahili as both written and spoken language, could be studied by historians.

The interior, however, did not have written records and so had to be studied by ethnologists. German administrators drew on many of the same stereotypes of the Orient to discuss and then attempt to eliminate the Arab and Indian presence in the region. At the same time, colonialists deployed stereotypes of Africa and Africans that justified German domination and plans to remake *Land* and *Leute* in the region.

Second, and more importantly, both the German representational space of East Africa and economic plans for it continued to evolve over the period I analyse in this study. The primary reason for this evolution was the actions of East Africans making space themselves. This occurred not only at the level of practice, where Lefebvre and Michel de Certeau see opportunities for people to challenge state power to make space, but in the other parts of the spatial triad as well. The shifts in the German spatial imaginary followed no neat rhythm. What pattern does emerge is one of failed German plans for economic development, followed by drastic shifts in the representational space of East Africa that justified new economic practices for the colony. German discourse and actions failed time and again to achieve their ends. Colonialists introduced new models reflecting the latest geographical and historical thinking in Germany, only to replace them with new ones after the models did not achieve their goals.

Because of the importance of cultural geography, we can observe a unique version of changes common to colonial empires in this period. All empires attempted to control both territory and labour in their colonies. German colonialists made use of a vocabulary established through cultural geography to do so. This lent a different character to the German discourse about race and space in Africa. As a discourse of progress, cultural geography provided a means for imagining colonial development as transformative at an earlier period than was the case in other colonial empires of the time. Germans imagined themselves in competition with Arab and Indian colonists not native to the region, and at times saw their competitors as tools with which to "improve" the colony's Black African population.

Representational space formed through several venues, from the writings and meetings of members of scientific and geographic societies to ideas about Africa circulating in working-class culture. To understand German representational space, I examine colonial propaganda, the writings of explorers and travellers, both published and unpublished, and public discussions of the German project in the region. I attend in particular to the discourses of cultural geographers and Germans in Ostafrika, who were the most powerful voices in shaping the future of the colony.

Of course, Africans developed representational space that did not match the German spatial imaginary. German colonialists intentionally obliterated African representational spaces in the establishment of the colonial state, so they are more difficult to find and played a smaller role in shaping Ostafrika's development than did German imaginaries. Nevertheless, African spatial imaginaries continued to play an important role and shaped German ideas. For local representational space, I rely on Swahili epic poetry and ethnographic works.[33] African representational space is also discoverable in the many ways in which Germans adopted local ideas about territory into their frameworks for understanding East Africa. Colonialists especially took on ideas about the interior and the people who lived there from coastal elites in their plans for economic development, but they also adopted aspects of the spatial imaginaries from intermediaries in the interior.

Cultural geography formed the primary bridge between representational space and representations of space. The ideological framework of *Kultur* as politics evolved out of a unique German intellectual tradition over the course of the nineteenth century.[34] Alexander von Humboldt wrote in his 1805 *Essay on the Geography of Plants* that "the civilization of peoples is almost always in inverse relation to the fertility of the soil they occupy," articulating the concept that societal progress could be read in *Kultur*.[35] From there, the idea of *Land* and *Leute* as interconnected developed from Wilhelm Riehl's *Land und Leute*, published in 1854. Riehl argued that there was a connection between people and landscape in Germany as part of an argument for German unification. Germans did not just live in the area to be unified, but had developed a deep relationship to the soil that called for a political union of people and territory.[36] Friedrich Ratzel laid out cultural geography's research program in the first volume of his *Anthropogeographie* (Human Geography), published in 1882. He argued that landscapes and societies were "inseparable," as everything people did happened in space. Cultural geography, also referred to as human geography, subsumed the study of history, as Ratzel claimed one could observe the different stages of human societies through the study of spaces they inhabited.[37] Ratzel's explanation of cultural geography's relationship with other disciplines explains why it became so powerful. Cultural geography was distinguished from other modes of studying space in its fusion of physical geography with ethnology and history. At a moment when Germans were looking for justifications for national chauvinism, cultural geography was appealing. It provided a means to think about German uniqueness from neighbours. The path to a stronger nation lay in strengthening the relationship between *Land* and *Leute*.

Representations of space encompass the various plans, models, and maps that colonialists made to depict East African space, both of an imagined future for the colony and its supposed present-day or past reality. Cultural geography's power was that it provided the means to think about Europe and the East African coast and interior as part of the same course of history and the interior as developing toward the model on the coast. Geography's power came from its combination of methods from different disciplines to study space and time together. Cultural geographers combined history and ethnography with physical geography to create a comprehensive explanation of social development that foregrounded the relationship between people and place. Friedrich Ratzel developed these ideas most famously in his *Lebensraum*, published in 1901, but his works from the 1880s on were attempts to form a picture of the entirety of human development. In a 1904 article in the *Historische Zeitschrift*, Germany's most influential historical journal, Ratzel admonished historians for their insistence on a separation between peoples to be studied with historical methods, meaning those with written records, and peoples without writing, who were left to anthropologists. What made history and anthropology "close relatives," he wrote, was that both dealt with societies' relationship with the earth. One could chart the progress of humanity in contemporary societies' transformations of the landscapes in which they lived. East African landscapes and societies could therefore be read as stages of the European past with which Germans were familiar. A comparison of an African ruler to a medieval German king thereby painted a larger picture of a society ruled through caprice and lands farmed inefficiently for the ruler's benefit alone. The history of all life, wrote Ratzel, was best understood as a competition over the limited space on the earth's surface. The culmination of human development and the eventual winner of this competition, befitting Ratzel's conservative agrarian politics, was a powerful state built on the work of farmers. Ratzel, and the myriad geographers he taught or influenced, believed that all human history could be united into one study through closer attention to space and the ways in which societies moved through it over time.[38]

To get at representations of space, I analyse maps of East Africa together with textual representations of space for how Germans and Africans depicted the colony's spaces and turned their representational space into something more tangible. The texts I use for this are descriptions of space found in travelogues, correspondence and, especially, in planning documents that lay out visions of territory and change. European travellers narrated and described what they saw to make spaces inhabitable and available for European exploitation. Though they claimed they were neutral observers, European explorers and geographers made the exploitation of colonial spaces possible.[39] African representations of space are filtered through European sources, whether missionaries, administrators, or ethnographers, but it is possible to access some aspects of them through both maps and words. Much like historians of colonial cartography have revealed that Indigenous knowledge entered European maps of extra-European spaces, East African ideas about space became part of German geographic knowledge beyond cartography. Germans relied on Africans as assistants – translators, guides, porters, research assistants, and soldiers – as they travelled and observed the colony. What they saw and understood depended on what their African intermediaries chose to show them and help them understand. The German interlocutors for such information, however, elided its origins and presented it as their own.[40]

The availability of representations of East African space to German audiences allowed Germans in the metropole to make space by imagining movement through Ostafrika, blurring the line between practice and other forms of making space. Cultural geography provided the means to move through time in text, map, and picture as a method of space-making. This discourse reached nearly every corner of German society. Not only readers of colonial newspapers but also people who consumed more popular publications would read about *Kultur* in Ostafrika. The prominence of such ideas is apparent even in textbooks for German schoolchildren. Although different kinds of sources addressed different audiences, I understand them as one discourse about the future of Ostafrika. For instance, a scientific explorer would usually write scientific papers about his journey but would also write for popular publications, publish a long travelogue with stories of adventure mixed with science, and contribute to discussions among government officials. Though each medium included and excluded different elements of an expedition, they together formed knowledge of East African space and people that both shaped government policy and popular conceptions of the region. Files from private colonial organizations involved in founding and administering the colony, personal papers, and missionary publications provide a window into how ideas circulated outside of official circles and show how tensions developed between private actors and the government. School textbooks, colonial periodicals, and Reichstag debates provide evidence for how knowledge from Ostafrika reached German audiences.

Finally, I also examine the practice of space. Michel de Certeau argued that spaces are created as people move through them and narrate their experience, linking them together "by 'modalities' that specify the kind of passage leading from one to the other" and make meaning out of place.[41] Here is where Africans were able to exercise greater control over the colony's spaces than in the other areas. With relatively few Germans in the colony, most of the people working through the German spatial project were local actors – Africans in the interior, and Africans, Arabs, and Indians on the coast. Through their actions, they shaped German plans for the colony into a reality that looked somewhat different. Germans could not simply conjure space out of nothing. They depended heavily on local knowledge, and non-European ideas about space entered the German discourse about the colony. Close readings of German and other European sources, as well as the relatively few available sources from non-European perspectives, reveal the ways in which non-European practices shaped the workings of the colonial state and remade European knowledge about East Africa. My understanding of spatial practice comes from reports by government officials and missionaries on their interactions with African societies. These sources reveal the ways in which African communities accommodated German plans for Ostafrika and reshaped them to fit their own needs. This book stresses the spatial aspect of such accommodation.

Ostafrika and German *Kultur*

Examining *Kultur* as ideology makes clear that the most important aspect of German colonialism was not violence but the creation of an economic, political, and biopolitical scheme for social engineering. Much of the historiography of German colonialism has focused solely on its violence. Historians have attempted to make connections between colonial violence and the National Socialist Holocaust. They have focused especially on the Herero-Nama genocide in Southwest Africa, the German dehumanization of victims there, and the elimination of limits on violence as a forerunner of genocide in Eastern Europe.[42] The centrality of violence has obscured other elements of German colonialism. While state violence is inescapable, an examination of the discourse around *Kultur* in Ostafrika brings to the fore the application of racial thinking in population management and economic planning. Studies of German imperialism in Eastern Europe have made clear the importance of concepts that combined ideas about race with ideas about landscapes to German expansion. In the conquest of Eastern Europe, Germans invented plans for economic development that depended on beliefs that there was a special German ability to remake landscapes.[43] It was in Ostafrika where such lines of thinking first became part of government policies. What Germans learned and knew about Ostafrika depended on a nexus of race and space in assessments of *Kultur*.

Contrary to longstanding assumptions, German colonialism was a significant phenomenon affecting metropolitan society.[44] Because Germany had few colonies compared to the United Kingdom or France, and the entire lifespan of the Germans overseas empire was barely over thirty years, the study of German colonialism was long minimized as unimportant.[45] But more recent work has shown that colonialism was an important factor in shaping German life in the metropole and German ideas about the world in this period, particularly in visual culture and in conceptions of human difference. Among Germany's African colonies, Ostafrika was also exceptional. Because of the coast's long interaction with other areas of the world, Germans were confronted with a more complicated racial hierarchy than they imagined elsewhere in Africa. Ostafrika dominated metropolitan discussions of the German colonies until at least 1900 as the nation's biggest, most populated, and most economically important colony.[46] Although historians have made the influence of agrarianism on Germany's settler colony in Southwest Africa apparent,[47] this book shows that colonialists extended the ideology of *Kultur* to non-European populations as well. It also draws attention to the importance of making space in the German colonial project.

In discussions of *Kultur* in Ostafrika, we can see an explicit transition from identification with Germanic tribes to an identity based in the cultivation of the soil in the early 1890s. From that point on, administrators and social scientists advocated for programs to further develop settled agriculture in the colony and break the military power of nomadic communities in order to force farming on them. The creation of settled agriculture through the reorientation of African politics and modes of living toward serving German economic priorities became administrators' main goal. Members of the German East Africa Company envisioned connections between Germans and nomadic peoples in Ostafrika, such as the Maasai, for the colony's first years. The colony's founders saw nomadic peoples as possessing the same spirit as early Germans and therefore as possessing the best potential for the creation of a new Germany in Africa. Over time, however, German colonialists came to believe that nomadism was a destructive influence and in conflict with *Kultur*.[48]

The German colonial period's importance to the development of racial thinking in East Africa has also been overlooked. This book shows that ideas about *Kultur* and colonial development depended on existing frames of knowledge and ideas about territory and its inhabitants that predated the German colonial period. Recent scholarship has clarified the ways in which colonial geographies set possibilities for African communities to themselves define ethnicity.[49] Scholarship on the formation of ideas about race and ethnicity in the region has largely ignored the German period and focused either on the precolonial past, on Belgian or British colonialism, or on the postcolonial period.[50] Histories of Semitic settlement and oppression that animated later politics in Rwanda and Zanzibar were first deployed politically as support for the German government's annexation of the Indian Ocean Coast and narratives about Arab settlement.[51] The German colonial period resulted in more continuities with the past than breaks from it, and Germans had to accommodate to local societies.[52] Attention to the decisions of German administrators makes clear the ways in which African knowledge became part of the ethnic divisions in the first place. German administrators turned racial divisions into governmental categories of difference and legislated markers of difference into the hierarchies of the colonial state. For instance, Germans adopted nearly wholesale the coastal view of Nyamwezi people, stereotyping them as porters and labourers to serve the colonial state. German administrators adopted coastal ideas about a binary between civilization and barbarism into plans for agriculture and their assessments of ethnic characteristics.[53] African geographies became the basis of European ones. African spatial practices became central elements in European discourses about African space.

Changing Ideas about Race and Space

Colonial Geography is organized chronologically. The first chapters introduce the ways in which the German East Africa Company argued for German conquest through its version of East Africa's history. Chapter 2 examines the precolonial spatialities of East Africa among societies in the East African interior, along the coast, and in Germany as the encounter between Germans and Africans began. The process of imagining East Africa as the embodiment of the past from which Europe had moved on began before formal colonialism. Scientific explorers depicted East African societies and landscapes as reminiscent of stories from Europe's past. Chapter 3 discusses Carl Peters and the Society for German Colonization's expeditions to the region and resultant treaties that they signed with a variety of local rulers in 1884 and 1885. It examines the emergence of a model of colonial development in which German political rule would suffice to produce economic improvement. The treaties and accompanying travelogues portrayed local politics as medieval to claim land for Germany. East African rulers, in this vision, signed treaties to become vassals of their new German king. Though Peters's goal of a private empire was common in this period of the "great land rush," his temporal-spatial argument was more explicit.[54] Building on the descriptions of precolonial travellers, Peters and his associates told the German public that Ostafrika's spaces resembled the Germany of centuries long before. Through rational administration and cultivation it could be made like contemporary Germany.

Chapters 4 through 6 examine an evolving debate over how to pursue colonialism that dominated Ostafrika's first years of existence. World-historical narratives became the dominant representation of East African space in Germany. The Society for German Colonization (*Gesellschaft für deutsche Kolonisation*, GfdK) argued that Ostafrika could be made into a profitable colony and an object of German glory through a focus on its *Land*. Mission societies and the Foreign Office disagreed. To missionaries, East African landscapes were underdeveloped primarily due to the practice of slavery, a practice they claimed belonged to Europe's past. They argued that Germany would first have to rebuild local societies to turn their people into productive labour and subjects before they could do the work of remaking East African space into what Germany needed. To win control of the coast, the company constructed a narrative of the history of Indian Ocean World as a unified space. In chapter 4, I analyse those narratives to create a model of social evolution in which African landscapes, evolving to be more like Germany, would produce changes in African communities and deliver economic benefits for Germany. Those narratives allowed Germans to create layers of sovereignty across Ostafrika and justify projecting power over the coast and narrow corridors inland.[55] Histories of difference became the means to define the coast as fundamentally distinct from the interior.

In chapters 5 and 6, I examine growing tensions over the relationship between race and space. In 1888, groups of coastal elites challenged private colonial authority to control networks linking the coast and interior through higher customs duties. German colonialism faced its first major crisis. The fighting on the coast, which is now known as the Bushiri War, intersected with contemporaneous concerns about Emin Pasha, the German explorer and provincial governor who had become stranded in Equatoria, in today's South Sudan. In responses to both issues, Germans proclaimed that Ostafrika was the site of a clash of civilizations between European Christianity and Oriental Islam, which had produced a moment of crisis for the colony.[56] Chapter 5 examines the agitation around Carl Peters's expedition to find Emin, based on his continued belief that what mattered was control of *Land*. Peters claimed that rescuing Emin would secure Equatoria for Germany and that Emin's experience there had shown his ability to implant *Kultur* in African landscapes. Chancellor Otto von Bismarck dispatched Hermann Wissmann to Ostafrika to fight the Bushiri War. German intervention thereafter aimed to strengthen connections to the land among people they believed were African and to exclude non-landed "foreigners" from being part of the colonial order to create absolute power over East African people and space. The government's response created a new basis for German claims to sovereignty, from the treaties of the mid-1880s to a humanitarian duty to protect of Black Africans from the Arab slave trade.

The final chapters of the book examine the formation of official German colonial policy in the 1890s, when race came to the fore of all discussions about Ostafrika's future. After Wissmann succeeded in his mission, the German government took control of the colony, and agrarianism became central to development planning. In what Juhani Koponen has defined as "development for exploitation," Germans made a long-term commitment to resource exploitation.[57] Scholarship on developmentalism dates most centralized planning to the postcolonial period, claiming the ideology of developmentalism was tied up with the politics of the Third World,[58] but German administrators clearly pursued projects that can be labelled as developmentalist in that they aimed to create both profits for Europeans and improvements in the living conditions for Africans.[59] Chapter 7 details the trial-and-error process from 1891 to 1893 through which new government administrators arrived at the doctrine of *Kultur*. Administrators adopted a division between the coast as semi-civilized and the interior as wilderness from precolonial coastal geographies. Their goal was to spread the semi-civilization of the coastal towns west. They began with a belief that technological solutions, most importantly railways, would allow them to do so while exploiting Ostafrika's resources, a form of technological modernity. Technological solutions, however, failed to achieve the desired effects. Administrators would then explicitly turn to cultural geography as the means of judging development and focus on settled agriculture as the colony's future. Thereafter, administrators would attempt to shape the social evolution of East Africans for Germany's economic benefit.

As I argue in chapter 8, Germans were unable to simply impose their own meanings on East African space, and local adaptations of German rule reveal the ways in which African meanings seeped into German practice. From 1893 to 1900, cultural geography became increasingly prominent, shaping not only administrative discussions but also missionary publications and textbooks for German youth. The field of

Heimatkunde (local history and geography), developed to better understand Germany itself, became the primary mode Germans used to consume information from Ostafrika. It provided new representations of East African space that were directly tied to belief in a modernity based in agrarianism. African communities did not fit neatly into the categories administrators imposed and were able to adapt to use the German system. African communities continued to shape the functioning of politics and spatial relationships into the colonial period. Ostafrika's inhabitants co-opted German geographies for their own ends. Though German rule was overwhelmingly destructive, some leaders and societies across the colony adapted the German presence to create their own political authority.

Chapter 9 analyses the concurrent development of attempts by the colony's administrators to create settled agriculture and the maturation of Friedrich Ratzel's geography to his later theories of *Lebensraum* (living space) and *biogeography*. The new geography provided the intellectual support to government intervention into Ostafrika's racial characteristics to create *Kultur*, yet a different kind of modernity. Under Gustav von Götzen, governor from 1901 to 1905, the administration adopted Ratzel's concepts of cultural diffusion wholesale. Götzen instituted state-controlled migration projects of agriculturalists from Sri Lanka and South Africa to deploy what he saw as positive racial characteristics. His eventual choice of Sinhalese farmers over Boer ranchers demonstrates the triumph of *Kultur* over even racial ideology. In managing race for *Kultur*, trying to change spatial practice, Ostafrika's administration reached its most intrusive stage. The end point of the study is the Maji Maji War of 1905–7. The war overturned much of the administrative system developed over the previous years and provided impetus for the German government to centralize authority over colonial affairs in Berlin, a decision which limited local initiative and after which Germany pursued colonial policies modelled more clearly after those of other nations.[60] This study, then, covers the period in which administrators on the ground in Ostafrika were the main decision makers about the colony's future and in which Berlin had little direct role in running Ostafrika, except when problems arose.

The debate over representations of space dominated the first two decades of German discussions of the colonization of East Africa. It dictated the possibilities for economics and politics in the region and shaped ideas about race and world geography. German colonialists evaluated Ostafrika's historical progress by *Kultur*, the cultivation of the soil. Different imaginaries of East African spaces and races became part of one discourse about the colony's future. Administrators mediated what entered administrative practice through cultural geographic theories of a relationship between landscape and racial character. Their geographies conflated race, space, and history and produced claims to territorial sovereignty based on a conception that societies' capacity for settled agriculture was the primary factor shaping history. Administrators adopted increasingly complex means of creating an idealized agrarian space in East Africa that would solve the problems of German modernity. In the German imaginary, time became space as colonialists tried to make a new progression of history on the landscapes of East Africa and bodies of East Africans.

2 Geographies of East Africa in the Second Half of the Nineteenth Century

In his 1877 travelogue, *Across Africa*, the English explorer Verney Lovett Cameron recounted a story that African porters had told him about Ugogo, in Central Tanganyika, on his expedition two years earlier. Some years before, an Arab merchant had undertaken a journey to Zanzibar from Unyanyembe, south of Lake Victoria, and was determined to extort supplies for his caravan with armed force rather than buying them. As he neared Ugogo, the people of the region "retreated into the jungle," destroying their houses and whatever supplies they could not carry and filling up their wells. According to Cameron's informants, 600–700 members of the caravan died.[1] Separately, the English missionary Tristram Pruen told readers in his travelogue about his porters relating the same incident, of a large caravan travelling through Ugogo using violence to get supplies and facing consequences for doing so.[2] The story served as a warning; subsequent travellers did not make the same mistake.

In the repeated story and Cameron's own journey through Ugogo, we can observe the workings of a representational space based in territoriality in late precolonial mainland Tanzania. The region's inhabitants clearly asserted their authority over natural and agricultural resources, as well as the right to travel through space. The members of Cameron's caravan clearly believed they were entering a region named Ugogo as they approached, defining precise borders for it. He reported that travellers "approach Ugogo in fear and trembling, apprehensive of being fleeced of half their stores in passing through."[3] Upon Cameron's arrival, the caravan was charged payments, called *hongo*, for food, water, and free passage. Cameron's caravan was not alone in paying hongo in Ugogo, or elsewhere in central Tanganyika; nearly all European travelogues from the second half of the nineteenth century describe such payments. These were not simply purchases of provisions. Hongo payments, though they contained ceremonial elements, were primarily economic exchanges. Travellers exchanged goods produced overseas such as cloth or industrial products for locally produced provisions and the right to travel through space.

Hongo payments and the caravan porters' naming of the regions through which they travelled belie claims that spatialities based around bordered territory in Africa date from European conquest. Following Jane Guyer, historians have argued that political power in precolonial Africa consisted of "wealth in people," the ability to mobilize large groups of followers, rather than the control of territory.[4] In relatively thinly populated regions, control of territory held far less value than control of people. People envisioned space as a collection of nodes and networks connecting them. Sovereignties overlapped, and rulers did not attempt to establish explicit control of space.[5] Such histories portray polities in which territory mattered little. It was only with colonial rule, Mahmood Mamdani has argued, that clear territorial borders became important.[6] I argue that the transition to the importance of territoriality to political power began before formal colonialism, as the region encountered global trade and capitalism.[7] New forms of political economy and spatial practice emerged in relation to the growth of trade from the 1830s. Older forms of authority broke down and ethnic loyalties eroded, allowing for the creation of new manifestations of political power.[8] Control of territory became increasingly important as a means of controlling networks and the trade goods that travelled along them.

The idea that African societies were not territorial developed out of Europe's own representational space of Africa, in which the lack of printed maps and border markers became a sign of Europe's progress and superiority over Africans. Modern European sovereignty became defined by territorial boundaries above all else. In the nineteenth century, "the territorial boundaries along which states claimed sovereignty became more sharply defined in both law and practice."[9] Europeans theorized territoriality as a stage in the evolution of human societies, combining history and space into one narrative of progress. Carl Schmitt, the twentieth-century German philosopher of law, theorized that sovereigns created authority by walling off space over which they could then exercise authority, which for him meant that true sovereignty developed in the creation of borders in Europe.[10] In cultural geography, which became the most common means for Europeans to describe East African spaces, the existence of clear territorial borders was a sign of societal evolution. The geographer Friedrich Ratzel claimed that more advanced races possessed clearer borders than less advanced ones.[11] They based these claims on a lack of Indigenous maps and the use of itineraries to describe space in Africa. Schmitt's and Ratzel's claims are indicative of a broader trend of claiming European advancement through the presence of recognizable political borders. European colonialists had a vested interest in imagining African polities as non-territorial, so their descriptions of polities express that interest. There are no maps created by the leaders of those polities to rebut colonial claims.

This chapter examines East African representational space and representations of space of three broad categories: people of interior societies, people of coastal societies, and European explorers and missionaries. These categories do not capture all the nuances of spatial conceptions, but they do capture general trends. Each category included many different communities, each with different ideas about East African space. The general trends here cannot capture all the shades of meaning in the conceptions of space at play in the region. Each group had its own preconceptions of the region's spaces that shaped all three aspects of the spatial triad. The general trends do, however, reflect how people developed spatial "imaginaries" out of their practices of space, the itineraries and networks that emerged as part of East Africa's integration into global capitalism.

Before the onset of formal German colonialism, multiple geographic imaginaries existed in East Africa. Each of these imaginaries developed out of adherents' experiences of the integration of the East African interior into the global commodities trade. Their experiences of integration produced different understandings of the relationship between political power and space. The three geographic imaginaries bore some similarities. For all three, itineraries rather than maps were the basis of spatial imaginaries. All three geographies included territorial authority and a conception of connections between people and place. There were, however, theoretical differences in models for understanding and managing space. Territoriality became more important in the interior than it was on the coast. Cultural geography became the primary means through which Europeans conceptualized East African spaces. With a cultural geographic framework, Europeans adopted African ideas about space and labelled them as primitive.

Spatialities of the East African Interior

People in the East African interior developed new spatial imaginaries over the course of the nineteenth century as the region integrated into the world economy and long-distance trade produced new representational space. As trade became increasingly important, networks that connected villages and people along trade routes became the foundation of economic and political power. The centrality of networks to conceptions of space is visible in maps drawn by participants in them, while showing at the same time the importance of those networks to power in political developments across mainland Tanzania. Though representational space still consisted largely of itineraries and networks, the increased importance of trade networks to power magnified the relevance of control of territory. Control of areas through which caravans travelled, particularly chokepoints or regions with scarce water or food, became the most important factor in political power. New political leaders, known as Big Men, established clear markers of territorial authority in order to assert control over such areas. In response, some communities adopted new practices to escape the power of these centralizing Big Men.

The conditions limiting long-distance trade meant that the East African interior integrated into the world economy later than did West Africa and that its integration and resultant political changes were thereafter more rapid. Two primary factors had limited long-distance trade in the region, which minimized the importance of territorial control prior to the nineteenth century. First, the region lacked navigable rivers, meaning that all trade had to travel overland. Second, the presence of the tsetse fly and sleeping sickness made it impossible to maintain pack animals, thus requiring that porters carried trade goods. At the turn of the nineteenth century, most of the area between the kingdoms of the Great Lakes and the coast was a region in which political borders overlapped or were nonexistent. Establishing political borders was not important compared to establishing control over people and their labour.[12]

Political arrangements and ideas about territory changed significantly beginning in the late eighteenth century with the arrival of global capitalism. Trade between coast and interior had been economically important for several centuries, but new trade routes developed beginning in the 1720s through the initiative of Nyamwezi porters, called *wapagazi*. Migrations, the distribution of natural resources, and the central position of the region of Unyamwezi, south of Lake Victoria and east of Lake Tanganyika, as well as a seasonal agriculture system that allowed young men to leave the area during the dry season in the trading system stimulated *wapagazi* to begin long-distance trade, a new practice of space. Porters from elsewhere expanded the trade over the eighteenth century, though the system continued to change until routes stabilized by around 1850.[13] Trade linked communities in East Africa with others and with global markets, rapidly bringing economic pressures from the slave trade, abolition, and commodity trades to bear.[14] The slave trade destroyed many existing political arrangements as political leaders proved unable to protect their followers. The slave trade increased people's value as commodities, meaning it often made more economic sense to sell one's followers than to maintain them as dependents.[15] Long-distance trade created new spatial imaginaries as East African societies encountered the Indian Ocean World and Europeans and their representational space.

Maps drawn by the Nyamwezi caravan headman Pesa Mbili in the early twentieth century are representations of space that remained oriented around itineraries and networks that persisted into the German

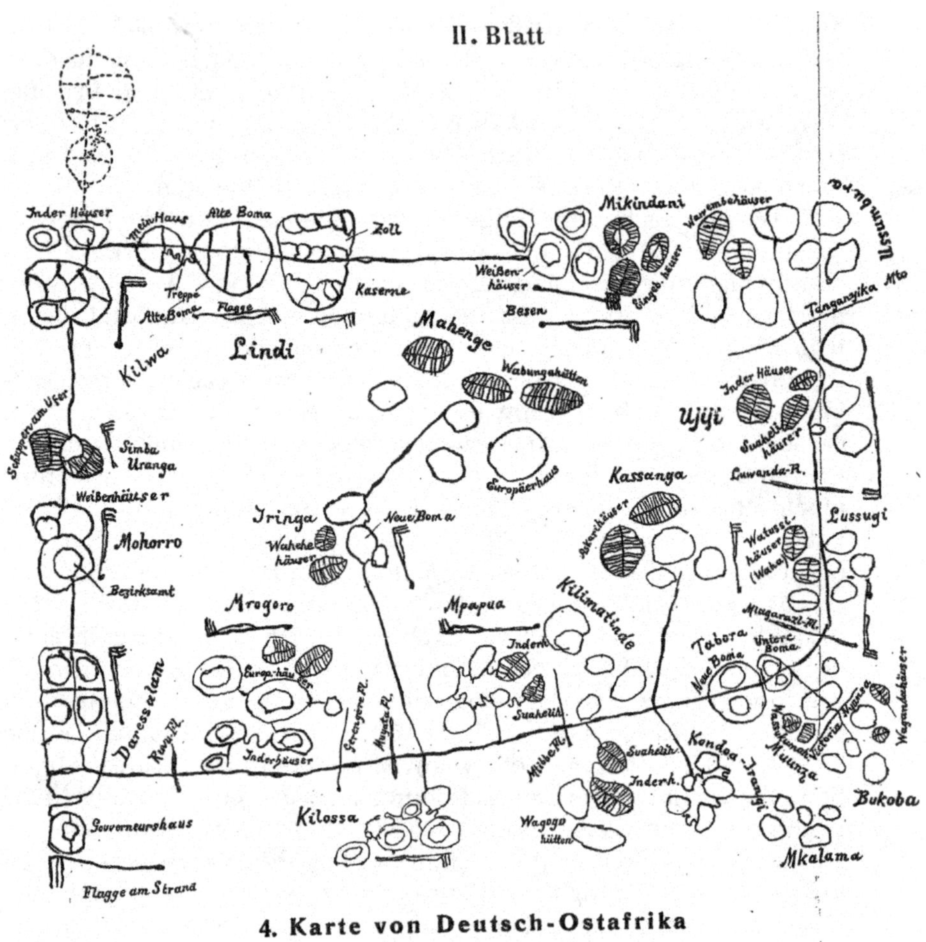

Figure 1. Pesa Mbili's map of the central caravan route. From Weule, "Zur Kartographie der Naturvölker"

period. The map turns a practice of space, the travel of a caravan between coast and interior, into a representation of space. That representation, in turn, reflects Pesa Mbili's representational space of the region depicted. Pesa Mbili's map of German East Africa shows not borders, but networks between places.[16] As Allen Howard has argued, precolonial Africans "perceived of places as interconnected, rather than as isolated, nodes because social reality and discourse involved linkages among people located in different places."[17] The caravan route in Pesa Mbili's map links villages and towns between Lindi and Ujiji, a straight distance of nearly 800 miles. For Pesa Mbili, the distances between the nodes of the network matter little; what matters is the connections between places and the people there. Each village appears as a collection of buildings, often radiating from one central building connected to the caravan route. Networks branch out from the main route to nearby places.[18] Long-distance trade made these places matter to Pesa Mbili. The villages on the route become oriented around their connections to the caravan route, which outweighs any need to accurately represent the distances between them.

Another map in the same collection, this one drawn by Sabatele, a Mambwe man from western Tanzania, makes the spatial imaginary's basis in itineraries even clearer. Sabatele's map represents the space between Dar es Salaam and the Great Lakes. His networks converge on Tabora, from which branches stretch out to Mwanza on Lake Victoria and Ujiji on Lake Tanganyika. Sabatele presented his route, too, as a straight line, though it involved turns. Sabatele marked some landscape features, such as the Makata plain and Kilimatinde Mountain, but otherwise the landscape is absent, the space through which the networks move empty. The map may appear simplistic, but it charts a complicated journey over hundreds of miles and marks the spatial elements crucial to the map's narrative. What matters here again are the nodes and networks, not the territory through which those networks run.[19]

The map in figure 3 presents space in absolute terms. One can see the caravan routes as they existed in terms of miles across the territory of mainland Tanzania. Pesa Mbili's map warps distances from their physical reality, but it better displays the representational space of Nyamwezi *wapagazi*. The warping of space is visible in the networks that connected people. Caravan porters' representations of space reflected their practices of space and representational spaces rather than an idealized mathematical vision of the world.

The transformation of space in Pesa Mbili's and Sabatele's maps was a global phenomenon in the second half of the nineteenth century reflecting the growth of global capitalism. Writing about railways in the western United States, Richard White draws a distinction between "absolute space," the number of miles covered by track, measured on the ground,

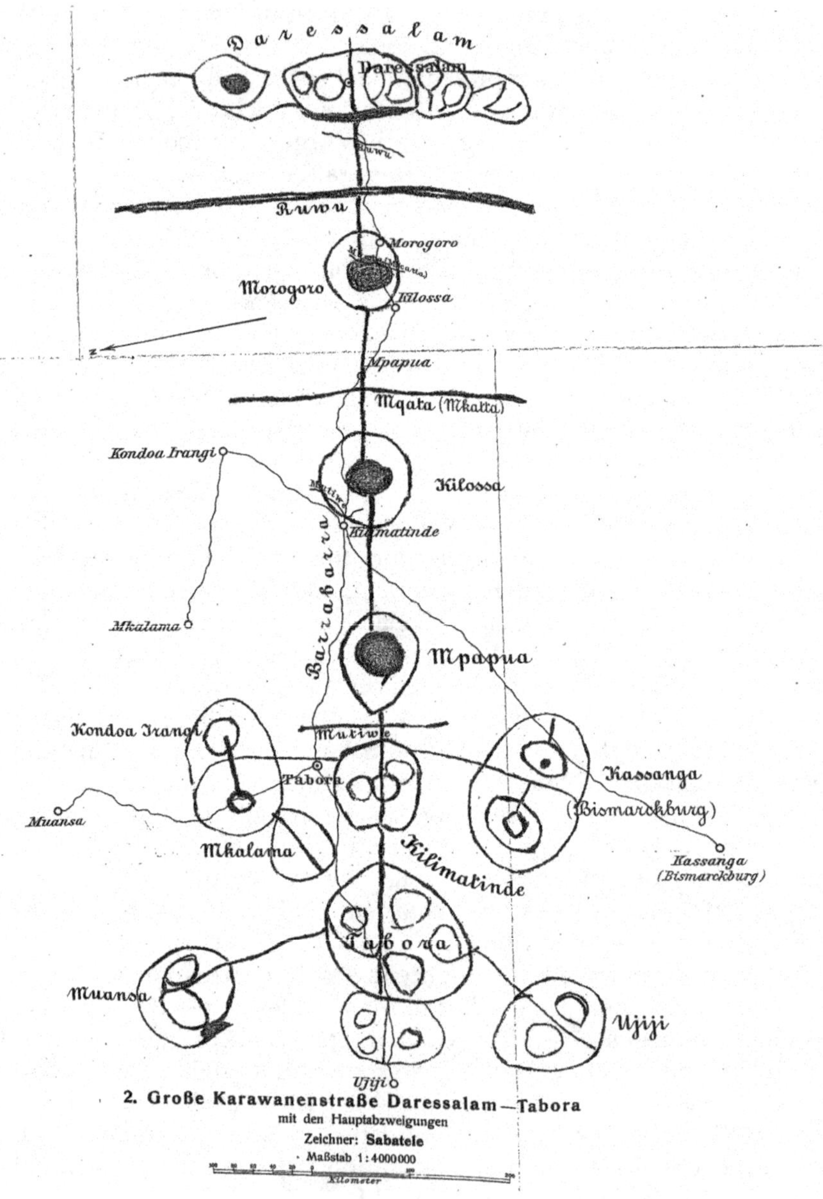

Figure 2. Sabatele's map of the central caravan route. From Weule, "Zur Kartographie der Naturvölker"

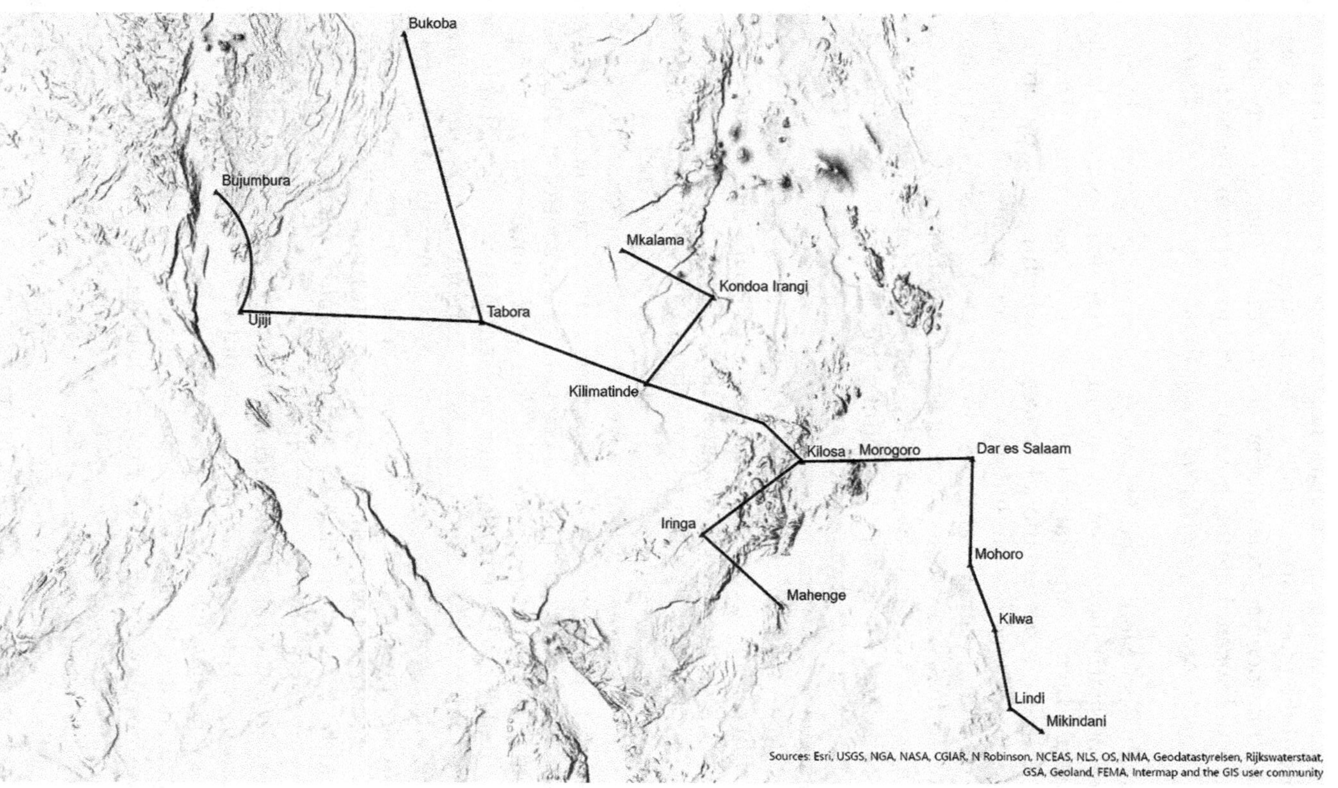

Figure 3. Pesa Mbili's route plotted on a map made with ArcGIS

and "relational space," space that emerged as people calculated distance in terms of the money to be made or lost through a particular journey.[20] A similar transition happened in mainland Tanzania. Though the people of the region did not attribute the same value in currency to movement across space, the networks that determined political and social power held similar value.

Though Pesa Mbili's and Sabatele's maps make apparent the persistence of spatialities based in networks; actions of political leaders indicate a growing importance for territoriality. New spatial practices produced new representational spaces in which control of territory and connections across the Indian Ocean were more important than before. The importation of firearms provided the means for local chiefs to increase their power and recruit more people to their sides. The community of interest between ruler and subject fractured as chiefs' interests became paramount.[21] Control of the caravan trade became the most important aspect of political and social power across the region. Charismatic war leaders copied chiefs in building their own standing armies and consolidated political power in the second half of the nineteenth century to become what some historians have termed "Big Men." Big Men attracted followers who hoped to build positions for themselves. Big Men used their connections with the coast as a tool for political power. Access to slaves, ivory, weapons, and cloth and other goods from the caravan trade, as well as cultural practices like Islam, tied people to Big Men.[22] Each of the villages represented in Pesa Mbili's and Sabatele's maps would have been under the control of a Big Man or a chief who operated like one. Although some historians have argued that the power of Big Men was not based in formal roles as territorial rulers, a closer look reveals the importance of territoriality.[23] Raiding villages and looting caravans served primarily to force trade to particular areas rather than to make money from stolen goods or people.[24] Caravan routes that passed through fruitful regions shifted to avoid high costs, political instability, or raids, meaning control of territory through which caravans had to pass because they had necessary supplies was valuable for demanding hongo.[25] Verney Cameron's experience in Ugogo, discussed at the outset of this chapter, makes apparent the control local rulers asserted over caravan routes. Johannes Rebmann, one of the first European travellers in the East African interior, was not allowed to travel onward without formal permission from a Big Man.[26] The Big Man had offered Rebmann assistance, but clearly expected payment for passage along the routes that he controlled. The explorer Joseph Thomson also saw the significance of territoriality and *hongo* as a payment for passage. In one case, villagers along his route made clear the "road was 'shut' by placing some green twigs across the pathway," barring the caravan's further progress.[27]

Hongo was the most important element in establishing and maintaining power. Big Men could set their own rates for passage, asserting their authority over what counted as currency in their territory. Thomson described a process of his porters restringing an enormous number of beads and sewing exact patterns on cloth to fit the specifications for hongo set in Maasai territory. For example, Maasai accepted only a fringed type of cloth "known as naibèrè" as payment.[28] Even in cases where Thomson did not himself meet the Big Man, he paid hongo in specific, exact forms for the right to travel. For instance, he sent Mandara of Moshi a gun, a "government sword-bayonet, a piece of cotton, two coloured cloths, and two flasks of gunpowder."[29] He also had to pay additional goods when a Maasai warrior killed one of his porters, as blood had been spilled on Maasai territory.[30] African practices of space set the terms on which Europeans could travel in East Africa.

The power of the Hehe, centred near today's Iringa, shows the development of complex practices to govern large territories. The Hehe people recognized the region south of Iringa as their place of origin. They conquered nearby areas but stood apart from the population as governors and garrison.[31] From there, they fought wars of territorial conquest through the 1870s and 1880s.[32] Mkwawa, the paramount Hehe chief, ruled over a large kingdom. He centralized power in his capital and distributed authority outwards through his subordinates, who were allotted specific territory according to their abilities. The most capable ones were posted on the frontier. All owed Mkwawa portions of their income from loot and trade, which he used to reward subordinates and ensure loyalty.[33] These outposts reported regularly on movements on the border and make clear that Hehe rule was territorial.[34] Observations from Mkwawa's border posts are evidence that he projected his power over a large territory. Georg von Prittwitz und Gaffron remarked that such posts were everywhere, such that the "entire mountainous region to the border of open land" was occupied.[35] To maintain control over territory as subordinates attempted to break away, one Ngoni ruler developed a new system of dividing his wives among four villages in the four cardinal directions, each with a "main wife." His nearest subordinates were assigned to these villages and communicated with the sultan through the main wife there.[36]

Big Men's authority over territory meant not only control of travel but also such authority asserted rights to resources. This became more important as overseas demand for ivory led to a boom in trade, and Big

Men attempted to control the valuable resource in their territory. The standard system was a claim to the "ground tusk" of any elephant killed, usually the larger of the two tusks.[37] A Catholic missionary observed claims to territory in a discussion with a village elder. The elder claimed ownership over the earth, the forest, and the water. They therefore had the right to take resources from it. If people travelled through their territory, it was on roads they created and drained, so they had a right to claim payment.[38] In one case, a Big Man diverted water flow upriver such that Thomson's caravan had no water until he paid hongo.[39]

Big Men who could control key provisions or chokepoints along the caravan routes could exert greater power, increasing the importance of territorial control. Gogo Big Men regulated trade through the control of water along dry sections of the central trade route, and the region had a reputation as especially expensive to traverse, as is clear from Cameron's story of the decimated caravan.[40] Even more apparent was the the role of control over trade in the power of Semboja of the Shambaa Kingdom. He founded the town of Masinde at a chokepoint for trade routes between the northern towns of the coast and the regions around Lake Victoria. According to an Austrian traveller, it was impossible to travel around Masinde. One had to travel through Semboja's town and pay hongo. The Austrian claimed the town was only thirty years old in the early 1890s, meaning it was built mid-century as caravan traffic through the region was increasing.[41] Tabora, as seen on Sabatele's map, formed the most consequential chokepoint for trade from the south and east of Lake Victoria and around Lake Tanganyika, which made it a key location for the consolidation of power.[42]

Some people attempted to escape Big Man authority, producing new spatial arrangements. Big Man rule was more successful on plains than in mountains, so many vulnerable agriculturalists moved to mountainous areas. They deserted large stretches of territory in the process.[43] The losers of struggles among Big Men and chiefs fled to other regions.[44] The town of Kilosa, along the central caravan route, formed as an agglomeration of people fleeing from farther south.[45] In the Uluguru Mountains, some communities settled in higher altitudes, which made it difficult for members of passing caravans to reach them. There, they were able to live in smaller villages than were people lower down the mountains and depended less on military force to maintain their livelihoods.[46]

By the 1880s, a spatial imaginary centred on trade routes connecting the Great Lakes and the Indian Ocean had become essential to political power in mainland Tanzania. Global trade produced new spatial practices, which in turn created new representational spaces of political power. People involved in the caravan trade conceptualized space in terms of networks linking people across the region, but control of territory was absolutely an element in political and economic power. Big Men established control over areas along trade routes, charging hongo for passage and building followings through their power over space. The consolidation of territorial authority was part of integration into the world economy and the movements of goods and people involved in trade. The long-distance movement of people and goods would also bring the people of the Tanzanian interior into contact with other spatial imaginaries, especially those of the Swahili Coast.

The Spatialities of the Swahili Coast

The integration of the Tanzanian interior into world trade over the course of the nineteenth century increased the importance of the towns along the Swahili Coast in regional and global economies. Zanzibar and parts of the coast developed economic specializations, made possible by international trade. The coast's representational space was centred on the Indian Ocean, but the interior remained important. Like societies farther west, the people of the coast conceptualized space in terms of economic and cultural networks. The important networks on the coast, however, stretched across the Indian Ocean to South Asia and the Arabian Peninsula. People were linked to space in terms of their economic and/or cultural value. The representational space of coastal society can be observed in Swahili epic poetry. This space was centred on the ocean as a world to itself, with power consisting of control of key points along trade routes both over the ocean and into the interior. Control of the nodes of the networks, the coastal towns and Zanzibar, were central to political and economic power. Territorial authority along the networks was less important than farther west.

The representational space of mainland Tanzania among members of coastal societies, like those of interior societies, centred on itineraries of travel. These imaginaries are apparent in Swahili epic poems. One poem by Mwalim Mbaraka bin Shomari, "Shairi la bwana mkubwa," is illustrative, though it is about a later German governor. The poem describes a journey of the governor's caravan around Ostafrika in itinerary form, listing places the caravan stopped and their attributes. The poet names the towns and villages to which the caravan travelled, describing their leaders and inhabitants in some detail. Except for a short anecdote about the governor shooting a lion that had killed one of his porters, the poem devotes no attention to the areas between towns. Though it is not a graphic presentation of space, the poem maps mainland Tanzania as sets of nodes and networks.[47]

Coastal communities were oriented toward the Indian Ocean but maintained important connections with the African interior. Networks along the Indian Ocean coast had first become heavily imbricated due to international trade in the eighth century, as Indian and Arabian slavers and merchants began travelling farther afield on their trading journeys. Swahili settlements, which began as Bantu farming communities, built trade connections across the ocean.[48] As the editors of a recent volume write, "the Swahili were simultaneously African and engaged in foreign trade, they were both farmers and traders, they had a range of relationships with groups both near and far, across land and sea."[49] The representational space of the coast included both the Indian Ocean World and the African mainland; individual circumstance might dictate which networks were more important. Connections to the Arabian peninsula, Iran, and South Asia were a source of power and prestige among coastal elites, whose representational space would have been centred on the Indian Ocean World. Those connections became even more prominent following Omani conquest at the end of the seventeenth century.[50] The people of the coast, who began calling themselves Swahili, after the Arabic word for "coast," became part of an Indian Ocean World, their lives organized as much by cultures and geographies of the ocean as by those on the African continent. Coastal people distinguished themselves from people inland through the concepts of *uungwana* and *ustaarabu*, denoting cosmopolitanism and urbanity.[51] They highlighted the aspects of their culture that connected them to Arabia, especially Islam and the Kiswahili language. Kiswahili was a literary language whereas the languages of the interior were not.[52] Freed slaves in particular adopted identities tied to the Indian Ocean World as a means of distinguishing themselves from those still enslaved.[53]

The importance of foreign merchants to the coast's society and economy strengthened those transoceanic connections as global trade increased. Indian merchants moved to the coast in significant numbers in the nineteenth century. One such merchant, Sewa Haji, dominated the trading community of Bagamoyo from the 1870s through the late 1890s. His firm supplied Arab, African, and European caravans with trade goods and porters and bought the ivory they brought back to the coast. He had representatives stationed at several towns in the interior: Kilimatinde, Tabora, and on Lake Victoria, as well as an agent on Unguja, the main island of Zanzibar.[54] Sewa's networks thus stretched from the Great Lakes to India, with his house in Bagamoyo as the primary node. *Ustaarabu* created networks between the members of coastal society and the other trading hubs of the Indian Ocean.

Contrasted with *uungwana* and *ustaarabu* was the concept of *ushenzi*, meaning "barbarism" or "savagery."[55] People to the west were *washenzi*, separated from the urbanity of the coast, and members of coastal society marked a border where the representational space of *ushenzi* began. One can observe this belief in the names of places. For example, *Bagamoyo* is a contraction of *buwaga mioyo*, meaning "calmed hearts," supposedly the feeling that one got approaching the town after travelling through the wilderness.[56] The author of "Shairi la bwana mkubwa," noted earlier, described people of the interior as "sorcerers" (*majusi*). On the southern route, caravans were named according to the political leader of the region from which they had come, dividing the origin points from coastal society.[57] Other Swahili epic poems refer to areas west of the coast as generic "forests" (*msitu*), fleeing events in the city to a generalized wilderness.[58] They demonstrate links between people and landscape in the spatial imaginary of the coast. The characteristics of *washenzi* were shared with the landscapes they inhabited, just as the *ustaarabu* and *uungwana* of the coast were associated with people there.

Poems reveal such conceptions of the interior as a space of danger and the exotic, a space apart from the society of coastal elites who consumed such poems, even as trade between the two increased. Sleman bin Mwenyi Tshande described his journey as an itinerary, a listing of places and the hongo paid in each, with little attention to the areas between places. He did note regions where "nothing human was to be seen." Anyone who had not themselves undertaken an expedition to the interior "does not know the worries of that world."[59] Mtoro bin Mwinyi Bakari described a need to travel to "the interior" as a young man. He had heard Uzigua was beautiful, so his "heart desired to get to know it." He seemed to find a blood-brothership ceremony conducted in one village as exotic.[60] Shorter allegorical poems also reveal such thinking about the interior as a space apart from the *uungwana* of the coast. The interior appears as a generic space for misadventure and possible danger. In "Der weite und der nahe Weg," one of the characters decided to take an unknown path and fell in a hole in the ground. Thus, the moral was that it was "better to take the far road that one knows."[61]

Coastal merchants imagined space in terms of networks, but these networks were different from those of the interior in that the spaces between nodes could not be controlled by one political authority, as they were maritime spaces. No political leader could build authority through the control of a chokepoint in an otherwise unimportant place. As Fahad Bishara has argued, there existed an Indian Ocean geography "constituted by shifting matrices of rights and obligations – one in which merchants could imagine the distance from Bombay to Zanzibar as a link in a network rather than an entire ocean."[62] Though the nodes in these networks were primarily in the Indian Ocean, the networks stretched as far as Salem, Massachusetts, as nineteenth-century globalization progressed. Agents in Zanzibar and caravan leaders were the most important actors in these networks.[63] Zanzibar was an entrepôt for trade, but it could not exercise complete control. Initially merchants build such connections, but other members of society copied them.

BuSaidi sultans in Zanzibar and their subjects created new kinds of economies and a political system based around control of the mainland nodes over the course of the nineteenth century. According to Abdul Sheriff, "Zanzibar was essentially a commercial intermediary between the African interior and the capitalist industrialising West, and it acted as a conveyor belt transmitting the demands of the latter for African luxuries and raw materials, and supplying in exchange imported manufactured goods." In 1840, the then Sultan of Oman, Said bin Sultan, transferred his capital to Zanzibar Stone Town from Muscat, as Zanzibar had become more important to his wealth than was Oman. Most of the arable land on the islands of Unguja and Pemba was planted with cloves, forcing Zanzibar to rely on food imports, as well as slave labourers, from the mainland. Networks stretching to the distant interior thus became a core part of the Zanzibari economy, necessary for the maintenance of a labour supply in the clove sector.[64] Demand from the interior and the rising value of ivory instigated the expansion of trade from far-flung places, particularly with Bombay.[65] Demand for luxury goods on the coast and increased access to credit from abroad allowed for increased consumer spending.[66]

The Zanzibari empire's focus on control of economic resources rather than explicit territoriality is similar to other non-European empires on the margins of Euro-American expansion in the nineteenth century. Like the Comanche empire of the American West, Zanzibari sultans did not try to centralize political control over their commercial empire. Rather than establishing control over the villages with which they came in contact, they used those villages as entry points into the networks of trade to which they were connected.[67] Like Khedival Egypt, Zanzibar attempted to extend its power into the African interior through informal means as a way of integrating into European-dominated international economies.[68] The primary concern was control of trade, particularly over who could charge tolls on goods travelling between the Indian Ocean and the interior.

The caravan system and its networks were dynamic in the face of European encroachment. In earlier centuries, the southern town of Kilwa Kisiwani was the main trading centre along the coast.[69] The BuSaidi conquest of Zanzibar shifted trade elsewhere, but more important were changes in response to European arrival. Kilwa Kisiwani possessed a deep harbour, making it an attractive station for British patrols against the slave trade. Coastal merchants therefore shifted to the neighbouring town of Kilwa Kivinje, which did not possess a harbour deep enough for European ships to operate.[70] Caravan leaders chose their routes based on the number of people who could be supported by the available provisions. Large caravans would frequently take less direct routes if they promised greater amounts of food and water.[71] Caravan merchants increasingly shifted trade to the northern and central caravan routes over the course of the nineteenth century as unrest near Lake Nyasa made the southern route more perilous. Trade with nearby regions operated separately from the long-distance caravan trade. Regional trade provided foodstuffs and less valuable items than the ivory of the long-distance trade and was understood as operating within a sphere of coastal society rather than in the wilderness farther west.[72]

Before German arrival, the people of the Swahili coast had developed a complex representational space that incorporated areas both across the Indian Ocean and in the East African interior. Those spaces were reckoned in terms of networks, but coastal people drew a clear divide between their own urbane society and spaces and the wild spaces and people of the interior. Connections to transoceanic networks were especially important for more recently arrived members of coastal society, former slaves and Indian merchants. Zanzibari power worked informally through those networks and did not involve territorially bordered sovereignty. Trade networks became increasingly important to Zanzibar and the people of the coast over the nineteenth century, especially through contacts with European merchants.

European Spatialities in East Africa

Beginning in the early nineteenth century, Germans and other Europeans became another set of actors in the spaces of East Africa. European explorers entered the region along existing trade routes, meaning that even the perspectives of scientific or religious travellers were dependent on East Africa's new connections to global trade. Europeans in East Africa attempted to rationalize the area's space according to models developed in Europe, but methods of understanding space based in cartography proved to be unable to create a complete picture of East African landscapes and societies. Rich descriptions in explorers' travelogues became the primary vehicle through which Europeans learned about the region. The ideas in those travelogues became the basis for conceptions of space that conflated the development of societies and landscapes in East Africa. By the 1880s, cultural geographic concepts had become the primary mode through which Germans produced representational space of the region. Germans imagined East Africa as a space for European adventure in which societies and landscapes were at a primeval level.

From the early nineteenth century, geography became one of the primary means through which Europeans conceptualized their place in the world and both people and space outside Europe. Felix Driver uses Joseph Conrad's term "geography militant" to identify a nineteenth-century European type, geographers actively promoting and working for imperial projects.[73] Geographical surveys served to mark and rationalize African landscapes, making them legible to European audiences and available for European exploitation.[74] Geography became institutionalized in Germany following unification in 1871. Publishers founded periodicals devoted specifically to geography.[75] *Petermanns geographische Mitteilungen*, published in Gotha, Germany, was the world's leading geographic journal.[76] Most German universities created positions in geography, and Ritter's cultural geography predominated in hiring for them.[77]

More than cartography, explorers' travelogues became the primary way that Europeans "knew" Africa. Whereas cartographers attempted to fill map spaces with political forms, travelogues depended on the depiction of landscapes as empty and available for European stories to take place without the interference of African peoples. Travelogues produced what the anthropologist Johannes Fabian defines as a *topos* of exploration, "a space without place ... (almost) without specifiable content," a set of spatial practices that made Africa into an empty setting for European adventure.[78] Without verified information about state borders, maps included large blank spaces in the centre of the African continent, spaces that appeared completely unknown to European readers. Even as exploration accelerated, maps lacked the feeling of veracity present in travelogues. For instance, *Petermanns geographische Mitteilungen* published the infamous "slug map" in 1856, a map which combined reports of the Great Lakes into one giant lake, which resembled a slug, snaking over millions of square miles.[79] Maps like this reflected the latest knowledge of physical geography, but readers believed they were less reliable for the truth about East Africa than written travelogues.

Though explorers were few in number, their ideas spread to a much wider audience through publication, inspiring increased engagement with Africa and becoming part of the German imaginary of the continent. German explorers became especially prominent in the exploration of East and Central Africa, where European inroads began relatively late in the century. Unlike other kinds of writing about the European presence in East Africa, travelogues appealed to both popular and elite audiences as providers of both adventure and scientific knowledge.[80] On the proximate level, many leading figures in Wilhelmine society supported exploration through the *Afrikanische Gesellschaft*, founded in 1878.[81] Kaiser Wilhelm I donated 50,000 marks to the society in 1879, and 75,000 in 1880.[82] Geographical societies in Leipzig, Dresden, Munich, and Frankfurt supported the work of the *Gesellschaft*.[83] As a region without existing European claims, East Africa became the part of the continent where German explorers and cultural geographers could most clearly establish the superiority of their techniques and knowledge over those of other Europeans.

The German explorer Heinrich Barth established cultural geography as the primary mode in which Europeans wrote about Africa. A student of Carl Ritter, Barth decided to become an explorer of Africa. His five-volume travelogue about his expedition with an Arab trading caravan in North and West Africa made him a celebrity in Germany and the United Kingdom, exposing a wide audience to Ritter's ideas as the means of understanding sub-Saharan Africa.[84] Barth's books inspired a generation of young men in Protestant Europe to explore and study Africa. British and German explorers became among the most famous people in Germany between the 1860s and 1890s.[85] Explorers, following Barth's template, travelled with African trading caravans and established concepts from cultural geography as the primary means that Germans used to understand African peoples and spaces.

Cultural geography enjoyed a reflexive relationship with the exploration of Africa. As explorers' and missionaries' writings depended on cultural geographic frameworks, explorers' and missionaries' observations worked their way back into geographers' works. The geographer Friedrich Ratzel described explorers as the "artists of an epoch." They had "provided a picture, whereas their predecessors had considered a dry, hollow word enough." This was the period of "detail painting" in representations of African space.[86] Ratzel provided additional support to the cultural geographic approach in exploration with the publication of the first volume of his *Anthropogeographie* in 1882.[87] Ratzel, drawing on his travels in Europe and North America, wrote that cultures could be understood by looking at the relationship between people and their environment. He maintained that people formed a spiritual bond with the landscapes they inhabited and that one could understand societies by looking at the ways in which they shaped and were shaped by physical geography. *Anthropogeographie* ensured the place of cultural geography in the German academy and influenced other disciplines. Ratzel was clearly influenced by the work of explorers and was among the early members of the *Afrikanische Gesellschaft*.[88] He believed that German explorers had achieved great things because of their own virtue and had provided important knowledge for Europe.

Cultural geography provided a framework to combine descriptions of spaces with ethnography. Spaces and the peoples who inhabited them appear in travelogues as inseparable parts of the same whole. Descriptions followed a regular pattern. Explorers began by describing a landscape, focusing on its physical features, before moving on to descriptions of the people who lived there, often described through the figure of a "typical" individual. Descriptions of people also focused on physical aspects: their build and their objects. Only then did explorers usually discuss culture or society, and then in limited ways. Space and people, *Land* and *Leute*, shaped and constituted one another in these accounts.[89] For Georg Schweinfurth, physical similarities among the Shilluk, Nuer, and Dinka in the Sudan showed the "remarkable law of Nature" that people would develop physical characteristics matching the landscapes in which they lived.[90] At another point, he remarked that cultural ideas about spirits could be traced to landscapes. While Europeans believed in ghosts and spectres because of their "dull leaden" skies, the Bongo believed spirits inhabited the forest, and people from the Orient, due to the blazing sun, feared the "evil eye."[91] Geographers rallied around cultural geography as a means of making geography relatable to students who had called it the "most sterile" subject before Germany acquired colonies.[92]

Explorers in East Africa plied the same routes pioneered by Nyamwezi and other caravans in the preceding decades to access areas to which Europeans had never before been. Although travelogues included maps, their narrative form and historical comparisons meant that itineraries became the primary representations of African spaces. Even when publications about explorers' journeys included maps, itinerary descriptions were also present, as publishers and readers did not believe that maps carried the "whole story."[93] For instance, Wilhelm Junker's map of his expedition was insufficient, so his travelogue included the itinerary, which lists each day's journey in terms of heading and the amount of time on the march.[94] Readers filled the blank spaces of African maps with explorers' narratives. These narratives rested on African intermediaries' representations of the spaces described.

What explorers could "discover" depended on what their intermediaries showed them, and upon the trade networks that existed outside of European control. Many explorers, however, attempted to

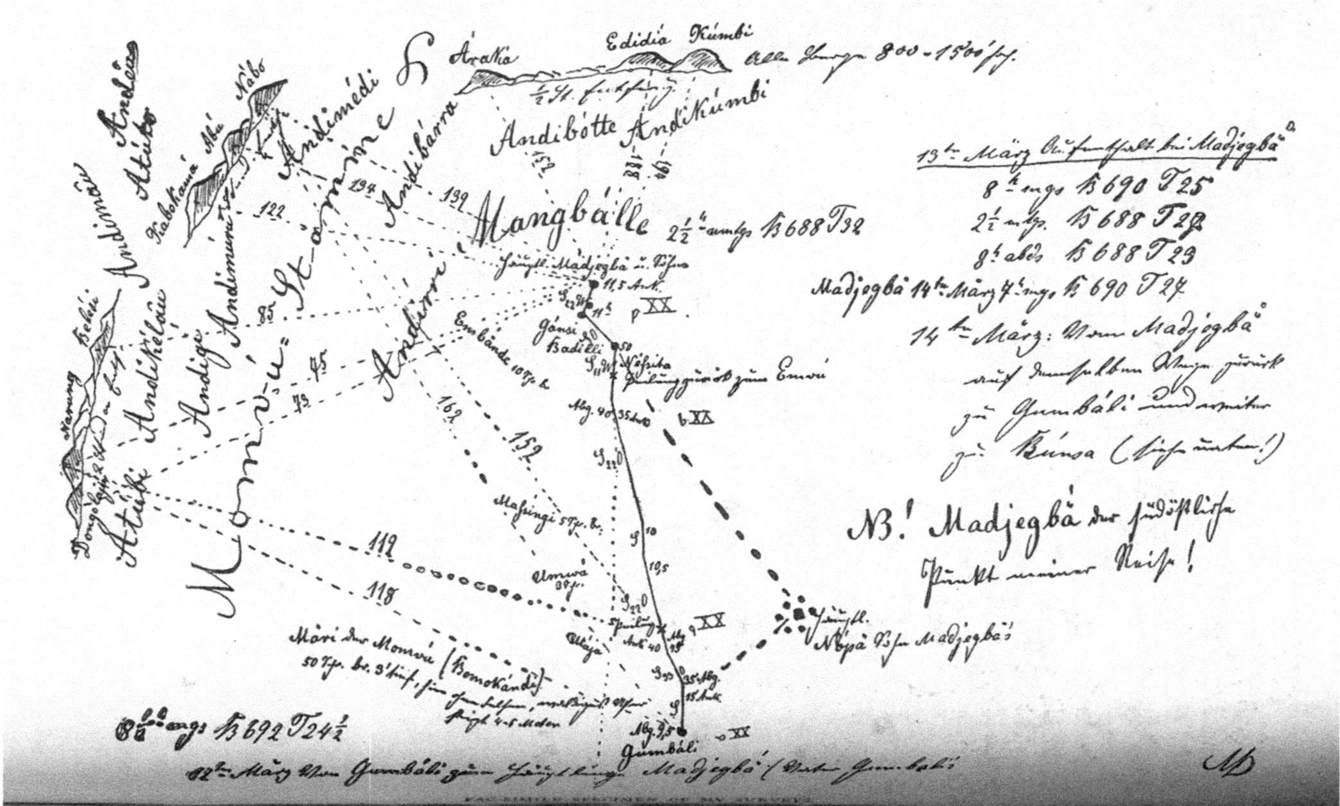

Figure 4.   Wilhelm Junker's map of his travels in northeast Africa. From Junker, *Travels in Africa*

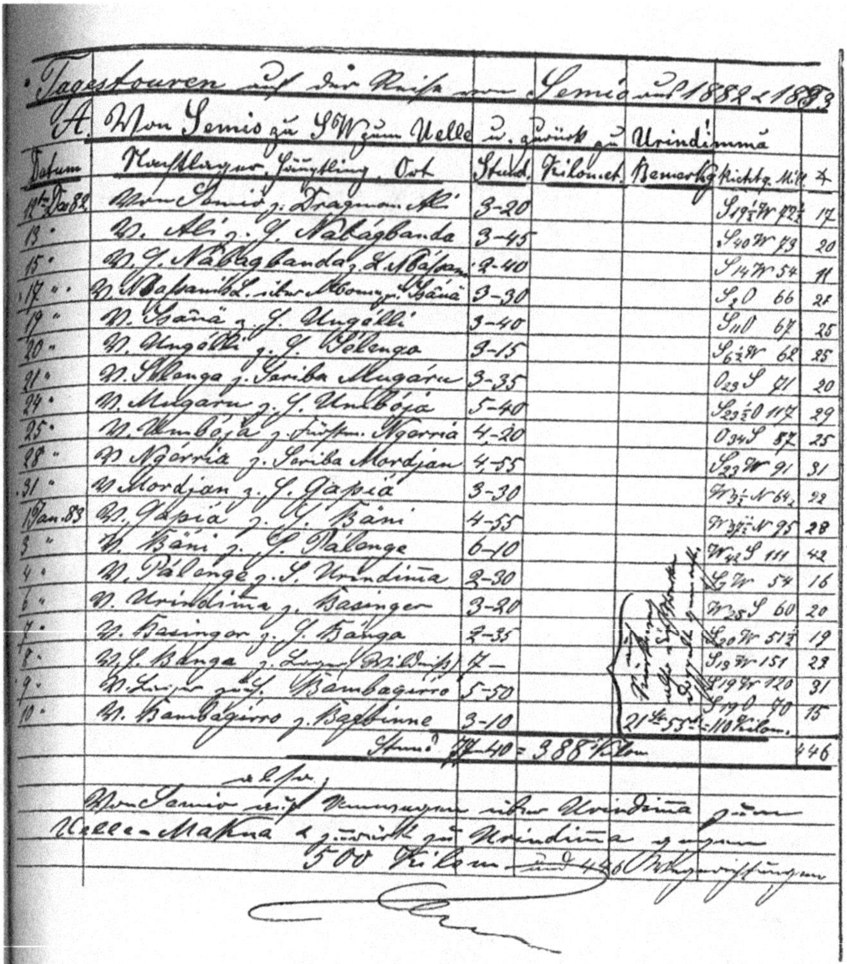

Tagestouren auf der Reise von Semio 1882–1883

A. Von Semio zu SW zum Uelle u. zurück zu Urindimma

Datum	Nachtlager, Tagesziel, Ort	Stund.	Kilomet.	Bemerkg	Richtg. Mill.	
11. Dez. 82	Von Semio z. Dragoman Ali	3-20			S 19½ W 72½	17
13.	V. Ali z. S. Nabagbanda	3-45			S 40 W 73	20
15.	V. S. Nabagbanda z. L. Nabassa	2-40			S 14 W 54	11
17.	V. Nabassa über Mbomu z. Sana	3-30			S 2 O 66	21
19.	V. Sana z. S. Ungolli	3-40			S 11 O 67	25
20.	V. Ungolli z. S. Belengo	3-15			S 6½ W 62	25
21.	V. Belenga z. Seriba Mugaru	3-35			O 23 S 71	20
24.	V. Mugaru z. S. Umboja	5-40			S 23½ O 117	29
25.	V. Umboja z. Fürsten Ngerria	4-20			O 34 S 87	25
28.	V. Ngerria z. Seriba Mordjan	4-55			S 23 W 91	31
31.	V. Mordjan z. S. Gapia	3-30			W 3½ N 64½	22
1. Jan. 83	V. Gapia z. S. Bani	4-55			W 3½ N 95	28
3.	V. Bani z. S. Palenge	6-10			W 42 S 111	42
4.	V. Palenge z. S. Urindima	2-30			S W 54	16
6.	V. Urindima z. Basinger	3-20			W 28 S 60	20
7.	V. Basinger z. S. Banga	2-35			S 20 W 51½	19
8.	V. S. Banga z. Lager (Bili-Durchfl.)	7–			S 19 W 151	23
9.	V. Lager z. S. Bambagirro	5-50			S 19 W 120	31
10.	V. Bambagirro z. Bassinne	3-10		21 St. 55 = 110 Kilom.	S 19 O 70	15
	Stund. 77-40 = 388 Kilom					446

Von Semio auf Umwegen über Urindima zum Uelle-Makua & zurück zu Urindima gegen 500 Kilom. und 446 Luftlinie

FAC-SIMILE II.

Figure 5. Junker's map rendered as itinerary. From Junker, *Travels in Africa*

hide this fact, and the explorer's authority depended on claims that he had discovered everything himself. Maps drawn by Europeans, and the itineraries they took on their expeditions, reflect precolonial African geographies, not only European knowledge.[95] Prominent in many European travel accounts was a border between "civilization" and "wilderness," mimicking the geographies of the coastal merchants who guided their journeys west. A German missionary stood on the Ndunguni hills that separated Wanikaland from the "wilderness" and looked back, thinking of the work, emergencies, and dangers that his expedition had gone through to spread the gospel there. Henry Morton Stanley described saying goodbye to "civilized life" when leaving Bagamoyo. An Austrian traveller wrote that the inhabitants of Taveta were the "last link with civilization" for travellers heading west from the coast. Joseph Thomson told readers that once past the coastal plantations, "with surprising abruptness," the traveller reached "a scene of desolation and sterility."[96] After crossing 800 miles of "dreary desert" and some steppe landscapes, Georg Schweinfurth entered the "true primeval forest."[97] Crossing into wilderness marked African space as exotic and dangerous.

To make the African wilderness recognizable to European readers, explorers placed its history within narratives with which those readers were familiar, which resulted in a picture of East Africa as merely a landscape for European adventure. Explorers frequently denied the people they encountered in the East African interior a history outside of natural processes and fell back on European history or literature to explain Africa. Schweinfurth described moving back in time to the period of the *Nibelungenlied* or the pharaohs as he moved south from Egypt, crossing cataracts that served as the "gates of the civilized world."[98] One German explorer described the Sultanate of Witu's history as one that put the Ghibelline-Guelph feud of medieval Europe to shame. A missionary from the British Church Missionary Society (CMS) described the Galla as the "Ishmaelites" of Africa, barbarous and ferocious perpetrators of deeds too horrible to describe. Two French missionaries compared one Big Man to the medieval French kings Charlemagne, Louis the Fair, and Henri IV. A British explorer thought the Great Lakes region resembled what Britain had looked like centuries before Christianity.[99] Another German cited Herodotus and the Egyptian pharaohs as predecessors in the project to "open" Africa.[100] One part of Schweinfurth's journey had "the character of an Odyssey," and he quoted the *Iliad*, Herodotus, and Aristotle in his descriptions of pygmies.[101] In addition to making spaces familiar, the use of historical comparisons placed African societies on a timeline of human development on which European societies were far more advanced. They adopted coastal society's division of the coast as a place of *ustaarabu* and the interior as a place of *ushenzi* into the European representational space of East Africa.

Though German missionaries in East Africa shared some of the attributes of scientific travellers, they focused on *Leute* instead of *Land*. Catholic missions promised to drag the "wild" people of Africa into civilization.[102] Missionaries adopted elements of local geographies of the interior in their writing, mimicking the focus on power through control of people. Although missionary societies were based on the coast, it was seen as merely a stepping-off point to reach the pagans of the interior, where missionaries believed Islam did not exist. By the 1870s, missionaries saw their work as a race to establish Christianity in the interior before Islam could spread farther.[103] Although missionaries did not explicitly work in geographical terms, mission societies divided work by language, thereby drawing borders around ethnic classifications. Adoption of one language as the language of instruction codified it as a "tribal" language, a marker of nationality.[104] The mission vision meant changing the people to change the land. Carl Büttner compared the sending of missionaries to Frederick the Great's decision to send schoolmasters to Poland with his armies. Remaking the land would be useless without remaking the people who lived there into farmers with interest in European culture.[105] This was a representational space of Africa as wilderness, but one in which its inhabitants mattered more than they did to the typical explorer.

Scientific explorers and missionaries were united in the view that Arabs and Islam were negative influences on the region. Arabs had made war cruel and barbarous in Central Africa. Worst of all was the slave trade. Human beings were "hunted as legitimate game" by Arab slave raiders. On the other hand, the European had to work with Arabs and other "Mohammedans" to accomplish his goals.[106] Following the largely successful and completed campaign to eliminate the slave trade on the Atlantic Ocean, European attention turned to the slave trade on the Indian Ocean as a humanitarian cause. European governments, explorers, merchants, and missionaries combined to paint a picture of the slave trade in East Africa as an imposition from Arabs on the coast and as a completely foreign element to the "natural" African societies they imagined in the interior. The cause became one of the main issues around which Europeans engaged with Africa in the mid- to late nineteenth century. Catholic missionaries used the spectre of an Arab slave trade to argue for the creation of missions in East Africa. The United Kingdom justified a series of interventions in Zanzibari affairs, beginning in 1822 and culminating in the abolition of the slave trade on the islands in 1873, in order to strengthen its strategic position on the basis of an Arab slave trade in East Africa it imagined as far beyond its actual dimensions.[107]

For empire in the late nineteenth century, this historical narrative of an Arab slave trade was important propaganda wrapped around a kernel of truth. It elided European and American participation in the Indian Ocean slave trade and cast the introduction of global trade to the region as a justification for European intervention.[108] European demands for "legitimate commerce" in place of commerce in people produced an increase in the use of slavery for production.[109] Zanzibar's booming clove sector produced an enormous demand for slave labour, both to pick cloves and to work plantations on the mainland to feed the larger population on the islands of Zanzibar.[110] The "enslavement frontier" pushed inland, away from Europeans, to areas not before subject to slave raiding.[111] But this trade had been in recession for over a decade before German colonization began, as the clove sector declined and British pressure forced the Sultan of Zanzibar to ban the slave trade in 1873.[112] Describing the "Arab slave trade" as such was a means to justify political interventions in the region that would increase European power over that of people from across the Indian Ocean. The idea that the African interior had been turned to desert pervaded accounts of the slave trade and the role of Islam among Europeans in this period. Articles associated Islam with barren landscapes, a projection of ideas about connections between Middle Eastern culture and Middle Eastern deserts being related through the actions of the people of the region to a faraway place.[113] Such descriptions were powerful representations of space that justified European intervention.

By the mid-1880s, Germans had developed a framework for understanding East Africa and its people based in cultural geography. Though there were differences between explorers and missionaries, they agreed that there was a close relationship between East African societies and landscapes. Their conceptions of East African space drew on cultural geographic conceptions of connections between societies and landscapes, work which had inspired the German role in the exploration of Africa from the middle of the century. The information that formed the basis of their concepts of East African space, however, came through African intermediaries who determined where Europeans went and what they saw. Missionaries and explorers were agreed that the primary impediment to East African *Kultur* was an Arab slave trade, which relegated both people and landscapes in the East African wilderness to a status reminiscent of Europe's distant past. The depictions of East Africa as wild, as behind Europe in time, and as besieged by Arab slavery would provide the impetus for colonization efforts.

Conclusion

Over the course of the nineteenth century, new geographies of East Africa emerged as the region integrated into the world economy. New practices of space led to new representations of space, which in turn formed the basis for changes in the representational space of the region. The growing international trade in commodities from the interior made trade networks increasingly important and control of trade the basis of political power. The interior developed a geography based around networks, conceptualizing space in terms of connections among people who lived in it. Territoriality became important as a means of controlling resources and parts of the trade routes. On the Swahili coast, space also worked in terms of networks, but networks that extended both into the interior and across the Indian Ocean. Those same trade routes became the means for European explorers and missionaries to access the interior. They developed their spatialities through participation in the long-distance trade system. Their new ideas about space, however, were based in cultural geographical conceptions that inhabitants shaped the territories in which they lived. Europeans developed ideas of East Africa as lagging far behind European modernity, its progress hindered by an Arab slave trade. Hidden within European spatialities, however, was a dependence on local forms of knowledge.

The three spatialities discussed in this chapter shaped one another before formal colonialism began in the mid-1880s. All had developed as part of the integration of the East African interior into the globalizing economy of the nineteenth century. In each, people attempted to make sense of the changing meaning of their new spatial practices of travel through the region and participation in long-distance trade. Cartography was less important in doing so than other forms of representing space, such as itineraries and the marking of networks through narration or exchange. European ideas about East African space were dependent on what they were told and directed to see by East Africans, meaning that African ideas about space became part of European ones. The arrival of a new actor in the region in 1884, the Society for German Colonization, upset existing frameworks for understanding space and attempted to fit East Africa into a different view of the world.

3 The Introduction of a Land-Based Approach to Colonization, 1884–1885

On 27 February 1885, the German government issued the German East Africa Company *Deutsch-Ostafrikanische Gesellschaft*, DOAG) a letter of protection (*Schutzbrief*) for territories the company claimed it had acquired from East African leaders. The *Schutzbrief*, signed by Chancellor Otto von Bismarck and Kaiser Wilhelm I, was the culmination of the DOAG's campaign to found a German overseas colony.[1] The company, under its previous iteration – the Society for German Colonization (*Gesellschaft für deutsche Kolonisation*, GfdK) – had undertaken an expedition to the Usagara region to acquire territory. The members of that expedition, Carl Peters, Joachim Graf von Pfeil, and Karl Jühlke, signed a series of treaties with local men and women they labelled "sultans," which ceded territorial rights to the German company. Peters had returned to Germany claiming sovereignty over 2,500 square miles of East African land.

Through the GfdK's travelogues, treaties, and campaign for government recognition, its varied representations of East African space – a conception that taking control of territory was sufficient for economic development – became the basis of German claims to East Africa. In this conception of colonialism, East African *Land* and *Leute* both would follow a natural course of evolution to become more like agrarian Germany. The *Schutzbrief* ratified the GfdK's "policy of rashness" (*Überstürzung*), in which it took direct action to create a German empire through private initiative. It provided government imprimatur to a radical political project that created a German colony through legal maneuvering and dubious treaties. The company threw itself into the existing European race for colonies and, with the *Schutzbrief* in hand, began an even more aggressive attempt to acquire territory, sending expedition after expedition to sign treaties further afield in East Africa to take hold of more land. Peters and Pfeil believed that the mere acquisition of land for Germany would lead to a powerful overseas empire.

Theories abound as to why Bismarck was willing to give in to the aggressive colonialist movement he had resisted through the 1870s and early 1880s, especially to an organization with so aggressive a program. Bismarck not only issued the *Schutzbrief*, but also used naval power to enforce it when the Sultan of Zanzibar, Said Barghash, protested from May through August 1885.[2] In the case of Ostafrika, it is clear that Bismarck did not direct events, but was rather pushed into action by public support for the creation of the colony. Peters and Pfeil built support in metropolitan Germany, despite misgivings about their enterprise, through a campaign to win public opinion. Central to that campaign was their organization's project to co-opt local geographies of the region into a historical narrative of the European past. Peters's and Pfeil's writings portrayed an East Africa redolent of nineteenth-century depictions of medieval Europe. The region's politics, in GfdK writings, looked like the Holy Roman Empire. More importantly, East African spaces appeared on a stage of evolutionary development leading to the agrarian economy of which many conservative Germans dreamed. East Africa, in these representations of space, was behind in time and available for European colonization.

In its colonial vision, the GfdK would build its empire on the pretense of a project to reimagine East Africa's and the world's historical and contemporary geographies. The first element of the GfdK's geographical project was repurposing explorers' travelogues into a new form that narrated an explicit physical conquest of land.[3] Peters and Pfeil knew about East Africa almost entirely through explorers' travelogues and adopted elements from them into their vision for the future of the region. They copied the tropes of precolonial travel writing, but their borrowing was a haphazard mix of ideas about how colonialism should work. At the heart of that mix was a focus on control of territory, that what mattered for colonization was control of *Land*, and that subsequent plans for making development were not nearly so important. They thus rejected the focus on remaking East African societies promoted by missionary organizations and many of the European explorers who had preceded them in East Africa. Though some Germans questioned the GfdK/DOAG's narrow focus on control of territory as the basis of colonial sovereignty, they did not challenge the representational space of Africa as a site for European imperial competition or the idea that East Africa was behind in time.

The second means by which Peters and Pfeil instituted their spatial vision was the treaties that they had signed with East Africans during their 1884 expedition.[4] It was obvious even to many contemporary German readers that the treaties were fabrications. The form of those fabrications, however, is important for understanding how the GfdK imagined East Africa and colonial rule. The treaties established a sovereignty based in land, construing East Africa's politics as similar to Peters's vision of medieval Europe. In the GfdK's imaginary, a bordering of territory was what made sovereignty. The GfdK was not the only private organization creating such treaties for colonialism in the period, and it fits the model of what Steven Press describes as "rogue empires" – private organizations disrupting the world order with the creation of private empires through fictitious treaties. Press, however, argues that these treaties depicted the territories in question as lacking any sovereignty.[5] The GfdK's treaties, I argue, depicted East Africa not as lacking in sovereignty, but as subject to medieval forms of sovereignty needing to be superseded by modern, European rule.

The treaties' African co-signers may not have understood exactly what the company was after, but likely saw Germany as a counterweight to Zanzibar in local politics and a source of the prestige goods so important to Big Man politics.[6] The treaties did not challenge Big Men's prerogatives, control of people or trade along the networks of precolonial geographies, meaning that the GfdK's geographies merely overlaid existing geographies of the East African interior; they did not replace precolonial practices of space. On the other hand, the organization's geographies did create conflict with Zanzibari and representational space in the interior. Peters and Pfeil attempted to exclude Zanzibari power from the territorial claims that they made through their treaties with statements that their African signers knew nothing of Zanzibar or Sultan Barghash. The result was a scramble for territory between the private German company and the Zanzibari sultan.

The GfdK Repurposes Travel Writing

When founding their new organization, the GfdK's leadership brought a mélange of ideas about African space, history, and politics together into one spatial imaginary to promote German colonialism. The GfdK's plans for colonization were wildly inconsistent and often incoherent, but through them all ran two central beliefs: great nations had colonial empires, and the most important step in creating those empires was claiming territory. The two primary figures in the organization at its founding, Carl Peters and Joachim von Pfeil, developed a new representational space that supported their colonial dreams. They borrowed heavily from explorers' travelogues and British colonialism in formulating their ideas about how German colonization should proceed. Their borrowing was haphazard, resulting in a helter-skelter approach to colonization and a lack of clear long-term plans. Peters and Pfeil depended on explorers' travelogues for what they knew about East Africa, but they stripped out the travelogues' ethnography, focusing only on descriptions of land and how that land could be useful for their vision of a German empire. Their vision would entail the centrality of territorial development, with an implicit assumption that East Africa's people would follow along. Through their descriptions of their expedition before and after the fact, the GfdK established the representational space for a future German East Africa.

The GfdK's founders declared that the organization's mission was to "secure an overseas *Heimat* for our nationality."[7] The usage of *Heimat*, a term that denotes both a place and cultural meanings that place has for its inhabitants, makes clear the organization's intention to seize land that would over time become an extension of metropolitan Germany. The GfdK formed at a Berlin meeting of colonial enthusiasts who were frustrated with existing German approaches to colonialism on 19 August 1884. The new organization explicitly rejected the approach and the liberalism of the *Deutsche Kolonialverein*, established two years earlier.[8] As Pfeil told it, available land was running out, and the *Kolonialverein* was doing no practical work. Three conditions mattered for choosing a place to make German. It could not be in another European nation's sphere of interest; it could not be in another nation's "language area"; and it must offer conditions for agriculture.[9] The emphasis on agriculture evokes a vision of settler colonialism, that German farmers could make overseas territory be more German. East Africa was an afterthought in the GfdK's early days; the organization's focus was on finding land that it could make German.

The expedition's members borrowed tropes from precolonial travel writing to justify their claims about East Africa. Peters emphasized his childhood in writing about his development as a colonialist figure. He grew up on the Elbe, which, he told readers of his memoirs, formed the historical border between Germans and Slavs. Peters's father was a friend of Carl Claus von der Decken, a German explorer of East Africa. The older Peters had told young Carl Peters about David Livingstone's travels and, pointing to East Africa on a map, told him "here lies the future of Africa."[10] He made it his mission to acquire a colonial empire for Germany in order to create opportunities for the post-unification generation of young *Bildungsbürger*, the educated middle class, like those made available to young British middle-class men because of the growth of the British Empire. Peters borrowed heavily from the British

Empire, but his conception of Empire was one of adventurers, like Stanley and Francis Drake, and not administrators.

The second key figure in the GfdK's founding, Joachim von Pfeil, also developed his ideas about African space from childhood readings of travelogues and later analysis of the British Empire. Pfeil, born in Prussian Silesia, was the only member of the initial GfdK expedition who had previously spent time in Africa. He had been inspired to go to Africa by an indirect connection to the explorer Gerhard Rohlfs, a relative of one of his *Gymnasium* (secondary school) teachers. Pfeil moved to British South Africa in 1873 and spent nine years there. He learned much from British and Boer colonists in a period he later described as his "colonial political schooling."[11] He returned to Germany in the early 1880s and joined the nascent colonial movement.

The GfdK settled on East Africa as the site of its ventures based on reading explorers' travelogues. Pfeil provided the impetus for the organization to look to East Africa. While laid up with an illness in Lorenzo Marques (now Maputo, the capital of Mozambique) before his return to Germany from his time living in South Africa, Pfeil had dreamt up a plan to acquire land in East Africa based on his reading of the travelogues of Henry Morton Stanley.[12] The GfdK approved the plan by unanimous vote on 16 September 1884, and its expedition to take land in East Africa departed on 1 October.[13] Accompanying Peters and Pfeil were Karl Jühlke, an old school friend of Peters, and August Otto, a young businessman travelling at his own expense to explore trade opportunities in the area. The qualifications of the expeditions' members did not fit those expected of European expeditions to Africa. None of the four had a scientific background. They were emblematic of the generation of *Bildungsbürger* who came of age following unification and sought outlets for their ambitions that did not appear available in Germany itself.[14]

A representational space based in a superficial reading of travelogues was apparent in everything about the expedition, from its initial planning through to Peters's and Pfeil's accounts of it afterwards. The GfdK's members had read travelogues avidly, and they had formed a vision of extra-European spaces as empty and in need of European settlement. Their actions on the expedition indicate conscious efforts to mimic the explorers about whom they had read. The expedition's accounts of its actions needed only those elements of precolonial travel literature that would support taking territory. Precolonial explorers had given short shrift to Zanzibar and the Swahili coast in their travelogues, so the GfdK paid little attention to the necessary work there and focused on the interior. GfdK members ignored the traditional moral and symbolic economies of exploration minimized in travelogues to do things their own way. When Justus Strandes, an official of the Hamburg trading house Hansing & Co., told Peters that it would take three to four months to prepare an expedition, Peters stormed ahead with the few porters he had managed to recruit.[15]

Peters's representational space of the East African interior is apparent in the tone of his post-expedition travelogue. Peters followed the model of earlier travelogues and marked a border between wilderness and civilization, but that border now marked areas available for German conquest. The tone of Peters's account changed when the expedition left Zanzibar for the mainland, from dry narrative to fantastical descriptions of the wilderness. The natural landscape became central to his retelling. The seas were blue, the foliage green and fresh. Peters declared that he would never forget the peculiar beauty of the first few days of the march. Before the expedition lay salvation; behind it lay fiasco and ruin. He felt as though he "were on another planet where life glowingly pulses through nature" and was overcome by yearning and melancholy. Like other travellers, Peters worried about the wilderness. He got lost hunting and feared being stuck alone for the night; at one point thought he would die and told Jühlke to rush on to the coast get the treaties to Germany. On his return to the coast, Peters again marked a border. He and Jühlke saw the cross of the French mission at Bagamoyo and took it as a sign that they were nearing the European "cultural world."[16] The bordering of wilderness and civilization was essential to claiming the treaty territory as available, as empty space for colonization.

In Peters's and Pfeil's descriptions of East African spaces and peoples, it is clear they believed that all that mattered for colonizing Africa was control of territory. Peters attached little to no importance to East African people, who were largely absent except when necessary to make a political point. Michael Pesek has pointed out that the Africans in Peters's account appear only in the context of the expedition, not as members of complex societies. Peters told readers that the Swahili were "invariably unwarlike and cowardly," like "always loyal dogs" who would follow one anywhere once one had earned their trust. They were "to a high degree dependent, where they believe they see their natural rulers" and "even in the Blacks rises the awareness that it would be better for them if whites lived among them as rulers of the land." If Germans could take control of East African space and make it more like Germany, the people of East Africa would necessarily change for the better as well. Most important for making a new Germany abroad was not German settlement, but taking territorial control of overseas areas and remaking them to be more like Germany, even if the inhabitants were not Germans. The GfdK thereby rejected the *Kolonialverein*'s focus on peacefully settling Germans abroad in favour of conquest.[17]

Peters and Pfeil copied descriptions of the slave trade from precolonial travelogues to argue that Arab slave raids had completely devastated the region's population, furthering their claims of empty and available land. East Africa was now lying like the "fruit trees of Mother Hulda," a Grimm's fairy tale in which an old woman needs help harvesting her abundant food. With the Mother Hulda comparison, Peters furthered claims that East African landscapes resembled fairy tales and made it appear stuck in the past. Peters made the outrageous claim that even Muinin Sagara, the "Sultan of Usagara," told him that the land would be better off ruled by Europeans.[18] They paid little attention to the destruction slave raids caused or their effects on East African communities. What mattered was that they had cleared *Land* of people, making it available for German conquest.

The GfdK's initial work in East Africa also drew inspiration from colonial enthusiasts in the other parts of the world. The model of drawing up territories granting sovereignty to private European actors began in Borneo in the 1840s. Private treaties like the GfdK's provided the framework for many other land grabs by private European organizations in Africa in the 1870s, most prominently Leopold II's Congo Free State.[19] Anglo-American models seem particularly influential in the GfdK's case. Peters used the term "adventurers" (in English) to describe the GfdK's initial fundraising and modelled his appearance after Stanley's. Like the GfdK, nineteenth-century American adventurers, known as "filibusters" from the Dutch *vrijbueter* (freebooter), were inspired by a vast travel literature encouraging expansion. Filibusters invaded several Central American and Caribbean countries, without support from the American government, to try to expand the realm of American republicanism in the 1840s. Filibusters too harboured grand dreams of national empire to open economic opportunities that were disappearing at home.[20] Similarly, Germany's expansion seemed to have stopped after unification in 1871, and those who wanted to create a greater Germany had to look beyond the European continent for territory into which the country could expand.[21] Peters entertained fantasies of going to the United States earlier in his life. He and Jühlke had discussed moving to Chicago to found a newspaper.[22] GfdK officials did not even have a charter from the German government; it would be the sovereign as a private organization with no state support. They thereby rejected liberal ideas of the state in favour of one in which the people living in the state would be subjects, not citizens. The GfdK's leaders could act as little kings in East Africa.

Through their descriptions of their expedition before and after the fact, the GfdK established the representational space for a future German East Africa. Though their borrowing was haphazard and Peters and Pfeil ignored established colonial practice, their colonial dreams were held together by the idea that a German colony could achieve greatness through the mere seizure of territory. The promise of the region lay entirely in taking control of territory, as the two adventurers eliminated ethnography and science from their travelogues. Control of *Land* would depend on German claims to territorial sovereignty. Narratives served to illustrate and explain the primary texts of their travelogues, which were the series of treaties that the expedition had produced. As legal documents, the treaties attempted to fix time to establish East African space as medieval, with the possibility of evolving to be more like modern agrarian Germany.

The Treaties

Between 23 November and 17 December 1884, the GfdK expedition signed ten treaties for East African lands. These treaties functioned as representations of space, as they included descriptions of the territory being ceded and the politics there. Reading them as such allows for a clearer idea of the GfdK's imaginary of African space and its plans for colonialism. In the GfdK's treaties and negotiations with the various "sultans" of East Africa, ideas from German cultural geography and Peters's visions of sovereignty came together with the local representational space to produce a space that advanced both the GfdK's goals and those of local leaders. The treaties drew on Peters's reading of medieval European sovereignty to certify the GfdK's control of territory, while also ensuring the continuation of existing spatial practices.[23] They constructed East Africa as similar to how nineteenth-century Europeans imagined medieval Europe, posing East African spaces as behind Europe in time and development.

The GfdK designed its treaties in Usagara as legal documents that would establish an official transfer of territory and created diplomatic ceremony to make them seem more legitimate to European readers. Peters and Pfeil attempted to create legal ceremonies that corresponded to prevailing European visions of African politics. They copied the ceremonies they had read about in travelogues, which they took to be accurate representations of African politics. According to Peters, the treaty ceremonies, which he called "diplomatic negotiations," followed a consistent form. The Germans and whomever they were signing treaties with exchanged gifts and performed ceremonies. Little negotiation actually took place. The treaties clearly anticipated European legal challenges to the GfdK's imagined East Africa, challenges to both their legality and their depictions of East African politics and space. Many treaties include a note that they were signed "according to the customary legal forms" in the region in which it was signed. At the end of each treaty was a list of witnesses, including both Europeans and Africans, meant to show that the treaty was legitimate according to the standards of both continents. The legal ceremonies were an element in the GfdK's attempts to shape representational space. Contemporary critics claimed the treaties had only been possible because Peters deceived the African signers with alcohol or tricks. The treaties granted a version of sovereignty to Africans for an instant, only to seize it and make it forever German. Each treaty labelled its African signer as a "sultan," a title that most would not have used, but which did connote supreme authority.[24] The treaties laid out the specific areas of state suzerainty (*Staatsoberhoheit*) that the GfdK members thought important. First was control of land, forests, and rivers. In exchange, the GfdK copied elements from previous treaties: protection of the sultans from any attacks, respect for their personal property, and payment of yearly presents.[25] The payment of yearly tribute was a normal aspect of local politics, as a European investigation of Zanzibari rule on the mainland in 1886 would show.[26]

The GfdK co-opted the sultans' resistance to Zanzibari rule to foreswear Zanzibari power on the mainland. To support its claims that Zanzibari Sultan Said Barghash did not have control over the territory it claimed, the expedition sought out corroborators. Salim bin Hamid, whom the GfdK named Barghash's "agent" in Nguru, declared that Barghash had no claim to the East African mainland, especially in Nguru and Usagara, and did not have sovereignty or protective rights.[27] Both the GfdK and the African sultans had an interest in blocking Zanzibari power and declaring the East African interior as land behind in time and without a territorial sovereign, as it allowed both parties to claim control of the tools of power in the region – land in the case of the GfdK and trade in the case of the sultans.

The sultans appear as sovereigns of small territories who are joining a larger political whole in exchange for promises of protection, a political constellation that Peters based on his reading of medieval European politics. During his brief academic career, Peters's reading of the 1177 Treaty of Venice challenged the orthodoxy that it had drastically reshaped relations between emperor and pope. Peters argued that Friedrich, rather than ceding political power to the papacy, ensured his temporal dominance over Germany.[28] He imagined sovereignty in East Africa as working along similar lines, placing the region's politics along a timeline of development with contemporary Europe as its endpoint. East African sovereignty, like its landscapes, was placed on a continuum of progress following an imagined European history.[29] In the GfdK's treaties, East African rulers would pledge loyalty and cede temporal authority to the GfdK, while maintaining their power over property and internal politics. The GfdK acquired economic privileges in exchange for promises of protection, with no promises of assistance in governance. The GfdK would act as the medieval emperor, the sovereign authority with the power of final decision. This element of the treaties made Usagara appear as a recognizable political constellation to German readers, one in need of rationalization and a strong central government.

Defining the power of the sultans in such a manner had the dual effect of not only creating terms for a German takeover, but of establishing a certain kind of law (which had not existed in the areas concerned), in which political and legal power was defined in terms of absolute control of territory. Fungo, a sultan in southern Usambara, agreed "for eternity," as the "absolute lord and master" to cede "all rights that following European terms are included in the sovereign rights of a prince." The "entire populace" of Mkondogwa saw the surrounding area ceded to the GfdK "for all time."[30] The focus on permanence and exclusivity was essential to the GfdK's claims to have assumed full territorial control over the areas in question. Those provisions were meaningless in the political context of the time and place in which the treaties' African signers agreed to them. Although contemporary European critics of the treaties claimed that Peters had tricked or forced the sultans into signing, their agreement makes sense given the circumstances. They reflect a two-way cultural illegibility where neither side could "hear one another" as both sides tried to find a "middle ground" on which to negotiate.[31] First, it was unlikely that the sultans were intimidated by the German expedition. Otto died of disease, and the other three travellers fell so ill that their porters had to carry them in litters.[32] Second, and more importantly, the basic form of the treaties and the ceremony around them were common to agreements in East Africa at the time; many leaders in the region had signed similar treaties with representatives of the Sultan of Zanzibar. To the African co-signers, the Germans appeared as just a new set of actors in an existing system and may have perceived the GfdK as a possible check to Zanzibari power. The treaties offered to help the sultans ensure the continuation of their control of *hongo* and the forms of authority they recognized while granting the GfdK territorial rights.

Notable with regards to the treaties' representations of space was the lack of clear boundaries to the territories that the GfdK acquired.

Though the primary reason for the treaties was to claim physical territory for the company, the treaties did not include cartographic coordinates for the territories they included, or even a description of their borders. The treaties aimed to make East African landscapes legible to German readers, but they ignored the standards of European treaties, which depended on a cartographic grid to mark one territory off from another. In fact, the German government would argue for the use of lines of longitude in defining colonial borders at the Berlin Conference, which would later provide the justification for recognizing the GfdK's treaties. The GfdK, however, could not find clear definitions of the sultans' territories because such definitions did not exist. The choice not to include clear boundaries, however, made the treaties' space conform more clearly to European representational space of Africa, as they made the sultans appear as premodern rulers who could not even clearly border their territory. Modern states had clear boundaries. Premodern ones did not.

Peters's and Pfeil's representations of space became the basis for the creation of a German colony. The GfdK's treaties, the founding documents of Ostafrika, depended on adaptations of precolonial spatialities of the East African interior. Peters and Pfeil produced representations of African space to conform to European expectations rather than attempting to convey the reality of what they encountered. They therefore formulated a method of taking political control in which they constructed the GfdK as a sort of medieval potentate and which left control over the people and economies of East Africa to the sultans of the treaties. The sultans signed the treaties because they did not infringe on their power and they could possibly use the Germans as a counterweight to Zanzibari power. The treaties secured the continuation of precolonial spatial practices in the region.

Fighting over East African Geographies in Metropolitan Germany

Building a German colony in East Africa required more than signing treaties to which most Europeans attached little value; Peters needed to win Chancellor Bismarck's approval for his venture and vision of East Africa's geography. Winning that approval was not easy, as several prominent Germans challenged the GfdK's representations of East Africa. In public campaigns, supporters and opponents of the organization debated the future of German colonialism, as well as the GfdK's East African representational space. At the centre of the debate was disagreement over whether taking territorial control could by itself be the basis of German colonialism. Even critics agreed with the GfdK, however, that African rulers practised a form of sovereignty that had been superseded in Europe's progress to modernity. Through Peters's campaign to get Bismarck's recognition, the GfdK's East African representational space became part of official German colonialism.

The most influential voices were those of previous German explorers in the region, who rejected the GfdK's representational space. Gustav Fischer, who had travelled in East Africa in the 1870s and early 1880s, wrote in a *Kolonialzeitung* article that the territory acquired was economically useless. Fischer believed the GfdK "dealt with fantasy pictures, not with reality" and that the land "in actuality" belonged to Said Barghash. What was worse, any failure of the Gfdk's part would doom future German ventures in East Africa. He proposed a different plan for colonization based on East African people rather than territory. The future was not in the exploitation of natural resources, but in the "methodical upbringing (*Erziehung*) of the Negro to work – herein alone lies the solution to the African problem." The GfdK had shown exactly how not to create a colonial empire. It had chosen a place with "virgin land," but the climate was unhealthy for Europeans and the people there did not want European wares. The best resource Africa had to offer was labour, which required missionaries, officials, merchants, and plantation leaders to become useful, and then only when those people had learned local languages.[33] Opponents challenged the very basis of the GfdK's ideas about possibilities for remaking East African space.

The GfdK expedition's members followed pre-colonial travellers' methods of representing space and published travelogues about their exploration and conquest to convince the German public. Like earlier travellers, Peters and Pfeil also went on lecture tours and presented their travels to metropolitan audiences.[34] They deployed historical narratives that posited colonialism as the highest stage of national development. Peters defended himself from attacks by citing the colonial histories of Australia and the United States to align himself with the proponents of settler colonialism that had been important to the founding of the GfdK. Arguing for his methods as a basis for settler colonialism placed the discussion of geography at the centre of the debate, particularly his belief that the control of territory was what mattered in colonialism. In Peters's telling of those colonial histories, representatives from Britain and the United States, respectively, had signed treaties based on their own laws to acquire territory. They had turned premodern societies and spaces into thriving agricultural spaces through the application of European methods.[35] Peters claimed both the British and Dutch had done the same thing, and that even Dido of Greek legend had done something similar to get land in North Africa millennia before.[36]

Some Germans protested the GfdK treaties on the grounds that their African signers did not understand the terms or that they undermined existing German trade in the region. But most of the protests accepted the GfdK's representational space. East Africa was a place where bands of robbers and cannibals "galumphed out of the earth." This form of protest accepted Peters's claims that East Africa was a space of premodern anarchy. Another newspaper criticized the GfdK for undermining the work that German merchants had been doing in Zanzibar since the 1840s. In West Africa, German merchants had led the push to acquire land; in East Africa, it was not in the nation's commercial interest.[37] Colonialists aligned with the Hamburg trading houses thus argued for a colonial model based in economic rather than territorial control.

In contrast to criticism from the government and much of the German press, some metropolitan circles celebrated the GfdK and supported Peters's colonial vision because of a vision of East Africa as a space of slavery. The famed explorer Georg Schweinfurth made a positive argument for the GfdK that contrasted with his anonymous argument that the GfdK misunderstood the region's politics. With German colonization, slavery would quickly disappear and the company could create new legal structures for Africans.[38] There was hope for a German colony in East Africa as blocking the power of Islam in the region. Schweinfurth connected the work of the GfdK to the Europewide anti-slavery efforts of the mid-1880s. While Schweinfurth argued that the Arabs of Zanzibar were more developed than were Black Africans, their development entailed slavery, an economic system European contemporaries classified as medieval. Islam was thus on a stage of development past Africa, but behind Christian Europe. It was open to question among contemporary Europeans whether the intermediate stage of Islamic civilization was helpful for Africa. This argument accepted Peters's version of East African rulers as medieval.

In the midst of the public debate, the GfdK reorganized as the German East Africa Company (DOAG: *Deutsch-Ostafrikanische Gesellschaft*) on 12 February 1885. The reorganization turned the society into something like the old trading companies that had established colonies in the early modern period, yet another mode of empire that diverged from its origins in the settler colonialist movement. The change made the organization more acceptable to the German state. Peters retained full control and pursued what he called the "policy of rashness" (*Überstürzung*), acting to accomplish his goals while everyone else was busy talking.[39]

Despite the public questions about the GfdK methods, Kaiser Wilhelm I and Bismarck ratified the DOAG's claims to sovereignty in East Africa through the *Schutzbrief* on 27 February 1885, the day after the Berlin Conference ended. The *Schutzbrief* meant official ratification of Peters's and Pfeil's representational space of Usagara as behind in time and in need of modern governance. The *Schutzbrief* cited the treaties from the GfdK expedition as the basis for protection. Germany agreed to take suzerainty over the territories in which the GfdK had acquired the "rights of sovereignty" through treaty. It also agreed to give the same protection to any future territories the DOAG acquired, leaving the DOAG as the sovereign power but with German protection. In issuing the *Schutzbrief*, Bismarck and the Kaiser recognized the organization's treaties as legitimate legal documents. Recognition entailed acceptance of the constellation of the region's politics contained therein, Peters's imagined medieval politics in which his organization could take control of territory while leaving many rights to other rulers. The *Schutzbrief*'s wording left questions about the DOAG's sovereignty unresolved. Later German commentators noted that it was short and contained relatively little detail, in contrast to later colonial pronouncements.[40]

Although other members of the colonial movement questioned the GfdK's methods, the German government recognized Peters's and Pfeil's geographies of East Africa and their treaties for territory. The DOAG's representational space of East Africa as medieval and depictions of Zanzibar as a weak colonial power were enshrined as policy for the future of Ostafrika. Though German colonialists disagreed about Zanzibari authority and the DOAG's focus, they did not challenge the idea that the company's treaties were valid in form. With the *Schutzbrief* in hand, Peters and Pfeil could continue to acquire *Land* for their private empire. The *Schutzbrief*'s recognition did not, however, settle territorial control in international politics, and Zanzibar would protest the German government's decision.

The Struggle with Zanzibar

The DOAG's "policy of rashness" brought it into conflict with the Sultan of Zanzibar over the workings of the Zanzibari empire, a conflict that entailed differences over how to understand spaces in the East African interior. The two sides held different visions of the area, which meant conflicting ideas about history and sovereignty there. Over the following months both the DOAG and Said Barghash would attempt to solidify their claims to control the interior. The DOAG would do so through a campaign to sign additional treaties for territory that would directly address the geographies of Zanzibari power on the mainland. Contained within these treaties were implicit histories of African politics. Those histories denied connections between the interior and the coast and claimed that the two were at different stages of historical development.

Despite issuing the *Schutzbrief*, Chancellor Bismarck was clearly not entirely convinced by the DOAG version of East African politics and took measures to balance the colonialist movement with international diplomacy. He moved to settle disputes over the acquisitions and prevent war with the United Kingdom or Zanzibar. He appointed the explorer Gerhard Rohlfs as his General Consul in Zanzibar on 27 January 1885, an appointment that favoured the aggressive arm of the colonial movement over the members of merchant houses who had previously held the position.[41] The appointment of Rohlfs makes clear Bismarck's acquiescence to political pressure from proponents of colonialism.

At the same time, *Schutzbrief* in hand, the DOAG set out to take advantage of the provision that the regime would protect its future acquisitions by claiming as much territory as possible. Future chancellor Leo von Caprivi later called this period the "period of flag raising." The DOAG spent nearly all its money sending out further expeditions to acquire land through treaty. Facing criticism of encroachment on Zanzibari territory, treaties signed on these expeditions had as their primary purpose a denial of historical links between Zanzibar and the interior. The company sent out a total of eighteen expeditions between 1884 and 1886, fourteen of those before political issues with Zanzibar were settled. All except one were designed to acquire territory.[42] There did not appear to be a system to the acquisitions, but recognition of all of them would have entailed an enormous colony, stretching as far north as Somalia and including the Comoros.

The race for territory became a dispute over how to understand Zanzibari power on the mainland. The DOAG drove on so frantically in its efforts in large part because Peters believed it was racing Barghash to acquire territory, posing a challenge to the DOAG's claims about Zanzibari weakness. Peters provided an analysis of Zanzibari claims to the DOAG treaty territory. The Sultan was not the ruler of the territory, he argued; rather, he was also a colonial power attempting to spread his influence there. The GfdK expedition had beaten him to the spot, and now he was trying to use other means to get the territory Peters had acquired fairly.[43] A central element in the DOAG's new treaties held that Barghash had no claim to territory. Each treaty included a specific statement that Zanzibar lacked any authority in the territory covered by the treaty. Sultan Hamolomo of southern Bondei, for example, declared that he was not a subject of Zanzibar, was in no way dependent on Barghash, and that Barghash had no rights to occupy any of his land.[44] Mwango of Taveta claimed that Lloyd Mathews had come through his village a few days before with 180 soldiers and 100 porters and asked to fly the Zanzibari flag. He knew that most of what the Arabs brought to the area really came from white people, so it would be better to have white people there to protect his people.[45] By expressly excluding the Sultan from the territories they were trying to acquire, Peters and Pfeil set a western edge for coastal society. Marking the western border of Zanzibari territory represented the area west of that border as "African," land to be exploited by European capital.

African political leaders signed with the DOAG if doing so would help their interests in creating a counterweight to Zanzibari power and would not affect their spatial practices. In a situation of mutual unintelligibility, the Germans arrived with ideas of territorial sovereignty that were not part of the East African political universe, demanding something intangible in exchange for anti-Zanzibari strength that was desirable to local rulers. Such motivations are apparent in the DOAG's treaty with Mandara, the "Sultan of Moshi." Mandara, according to Jühlke's transcription, claimed that he would be happy if the Germans kicked the Arabs out of the area. He himself was "a free independent prince, just like the Sultan of Zanzibar, and holds perhaps the same power as he." Mathews had shown up just ten to twelve days before, offering money to raise flags and asking not to let the Germans through. Mandara had refused, saying he was a fully independent sultan and had given a gift to Mathews to show his independence. Mandara claimed "all rights" of "state sovereignty." He agreed to "personal friendship," but asked Jühlke to protest the Sultan's claims. The treaty included explicit language that Jühlke would raise such protests if necessary. The treaty also obligated the company to educated Mandara's sons "in the German manner," showing Mandara's larger ambitions.[46] Mandara saw himself as the equal of Barghash, and likely hoped that educating his sons in German institutions would give them standing to challenge Zanzibari authority in European eyes.

The case of Semboja of the Shambaa Kingdom illustrates alternative representational space in the face of German encroachment. Semboja claimed that he was a subject of Barghash when the DOAG tried to get him to sign a treaty. Semboja's alliance with Barghash was crucial to his ability to control the wealth necessary to manage his social and economic networks. An alliance with the DOAG did not offer the same advantages. Semboja, posted by his father to a remote part of his kingdom, built a trading centre at Mazinde, and took advantage of demand for slaves in Zanzibar to build a powerful position through trade in Shambaa subjects. With his profits from the slave trade, Semboja pursued a war against his father's successor, Shekulwavu. Shekulwavu pursued an alliance with one of Barghash's enemies, the former *liwali* (governor) of Mombasa, and Semboja used that alliance as a pretense to ask for Barghash's support in the war. Barghash agreed, and his soldiers solidified Semboja's position as the most powerful figure in the region.[47] Semboja's decision to side with Barghash makes perfect sense within the context of local politics. He was more powerful than the treaty sultans, and his alliance with Barghash offered more benefits than the DOAG could provide. He did not count on the power of the German state arriving in support of the DOAG.

Barghash filed a formal protest against the DOAG's actions with the German government on 27 April 1885, a protest that required the reorientation of the spatial practices of the Zanzibari empire, from the control of trade the Omani sultans had pursued until that point to the control of territory along European lines. He wrote to Bismarck on 21 May that his realm stretched in an unbroken line along the coast from Warsheikh in the north to the Tong Bay in the south.[48] Barghash adopted territorial authority as the basis of his power in the face of the German threat to his informal empire. Barghash also attempted to make the DOAG's further progress impossible by adopting its methods of acquiring territory. He made it difficult for German expeditions to travel. Barghash's "police chief," Kari Haji, paid off the porters hired by one DOAG expedition so that they would not proceed farther. Another expedition had to turn back because too many of its porters deserted for it to progress further. Barghash sent a force under British officer Lloyd Mathews to protest the GfdK's claims and assert Barghash's own. Mathews returned with a treaty signed by twenty-five "sultans" in the Kilimanjaro and Chaga regions, saying they were subjects of Barghash. Mathews wrote that he had placed the entire Kilimanjaro region under Barghash's protection, and that the ideas that the DOAG attested to Mandara were too high-flown for an African chief, anyway.[49] Efforts to impede the DOAG's advance were clearly directed just at the company and not at Germans in general. At the height of the tension, in late June, German merchants were still able to organize trade caravans.[50] Barghash's efforts to block the DOAG with force ultimately failed, meaning the organization could continue to sign treaties and build its case.

Barghash's protests unified the colonialist movement in support of the DOAG in Germany by turning the dispute into a conflict over territory. Although the press had been dubious about the DOAG's claims about sovereignty in the territories concerned, many newspapers rallied around the company when it came to the nationalist issue of preventing foreign incursion. Sections of the German press reacted vociferously to Barghash's and Mathews's actions. The *National Zeitung* called it a "violent attack on foreign rights," likely because of the British Mathews' involvement. Focusing on one treaty, the *Deutsche Tageblatt* called Zanzibari actions an attack on Mandara of Moshi's sovereignty. Mandara's claims were stronger than Barghash's.[51] German public opinion supported the DOAG's representations of East African space and the idea that territorial control was what mattered for empire. Barghash's actions quieted alternative German geographies of East Africa. Peters lobbied the Foreign Office for support in the border dispute. He claimed that he had had Barghash's territorial claims in mind when signing the treaties, and so worked to get free of the "coastal belt" to where he determined Barghash's power ended, two to three days' march beyond Saadani. He asked Bismarck to determine the exact borders of the Zanzibari Sultanate.[52]

The well-known explorer Ernst Vohsen became the most prominent voice in support of the DOAG, combining claims about Zanzibari territoriality with concerns about Islam in a representation of East African space as lawless and stuck in the past. Vohsen described any attempts to stop the DOAG's progress as part of an Arab conspiracy. King Isike of Unyanyembe, who had claimed allegiance to Barghash, was just a "straw puppet in the hands of the Arabs of Tabora," led by Seth bin Juma, Barghash's unofficial governor. Arabs had put Isike on the throne; he owed them everything. Barghash exercised power, according to Vohsen, but through trickery and force rather than legal agreement. Vohsen suggested building stations along the caravan routes to protect German passage.[53] He cast Zanzibari power as illusory and in need of replacement with a German protective force.

Vohsen also tapped into a set of local grievances against Isike and his relationship with the Arabs of Tabora for German aggrandizement. The contrast with Peters's and Pfeil's spatial imaginaries could not be clearer. Vohsen had travelled in the region and had obtained an understanding of local politics that he thought could be instrumentalized in German favour. Unyanyembe had been engaged in a civil war since the 1870s, with Isike and Arab merchants aligned on one side. Some Wanyamwezi had instead aligned themselves with Mirambo or Nyungu, Isike's rivals.[54] Isike's goal was to reshape spatial practices by taking control of trade and ensuring that Tabora was the entrepôt for slaves and ivory shipped to the coast.[55] Using his armies and his alliances,

Isike built a powerful position in Unyanyembe by the middle of the 1880s.[56] Isike's power had created resentment, and the German presence offered a counterweight to it. Vohsen suggested working with Mirambo or Nyungu instead of Isike as a means of establishing more legitimate rule in the region. Whereas Peters and Pfeil had invented a version of East Africa's politics as medieval to fit their vision for German empire, Vohsen suggested manipulating local politics to German advantage.

At the same time, members of the DOAG adjusted their representations of space to win over German audiences. Pfeil proposed that writing about East Africa ought to change to fit the needs of German colonialism, which meant transforming exotic African spaces into familiar German ones. The time for tales of "adventures of travel" was over. The colonial project's success depended on physical labour, the shipment of goods, and cultivation of fields.[57] In Pfeil's immediate account of his time in East Africa, from 1886, he disclaimed the importance of continued travel writing about the region between the coast and the DOAG's first station, at Sima.[58] The narrative structures of the new writing would shape Ostafrika as a space of German economic activity resembling Europe rather than the explorers' travelogue space of adventure and science. The DOAG's supporters frequently employed new representations of space, maps of the territory that the organization claimed. As noted above, the treaties did not contain explicit borders for the land included. Turning the treaties into map form required the invention of borders for Usagara to make the treaties legible for German audiences with little understanding of the politics or landscapes of East Africa. Mapping meant mastery of East African spaces. The task of making recognizable, rational borders out of the DOAG territory, however, is not possible. Usagara appears with rounded borders, unknown anywhere in Europe. The DOAG and its supporters, even with their claims to land, could not articulate those claims in a way that met European expectations for sovereignty.[59] By making the borders appear somewhat arbitrary, the maps supported representations of the region as lacking the markers of modernity.

Bismarck adopted the DOAG's spatial imaginary in his response to Barghash's protests. The Foreign Office instructed Gerhard Rohlfs to declare Barghash's protest ungrounded and to raise a counter protest in the name of the Kaiser against Barghash's after-the-fact occupation of territory in a German protectorate. Barghash's protest was an attack on the rights of German subjects, as the DOAG had acquired those lands through treaties with independent princes. Barghash had recognized their independence when he told earlier travellers that

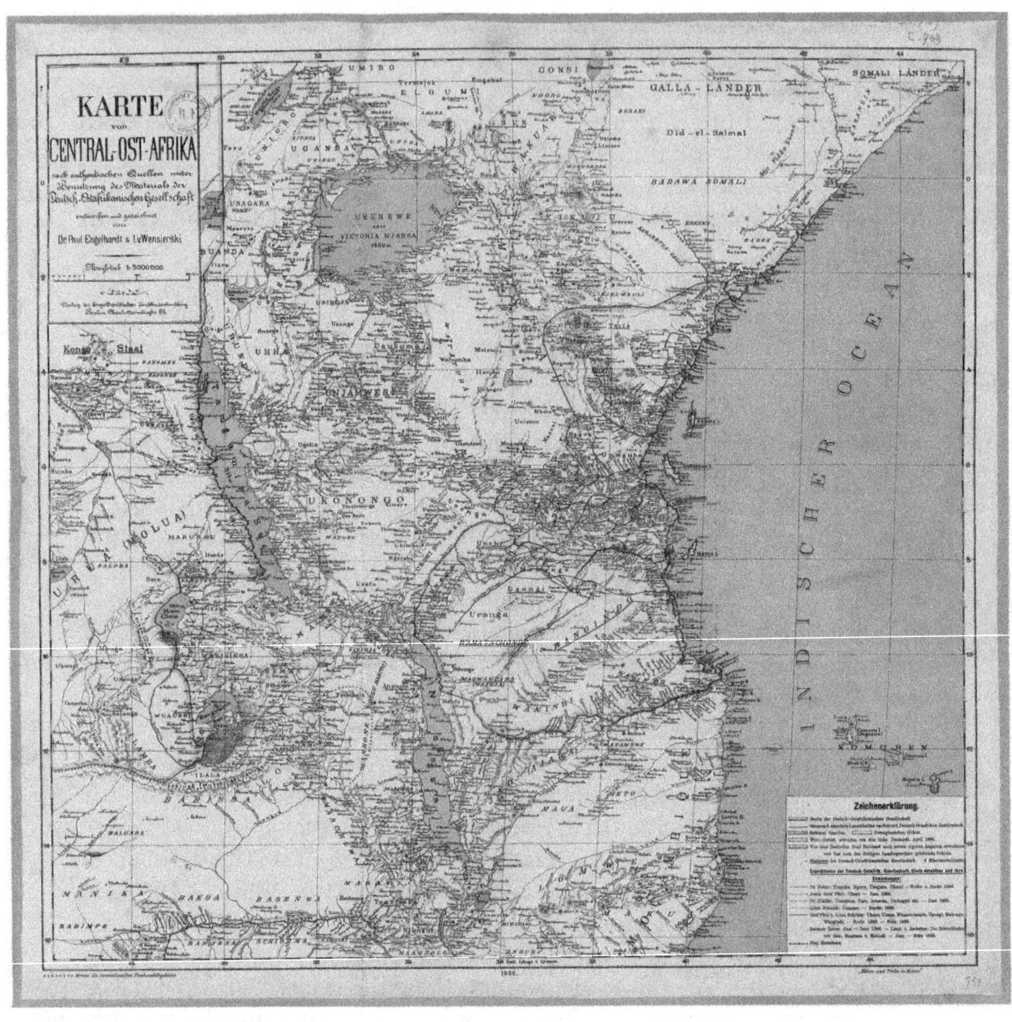

Figure 6. Engelhardt and Wenzierski, "Karte von Central-Ost-Afrika"

his sovereignty did not extend beyond the coast. Rohlfs confronted Barghash with the free trade provisions in the Congo Act as support for the DOAG's claims that it had a right to recruit porters and to travel. Bismarck dispatched four naval ships to Zanzibar on 27 May to back up his claims.[60] Bismarck hoped to use an overseas empire as a bargaining chip to split Britain and France. The agreement ratified the DOAG vision of East Africa's geography and the Zanzibari Empire.

By late May of 1885, other European nations had, for the most part, ceded what was to become Ostafrika to Germany. The period of flag raising had been successful in that it had created a basis for the international community to recognize the DOAG's claims to East African territory. African participation in the DOAG project depended on whether rulers thought they could use the DOAG for their own political advantage. Included in the recognition of territorial claims was an acceptance of Peters's and Pfeil's East African representational space, of political authorities as territorially premodern, as well as the idea that a private company could serve as a sovereign power in Africa. Bismarck's support of the plan ensured that the DOAG's *Land*-based approach to colonization would form the basis of German work in Ostafrika moving forward.

Conclusion

Barghash backed down under threat from the German navy and began pulling his troops out by 24 June. The German ships arrived on 7 August and threatened to shell Zanzibar if he did not agree to recognize the DOAG's treaties.[61] Under pressure from Britain, Barghash recognized the DOAG's claims on 13 August, and agreed to let the society use Dar es Salaam's harbour for its transportation needs. The agreement between Zanzibar and Germany, signed on 20 December 1885, treated Zanzibar as a less-sovereign nation. On one hand, the agreement treated Barghash's commercial empire as another player in the European Scramble for Africa by forcing him to agree to the free-trade provisions that European powers had decided. On the other hand, it subsumed Zanzibari law to German by exempting German subjects from Barghash's authority. It promised free trade for both sides and obligated Barghash to prevent the monopolization of trade. Where the treaty was particularly onerous was on questions about the new DOAG acquisitions and the legal status of Germans. Goods from west of the Sultan's dominions would be exempt from duties. The sultanate's authorities could not intervene in disputes between Germans and "members of other christian [*sic*] nations," a restriction extended to his subjects who were working for German companies or the German state.[62]

The agreement marked the end of the first period of the German colonization of Ostafrika. The DOAG had established German sovereignty and control over thousands of square miles in East Africa through treaties and diplomatic negotiations between the German government and the Sultan of Zanzibar. It had established a basis for the colony in the control of territory. Its haphazard approach to empire had succeeded in creating a colony based on territorial conquest. The treaties that it had signed with the East African sultans had established a legal basis for Bismarck to protect Ostafrika, while not challenging existing spatial practice. Not all colonialist Germans had accepted the geographies contained in the treaties, and the organization had to wage a public campaign to win over the government. Despite that campaign's success in Germany, it brought the DOAG into conflict with Said Barghash over territory. Barghash adopted territorial claims as part of his formerly informal economic empire. Settling the territorial dispute between Zanzibar and the DOAG was the next step in making East African space.

4 Inventing *Hinterland*: Historiography and Cultural Geography in the DOAG's Takeover of the Indian Ocean Coast, 1886–1888

The *Deutsche Kolonialzeitung*, the primary periodical of Germany's colonial movement, published a series of articles on the history of East Africa's coast in early 1888. The history of the town of Kilwa, one article claimed, demonstrated that the entire history of East Africa "in truth constitutes only a battle of two continents – Europe and Asia – over sovereign power in the third continent – Africa."[1] The colonization of Africa thus became the latest stage in a world historical struggle. Asia had won in Egypt, and had established itself on the East African coast, especially at Kilwa.[2] Another article declared the "end goal" of the "Asiatic settlers" had always been the products of the "East African and Central African *hinterlands*," envisioning a history of an Indian Ocean World linked by trade in commodities. The products of the *hinterlands* needed to also be the goal for the Germans, who would take up the mantle from Arabs as colonists in the region.[3] German colonialism was thus posed as the culmination of a long intercontinental struggle over East African space.

The *hinterlands* of East Africa had become the central concern of colonial geographies over the previous three years. The GfdK's "policy of rashness" had been confirmed in early 1885, as the German government recognized Carl Peters's treaties and promised to protect the territory he had acquired. Where exactly that territory lay and how development would proceed, however, were open questions. Peters claimed the goal was to reach the "natural borders" of East Africa, the Great Lakes and the Indian Ocean. Acquiring international recognition for those borders required the deployment of historical narratives and cultural geographies of the region that made German sovereignty appear to be part of the natural progress of history and part of a pan-European struggle against the influence of Islam in Africa. Over the three years from 1885 to 1888, German representations of East African space went through two changes. First, maps replaced travelogues as the primary representations of Ostafrika, highlighting borders and territorial sovereignty. And as the DOAG changed its methods, historical writing took the place of maps, as they enabled claims to territory allotted to Zanzibar. These narratives established the Indian Ocean World as a unitary space for historical analysis. German narratives focused on the blooming *Kultur* of premodern East Africa. Arab colonialism had brought this to an end, and it was only Germany that could restore the region's former glory.

Though the term was not used at the time, many histories of the Berlin Conference in 1884–5 credit the German delegation with promoting settlement based on the "Hinterland Theory," meaning that a European nation that claimed a piece of Africa's coast should reserve the areas inland from that coast along lines of latitude as its sphere of interest.[4] *Hinterland*, as a simple relationship in space, provided a cheap and easy means to claim territory without reference to historical claims to it. Although Germany was the power that promoted the Hinterland Theory, Germans began a process of undermining the agreement almost immediately through the mobilization of representations of East African space that supported German territorial claims to areas not allotted by the conference's *hinterland*. Ostafrika posed a clear contradiction to the physical definition of *hinterland*. The DOAG had agreed that the Indian Ocean coast was part of the realm of the Sultan of Zanzibar, Said Barghash, who had the full support of the British government as a proxy power in the region. According to the Hinterland Theory, Barghash's control of the coast would make the Sultan the rightful ruler of the territory west to which the DOAG had laid claim. Creating a German territory, then, meant defining Zanzibar as a failed colonial power that had not successfully established sovereignty in its *hinterland*. Through an international commission to delimit Zanzibar's borders, the German government successfully created a narrative of failed Arab colonization in which Zanzibari influence had never spread inland from the coast and where the interior remained in the medieval past. In this narrative, a Muslim state such as Zanzibar was unable to modernize space, and thus was exempt from the *hinterland* theory.

From 1886 to 1888, the DOAG tried and failed to create new spatial practices in East Africa for economic development. Its attempts to create new economic networks that would make the colony profitable were unsuccessful, and the company faced near bankruptcy not long after its grand fantasies seemed within reach. To justify this failure and chart a new direction forward, they developed a new representational space of the region in which they redefined *hinterland* to mean economic connections rather than lines of latitude. Histories of Arab colonization in East Africa as part of an Indian Ocean World became increasingly important to the German project. The DOAG and its supporters claimed the region had been linked with the Mediterranean and Middle East since antiquity. They recast Zanzibar, and Arabs more generally, as colonists who had successfully created economic connections with the interior but who had also brought slavery to the region. Those histories served to make new forms of space in East Africa seem part of the natural progress of history. German sovereignty over East Africa was justified as a humanitarian presence, protecting Black Africans from Arab and Islamic slavery.

This avaricious imperial history of *hinterland* has been obscured in the historiography of the Indian Ocean World and other maritime cultures. In today's usage, *hinterland* serves as shorthand to describe the broader geographical area connected to a port city through imports and exports: a *hinterland* is territory that receives imports and sends exports through a particular port or area of coast. *Hinterland* has become convenient shorthand in western European languages for the economic catchment areas tied to a coastal city, as a means for describing areas that are economically connected even if divided by political boundaries. *Hinterland* as an idea thus holds promise, some scholars argue, as a means of decolonizing imperial geographies. Drawing on Fernand Braudel's work on the Mediterranean, synthesizers of Indian Ocean history have embraced the concept of *hinterland* as one with the power to undermine nineteenth-century European geographical categories created through empire with *longue-durée* historical narratives of connections across the ocean. *Hinterlands* cross political boundaries to show the ways in which people, societies, and cultures interacted. The idea of an Indian Ocean World as a cosmopolitan space pre-dating European expansion and ended by colonialism has offered an antidote to the geographical projects of European imperialism.[5]

Though *hinterland* appears to offer a corrective to other European geographical categories and their problems from the colonial era, its history in German East Africa complicates that narrative due to its origins in the German project of claiming East African territory. Today's meaning of *hinterland* was the result of the processes described in this chapter: The DOAG and its allies framed the concept of *hinterland* as economic connections in order to define the Arab and Muslim presence in East Africa as slavery and to take control of territory. They invented the concept in its current iteration to make a new representational space of East Africa as the site of a historical European struggle against Islam.

The DOAG used its new historical narratives and new meaning of *hinterland* to demand and win control of the coast from Zanzibar.

Hinterland at the Berlin Conference

At the Berlin Conference of 1884–5, twelve European nations, the Ottoman Empire, and the United States agreed to a document, typically called the "Congo Act," that set rules for drawing borders in Africa.[6] Contrary to the standard narrative of the conference, it did not set fixed borders for European colonies; it merely laid out rules for determining borders in the future.[7] European theories of sovereignty depended on bordered territory, so the conference set rules for its creation. The resulting decision was that latitude should determine borders in parts of the Africa about which Europeans knew little. European states would control areas inland from their coastal territories. This would be the case in Germany's other colonies, Togo, Southwest Africa, and Cameroon, as well as for most European claims. But in East Africa, international agreement allotted the coast to Zanzibar. The protective letter that Chancellor Bismarck issued the DOAG showed that European nations believed the Hinterland Theory only applied to European states and not to Zanzibar.

At the conference, the German delegation proposed future borders based on the concept of *hinterland*. Without having to establish "effective occupation," as indicated by administrative or military posts, European states could claim large spheres of influence. It ensured their future power to block rival colonial projects without the prospect of war. Such a concept was not new. It undergirded the United States' expansion west from the original thirteen colonies, for example, as states on the Atlantic Ocean claimed territory stretching to their west, seemingly without limit.[8] As a basis for borders, *hinterland* would prevent conflicts among European states attempting to grab more territory. The concept was the ultimate idea of a neat territorial sovereignty, marked cleanly on a map along scientific lines.

It is clear that the 1884 definition of *hinterland* did not contain its present-day meaning of economic connections when we look at the presence of the prefix *hinter-* in other German geographies of the time. *Hinter* means "behind" in German and does not denote connections beyond a simple relationship in space. The prefix was used in geography to denote such spatial relationships. For instance, there is an area in the Rhineland called *Hintertaunus*, so named because it is farther from the cities to its north than the other parts of the Taunus mountain range. *Hinterindien* was the term nineteenth-century Germans used for

Southeast Asia, as it was beyond India. These relationships are all in terms of distance in relation to the *Vor-*, "forward" regions, not connections between the areas to which they refer.[9]

At the same time, the Society for German Colonization had been making its own attempt at colonialism that did not fit with the framework laid out in Berlin. The treaties with which Carl Peters returned to Germany did not claim any territory on the coast. The GfdK/DOAG claims were inland from the Indian Ocean, which Europeans generally allotted to the BuSaidi Sultanate of Zanzibar. Thus, there was no German coast *hinter* to which Germany could lay claim according to the *hinterland* theory. German East Africa appeared to contradict the vision of colonial borders that European powers agreed to at the Berlin Conference. Though the sultanate claimed territory on the African continent, Bismarck did not invite its representatives to attend the conference. What the DOAG and the German government worked with in 1885 was a mix of concepts of colonial sovereignty, one requiring recognition and development, the other merely cartography.

Historical Narratives in the Debate over Zanzibar's *Hinterland*

Sultan Said Barghash would not cede his territorial claims without a fight. He had sent Lloyd Mathews with a detachment of soldiers to sign treaties disproving the DOAG's claims and had formally protested the Company's claims with the German government. The result was conflicting claims to areas near the Indian Ocean and as far west as Mount Kilimanjaro. Over the next several months, the German and British Foreign Offices debated the extent of Barghash's territory on the mainland. At the core of the debate was a disagreement over whether European colonial geographies applied to a non-European state, specifically Zanzibar, a disagreement in which the two sides deployed conflicting representations of the region's spaces to solve. The Zanzibari, British, and German commissioners all utilized historical narratives of the region to argue for their version of East African mainland spaces. The discussions demonstrate a new European interest in the expansion of territorial sovereignty in the 1880s informed by the ideologies of the New Imperialism.

Barghash's authority lacked the clear markers of territorial authority that European observers expected, but he cited other factors as proof of his sovereignty. The Sultan claimed that his rule stretched all the way to the Great Lakes. In a letter to Otto von Bismarck, Barghash wrote that his authority was "firmly established" in Ujiji and Unyanyembe, as well as other places along the caravan routes. His rule throughout the region was "clear and indisputable."[10] Barghash paid the Bombay and Zanzibar firm of Jairam Sewji to maintain a staff of customs officials in the towns of the coast. Some of the firm's officials bore the titles of *akida*, *diwan*, or *liwali*, which denoted the power to enforce Barghash's justice and command soldiers.[11] Barghash had also named a *jumbe* (chief) under his authority on Lake Nyasa and had *liwalis* in Unyanyembe, Ujiji, and Manyema. His flag flew along the northern caravan route. If observers doubted the existence of these officials, Barghash's authority had been proven by the use of his letters by European explorers across all of East Africa.[12] Barghash thus represented most of mainland Tanzania and southern Kenya as part of a cohesive governmental system based in Zanzibar. Barghash described his mainland presence in a way that would most clearly conform to the European expectations of "effective occupation" discussed at the Berlin Conference.

Previous European observers had disagreed over Zanzibar's borders; they had devoted little attention to accurately representing the extent of Zanzibar's authority in the interior. Europeans attributed racial markers to spaces and allotted those spaces to Zanzibar or African rulers based on their perceived racial characteristics. A British missionary had written that though Zanzibar held nominal sovereignty, chiefs in Usagara, Useguha, and Nguru were in reality completely independent and did not recognize Zanzibari claims.[13] The British government had maintained a presence in Zanzibar for decades, treating the sultanate as a proxy power in the region, but it had done little to clarify borders because its focus had been on trade. The 1861 agreement between Zanzibar and Oman that split the two had said only that Zanzibar's mainland possessions stretched from Mogadishu to Cape Delgado, but not how far west. A House of Commons select committee on the slave trade from 1871 had determined Zanzibari authority stretched from the equator to ten degrees south latitude, roughly to Lindi, but again ignored the western border.[14] These interpretations only confused the situation.

In 1885, the British government produced a textual representation of the East African mainland as a space long colonized by Zanzibar. This representation appeared as part of a historical narrative in which Zanzibar was the centre of an Arab ruling race that had come to East Africa and brought a degree of civilization to African societies. John Kirk, the British consul general in Zanzibar, wrote that Zanzibar had "been ruled by successive waves of northern races, of Arab and Persian origin," since the outset of the "Mahommedan era." There was no "trace of a native ruling race," though "barbarian" Africans had blocked the progress of civilization.[15] The British Foreign Office claimed Zanzibar's borders stretched at least sixty miles inland from Cape Delgado in the South to Port Dunford in the north, and that there was evidence of Barghash's sovereignty farther west, among the Chaga near Mount Kilimanjaro.[16] Evidence for this claim came from the statements collected by Lloyd Mathews as part of Barghash's rebuttal to the GfdK treaties, signed by "sultans" in the region who declared they were Zanzibari subjects, a form that looked much like the GfdK treaties to acquire territory.[17]

The British and German governments deployed concepts from European international law to support their positions on Zanzibar's borders. The UK's legal position relied on the natural environment to define borders. Its Foreign Office decided that control of a coast should entail the control of the area through which rivers flowed that emptied on a coast where it exercised sovereignty. These rivers' watersheds formed internal borders. It attributed this theory to the Swiss jurist Johann Kaspar Bluntschli, who had written earlier in the nineteenth century.[18] Like the Hinterland Theory of the Berlin Conference, this theory of sovereignty depended entirely on physical geography with no concern for how Zanzibari authority worked. Germans depicted Zanzibar as a "coastal state," different in character from European states and therefore in need of a different kind of *hinterland* from that used at the Berlin Conference. The German Foreign Office cited the eighteenth-century Dutch theorist Cornelius von Bynkershoek's theory of the "cannon shot rule" to argue that a coastal state, such as Zanzibar, only maintained sovereignty so far as its weapons could reach from the sea. "Effective control" meant the area that could be effectively patrolled by ships.[19] Such precedents supported a limitation of the Zanzibari *hinterland*.

The German Foreign Office threw its weight behind the DOAG based on a representation of the East African mainland as a space where no real colonization took place. The Foreign Office set historical terms for understanding Zanzibari sovereignty in an *aide-mémoire* sent to Earl Granville, the British Secretary of State for Foreign Affairs. The letter claimed that East Africa's history had been a struggle between "Arabian pirates and slave-dealers" and Africans. Zanzibari power lacked "a proper territorial character," and the sultanate only controlled "twenty-five to thirty widely distant" customs houses. Areas between them were "completely independent," full of "numerous barbarous tribes" hostile to and stronger militarily than Zanzibar. West of the coast, no Zanzibari sultan had ever "exercised any of the rights of sovereignty."[20] In the interior lived "heathen peoples," who did not recognize Zanzibari power at all. The UK's support for Zanzibar was spreading "bloody Mahommedan propaganda" in Africa.[21] The DOAG had first articulated this representation through a historical narrative of East Africa as racial struggle in which Zanzibar oppressed Africans and did not exercise territorial control. A DOAG member who had recently travelled to East Africa and signed land treaties for the Company wrote that both Barghash and the UK had only ever bought land, never acquired sovereignty. The Sultan controlled only isolated points on the coast, and Arab power was almost gone now that the slave trade was on the way out.[22] Such claims about Zanzibar made it impossible to treat the sultanate as a fellow member of the community of nations involved in colonizing Africa.

Existing concepts from international law proved unable to resolve the disagreement over how to define the end of the Zanzibari *hinterland* and the start of the African interior. Both British and German governments called upon concepts from international law that had been used to settle disagreements over colonial borders in the past. The presence of Zanzibar, however, made the situation more complicated. This was not simply a disagreement over borders between European states, but over how to define the borders of a non-European state that was not part of the European Concert of Nations. Both European governments had turned to history to represent Zanzibari rule on the African mainland. Their conflicting historical narratives of Zanzibar as either a colonial power or a slave-raiding interloper entailed different East African representational spaces: the British as a long-colonized space, the Germans as a region of savagery untouched by civilization. To resolve the dispute, Germany and the United Kingdom invited France to create a three-power commission that would delimit Zanzibar's *hinterland* on the African mainland and prevent conflict between Germany and Zanzibar.

Delimiting Zanzibar's *Hinterland*

The Zanzibar Delimitation Commission's eventual decision resulted in borders for the Zanzibari empire that conflicted with the *hinterland* of the Berlin Conference. The issue at the heart of its debates was whether the German or British representational space of the East African mainland was more correct. In the discussion, the commissioners and government officials combined concepts from European international law with late nineteenth-century racial thinking to make sense of East African politics and create territorial sovereignty. The commission's final agreement imposed territorial sovereignty following European law but based that sovereignty on definitions of authority provided by local intermediaries. This agreement was based in representations of mainland geographies that reflected coastal communities' divide between coast and interior. The coast was to be treated as semi-sovereign, similarly to Europe's incursions into China, while the interior was to be African wilderness, available for European colonialism.

How each nation imagined the work of the commission differed, but the German perspective was favoured by the commissioners' decisions about how to structure their process. The three commissioners were all diplomats: Otto Schmidt-Leda, Herbert Kitchener, and Achille Raffray (Victor-Gabriel Lemaire would replace Raffray after Schmidt complained he was a "German hater" and irreversibly biased).[23] Schmidt, Kitchener, and Raffray were unanimous in a decision to exclude Barghash from participation in the commission, participation that was "neither in the interest of their work nor the spirit of their instructions."[24] The choice to deny Barghash's request to participate repeated the claims that Zanzibar was not a real nation that had been used at the Berlin Conference. They decided to interview people about Zanzibari authority rather than relying on physical geography. The German Foreign Office ensured that all decisions on borders would require unanimity among the commissioners, so intransigence would be a viable negotiating tactic.[25]

The commission decided to limit its work to the coast because of the travel difficulties and expense of inquiries farther west, a decision that limited the commissioners to the representational spaces of the coastal cities discussed in chapter 2 and which ignored the opinions of the people their decisions would most directly affect.[26] The commissioners began their examination of the coast in the south and worked their way north. They questioned what officials they could find in each town. These were typically Barghash's *liwalis* or *akidas*, the officers he paid to manage justice and customs in each town. The choice of those officials, of course, meant that the voices they heard were people attached to Barghash for their authority. Questioning followed a standard format: title, ethnic origin, role in Islamic legal system, number of soldiers, borders of authority, names of neighbouring tribes, villages, and their leaders.[27] Most of the answers were extremely similar. The officials had been appointed by Barghash or one of his predecessors, had Zanzibari soldiers at their command, and did not provide specific information about western borders.

The instructions that the governments provided for their commissioners allow us to see how their representational spaces of the region differed. In Germany's framing, recognition of Zanzibar's colonization required occupation and taxation. The test would be payment of tributes or customs to Barghash. Herbert von Bismarck instructed Schmidt to test how far dependence on Zanzibar stretched in order to "geographically border who recognizes the Sultan's sovereignty" in the regions "behind the coastal strip."[28] Further instructions to Schmidt declared he was to recognize no Zanzibari sovereignty on the mainland. "Native rulers" in Bagamoyo, Mombasa, and Lamu did not recognize Zanzibar, and there was no central administration for the mainland. In the interior, Barghash exercised no sovereignty whatsoever. The "hinterlands of the coast of the Indian Ocean to the Great Lakes" were inhabited by many different tribes (*Völkerschaften*) ruled by "native princes." The commission should therefore only look for coastal points the Sultan occupied and "how wide the small coastal strip" over which Barghash exercised "nominal sovereignty" stretched.[29]

The British instructions included the elements of occupation and taxation of the German ones but left Kitchener the leeway to look for other "indicia" of Barghash's power, a much broader definition that also abandoned physical geography as the means of making African borders and supported British claims that Zanzibar had colonized the mainland. These discussions predate Frederick Lugard's articulation of "indirect rule" in British Africa, though British instructions bear some markings of what later became official doctrine. The Foreign Office told Kitchener that there was "much uncertainty" about the extent of Barghash's territory and authority over "tribes" in the interior. Kitchener was to ask people along the coast whether they paid tribute, taxes, or customs to Barghash, or practised any other "acts of homage or fealty."[30] The instructions recognized that previous British understandings of Zanzibari power on the mainland may have been mistaken, but they also reflect a British desire not to use the Hinterland Theory for non-European states.

The commissioners selectively read the testimony they collected to frame local representations of space favourably for their positions. Schmidt stressed the fact that several of the interviewees could not name the regions (*pays*) over which they claimed authority, something he took to mean a lack of authority there. He cast the commonalities among testimonies as a reason to doubt their veracity. The officials were all ready for the commission's arrival, likely prompted by communication from Barghash. Fear of Barghash prevented free expression. Barghash's authority over the officials interviewed could not be doubted, but it was unclear whether "the Sultan's sovereignty is recognized by the native population overall on the coast and how far into the interior his sovereignty stretches."[31] What would have been enough for Schmidt can be seen in the testimony of the single official who was able to provide the kind of evidence the commissioners thought sufficient. Pangani's *liwali*, Salem bin Ali, stood out for his ability to identify the people subject to his authority. He declared that all Usambara obeyed him. People came to him for justice, and he sent soldiers if people did not obey. In contrast with most of the other officials interviewed, Salem was able to name three chiefs in the interior: Musamboya, Kisnwéré, and Kitwanga-Chayo, all of whom he claimed had come through Pangani to pay homage to Barghash. Salem also knew where at least one of the chiefs lived; Musamboya was in Mzindi, seven to ten days distant.[32] Salem's testimony stuck out among the vague statements of his contemporaries as the exception that proved the rule for Schmidt.

Otherwise, the testimony of the various *liwalis* and *akidas* reflect a conception of the people of the interior as indistinguishable *washenzi* ("wild people"), contrasted with the urbane people of the coast. This was clearest in the testimony of Tanga's *liwali*, Seliman ben Naceur el Karoni, who claimed that he regulated affairs among the *washenzi* to the west of the town, whom he could call to arms if necessary.[33] Other officials simply provided general statements about their territories' extent to the west. An elder from near Samanga, on the Rufiji Delta, claimed Zanzibari power existed as far away as three to four months' march.[34] Dar es Salaam's *liwali*, Mohammed ben Seliman el Karoumi, claimed he sent soldiers at any report of a crime within five days' march. Chiefs came to him, but upon questioning, he could not name a single one. He did know the names of coastal villages whence people came for justice. These descriptions provided evidence for the German argument that Zanzibari authority existed only along the coast.[35]

In sorting through a political situation they did not understand, the European commissioners settled on the minimalist position with regards to Zanzibar's borders. They decided on a representational space of the interior as uncolonized. The agreement set Ostafrika's initial eastern border and cast the Zanzibari empire as a nation-state with set borders. The commission decided that the Sultan of Zanzibar's "sovereign rights stop 10 English miles from the coast."[36] Both sides agreed that the initial agreement was provisional and that they would work to create permanent borders.[37] The committee's decision reduced complicated political situations to a simple border. The commissioners did not consider economic connections, only political sovereignty.

The commission's decision clarified how European nations would define sovereignty in Africa: by taxation and legal structures, establishing international agreement on an issue that might otherwise have been subject to dispute. The agreement was extremely favourable to the DOAG. It created territorial borders for Zanzibar that would allow the DOAG to expand its colony without concern for Zanzibari claims. The DOAG would be able to place representatives in two ports, Dar es Salaam and Pangani. This followed the model of treaty ports that Germany had established in China.[38] Germany agreed to recognize Barghash's sovereignty over the islands of Unguja, Pemba, Lamu, and Mafia, as well as a coastal strip, ten miles wide.[39] British officials agreed to support Germany in negotiations with Barghash over customs on goods travelling from the German territory to the coast, in recognition that trade was important for the colony's future.[40]

The agreement provided international recognition for the DOAG's "policy of rashness," its model of private empire. The company was now free to continue its acquisition of territory in the East African interior. The explorer Gerhard Rohlfs described what was left to Zanzibar as a "small coast space, the *Vorland* of the German and English *Hinterland*."[41] *Kolonial-Politische Korrespondenz* declared "a vast land stands open to exploitation and cultivation."[42] Many of the more strident German colonialists, however, were displeased with the agreement and would attempt to undermine it over the next few years. The *Kolonialzeitung* lamented the agreement as giving the "lion's share" of the "virgin interior" to the British. Though Germany had taken an East African colony, the DOAG's plans for the future of that colony remained unclear.

Empire as Control of *Land*

In the first two years of Ostafrika's existence, the DOAG treated colonialism as a matter of simple territorial control, meaning it paid little attention to East African societies. After its successes in winning territory, the DOAG struggled to shift to a new model of colonialism and continued to sign treaties for land. The focus of DOAG publications and public speaking tours was almost entirely on territory, and little time was spent attending to the people who lived in the colony. The company articulated a representational space of East Africa as untouched by *Kultur* or the global economy. This representational space was used to support an argument that remaking East African territory would create a productive colony without having to work with the region's societies. Its leaders declared that their new colony contained great riches that only needed small German investments and settlements to tap. One of the main effects of this model of colonialism, however, was to increase the power of local intermediaries, on whom Germans became dependent for their control of existing networks.

The Company's plans continued its earlier methods, ignoring African spatial practices as Germans attempted to create the historical evolution of East African spaces that Peters and Pfeil had claimed were possible. The DOAG's first priority for development would be the establishment of infrastructure, which it set out to accomplish through the foundation of stations, which would serve as economic and political outposts for the DOAG to extend its power west.[43] Stations would serve not just to support trade but to create "real occupation" of territory.[44] In the DOAG's estimation, these outposts would create a spatiality of East Africa as a network of points, each radiating Germanness to the surrounding territory. The DOAG displayed a lack of long-term thinking about the stations. Some were clustered together, making many of them redundant, and paid no attention to surrounding societies. Officials put little thought into how outpost leaders would either create alliances with local leaders or stifle their dissent.[45] Rail would link the network's points in the long run, the DOAG thought, indicating the importance of networks to the company's future colonial vision and a belief that it could excise the caravan system and associated spatial practices from the colony. Railways would "open the *hinterland*."[46] The DOAG planned to build railways to open new districts, farms would spring up along the railways, and "where until then there was wilderness, the life of civilization would evolve," making East Africa more like Germany.[47]

In contrast to German plans, the stations created a permanence for Big Men as elements of the built environment, making spatial practices more difficult to change. For local communities, the German presence became merely another, if violent, factor in patronage politics. DOAG officials became dependents of local Big Men, thereby increasing those Big Men's political influence. Though nominally under DOAG control, politics continued as usual in the colony.[48] Stations were meant to create a more permanent presence for German travellers, but they had the effect of binding their inhabitants to local political conditions because they lacked the power to compel local leaders to do their bidding. Station officials had to work with local leaders to get labour to build stations and agriculture.[49] The station brought trade in goods from the coast, and people from the surrounding region, the basis of Big-Man power. Initial German colonization thus increased the role of Big Men in local practices of space, making their residences even more the centre of spatial networks.

Even on the coast, the DOAG was dependent on local assistance and local politics. Peters rented buildings in Dar es Salaam from the Indian merchant Sewa Haji, with whom he also contracted for porters for expeditions to Arusha and Mpwapwa. When the town's *liwali* later tried to throw the Germans out, Sewa stood by the DOAG and ensured they would not become the "laughingstock" of the town.[50] Sewa thus served as the provider of a safe urban space for the new German colonists, whose position with regards to Zanzibar, the United Kingdom, and the German state itself was tenuous. Sewa was more expensive than other contractors, but he guaranteed that his porters would work and not desert. The company's patronage increased Sewa's profits and his access to trade goods that were necessary to maintain and build his power in Bagamoyo and along his networks.

The DOAG model of empire continued to be based entirely around the control of territory. Building stations and constructing railways, the company's leaders thought, would transform spatial practices in the surrounding areas to produce economic productivity for German benefit. One of their main effects, however, was to increase the power of local African authorities, who became more important for longer-distance trade. The precolonial networks that had linked distant communities remained in place and more important to trade than anything the DOAG had done. The stations were clearly not a success in creating economic development, and money was not forthcoming for larger infrastructure projects. Critics of the DOAG model would become more vocal and demand changes in how Ostafrika operated.

Criticisms of Plans for German Settlement and a Shift to Trade

As discussed in chapter 3, some members of the colonial movement had been dubious of Carl Peters's and Joachim von Pfeil's project to found Ostafrika from its outset. Those critics became increasingly strident as the DOAG project stalled in 1885 and 1886. The focus on territorial control alone proved unpopular, especially with missionary organizations interested in colonialism for Christianization. The most influential of the criticisms was over the DOAG's economic model, which had not produced enough money to even pay for itself. Critics articulated a different representational space of East Africa to demand the company shift its focus to control of the caravan routes that had attracted German interest to the region in the first place instead of building railways for which few Germans wanted to provide capital. African trade networks, a form of spatial practice, and the attendant representational space of the long-distance caravan trade, won out over Peters's and Pfeil's spatial imaginaries. Ascendant members of the colonialist movement attempted to adopt elements of the caravan trade's spaces into a new representational space. Their new spatial imaginary was an East Africa of flourishing trade routes directed by African communities that Germans could utilize rather than the DOAG's representational space of transformative settlement by German farmers.

Evangelical mission leaders were especially critical of the DOAG's geographies and began to challenge the DOAG's approach. Pfeil expressed the DOAG's hopes in a speech at the General German Congress for the Progress of Overseas Interests in 1886. He told his audience that the area's trade had already shown for a millennium that it could produce riches for a European power, as seen by Portuguese conquest.[51] The company was in the position to create an "organic relationship" to native populations in order to get natives to work. People were settling near German stations because they trusted Germans and were willing to subject themselves to German control.[52] The evangelical missionary Carl Büttner stood up and told Pfeil the plan "made his blood run cold" with its total dehumanization of Africans whom missionaries hoped to convert to Christianity.[53] The Leipzig Mission Society was so turned off by the DOAG approach that it built missions in the British sphere of interest to the north rather than in the German colony.[54] Evangelical mission societies feared mission work could be compromised if missions became too closely tied to the German state and wanted to avoid becoming a purely national undertaking.[55] Other German colonialists, more politically powerful than missionary organizations, also criticized the DOAG's insistence that East African territory could easily be made more German.[56]

To avert the issue of missionary criticism, Peters and his allies created their own missionary society that would pursue Christianization along the nationalist lines they envisioned, attempting to instrumentalize missions as tools to make Ostafrika more German. Karl Jühlke published an article promoting the society's work. No longer would German missionaries have to work in foreign territories, but could devote their work to the glory of their fatherland.[57] A missionary's work would "deepen," as he felt called to "build new bridges between the old *Heimat* and the newly acquired territories."[58] The company's vision for missionary work was purely national; missionaries were to convert Africans to serve Germany.

The focus on territory alone meant a shift to maps as texts to communicate visions for the future of East Africa, replacing travelogues as the primary German representations of East African space. Maps made Ostafrika appear a unified space available for German exploitation, in contrast to the broken itineraries of travelogues. Maps, as was the case with the Delimitation Commission, became increasingly common in publications about Ostafrika in this period. Cultural geographers, however, challenged the veracity of maps as representations of space, and demanded alternatives. Friedrich Ratzel critiqued the colonialist obsession with cartography, pointing out that maps showed only physical geography, omitting the societies who inhabited the regions shown. Ratzel noted that it was extremely difficult to map Africa's ethnographic and political relations, so the map included supplementary pages to explain what could not be shown in map form. While the map did provide clear political boundaries for European colonies and a clear vision of the region's *Land*, they captured little information about the region's *Leute*. For a cultural geographer like Ratzel, interested in the relationship between *Land* and *Leute*, the cartography of German East Africa fell short for explaining the region. It showed the DOAG's vision of the colony's future, but not the reality on the ground, with the messiness of local politics.[59] He thus articulated a cultural geography argument against the DOAG's focus on territory alone, one based both in alternative representations of space and an alternative representational space in which African communities played an important role.

Despite the DOAG's belief that territorial control was what mattered, much of the German colonialist movement focused on trade. As a clear indication of their importance, the *Kolonialzeitung* focused on the Indian Ocean ports. An article from the beginning of 1886 described Dar es Salaam and its trade (including a map of the depth of different parts of the harbour). The city was full of empty buildings, giving the "impression, as though an imposing residence was supposed to have been built overnight through the will of a despot," but a more powerful enemy had stopped the building before it could begin.[60] The Foreign Office, it is clear, thought that control of trade would be what mattered for the future of Ostafrika, supporting a vision of empire through the trading companies that had established Germany's other colonies.[61] Such descriptions represented East Africa differently, as a space where methods conforming more to the European standard for colonizing Africa was necessary. Above all, the Indian Ocean ports, and the ability to charge duties on ivory and copal, became the most important elements in German visions for the colony, prompting the DOAG to change its rhetoric about Ostafrika's future.

Critics forced the DOAG to adopt trade and economic connections into representations of the colony's spaces. The lack of funding for railways meant that a short-term solution would need to be found to make the colony profitable. The DOAG described a unified "trade region" with its centre at Zanzibar, which had rapidly developed with a pace "reminiscent of America," the best evidence for the untapped latent economic strength of the continental hinterland.[62] The colony's value at that moment was in its control over trade routes. The "great goal" on which the DOAG claimed it always kept an eye was "the connection of the coast with the Great Lakes of the interior."[63] The company would strike out to reach Lake Nyasa/Malawi "and therewith open for trade all of Southeast Central Africa." Control of trade was more important than territorial sovereignty. In a report from 1887, the DOAG's leadership claimed "the particular advantages of our undertaking are the geographic location of the acquired territories, the quality of their soil, and the peculiar relations there." All the caravan routes from the interior of Africa led to the German territory. Additionally, the area's rivers could be used to further trade.[64]

Peters remained convinced that it would be "child's play" for Barghash not only to keep German occupation of the hinterland "in check," but to completely ruin the company by occupying the coast. The owner of the coast was "naturally sovereign of the entire hinterland in terms of trade,"[65] as it was much cheaper to get goods to world markets with control of the coast. In "uncultivated areas the owner of the coast was the natural master of the Hinterland as well."[66] Critics of the DOAG had forced Peters to abandon the "policy of rashness" for something more practical: a colony based in trade. The shift to trade necessitated new geographies to replace the sole focus on the acquisition of territory. The company told its shareholders at the end of 1887 that Peters was negotiating for Dar es Salaam as "entrance and exit gates" for the coastal strip. The grand dreams of empire along North American lines had disappeared, for which Peters and his allies blamed Zanzibar. Their dismissal of Zanzibari power from a short time earlier had been replaced by a conception of Zanzibar as a rival empire, visible on maps that showed Ostafrika's eastern border. Supporters of aggressive colonialism would adopt a new representation of East African space based in historical writing to remove that border.

Defining *Hinterland* Anew through Indian Ocean Histories

Before moving to control the coast, however, German colonialists had to change the meaning of *hinterland* and adjust their rhetoric to justify control of the coast. Histories of the region, written by colonialists and social scientists, provided arguments for a new East African representational space, one that combined coast and interior into one economic unit – a coast and its trade *hinterland*. The histories that DOAG members and their allies wrote posited deep economic connections between the coast and the interior, dating back centuries, and made those economic connections out to be the most important factor in the region's history. Such histories cast the German enterprise in East Africa as a mission shared with other European nations. They depicted the DOAG's work as advancing European civilization, not only to improve Africa, but to defeat Middle Eastern civilization, as well. That new bordering provided backing for the DOAG's move to take control of the coast from

Zanzibar. German colonialists modified the meaning of *hinterland* to one more closely matching today's – economic connections to a port – to push for, then win, control of the coast. Doing so meant reimagining the history of East Africa to make German claims to the coast appear part of historical progress to defeat Arab colonialism.

To argue for a change of focus in Ostafrika's orientation, colonialists increasingly deployed histories of the Swahili coast's connections to ancient and medieval civilizations to make their argument. These histories represented the East African mainland differently from earlier European writing about the region. East Africa appeared now as part of an Indian Ocean World that connected it to a golden age dating to antiquity. DOAG member and former Reichstag deputy Carl von Grimm wrote a history, *Die Pharaonen in Ostafrika*, meant to be the first in a series connecting East Africa to the Egyptian and classical worlds to establish it as eternally colonial space. Germany was thus not a destructive interloper, but promised rather to restore the glory of East Africa held in the ancient world by eliminating the negative influences of Arab colonization.[67] Coastal towns had been walled and had garrisons of several thousand men, but when Mombasa fell to an Arab invasion in 1698, political unity had disappeared and with it the level of culture collapsed.[68] Grimm argued that Vasco da Gama's descriptions showed the coast's capacity for development if Arab control could be ended, as the region had once been rich and could be again with the right kind of development.

Germans' evolutionary view of history and adoration of Ancient Greece enabled Germans to imagine a Golden Age of East Africa connected to Indian Ocean trade in antiquity. That ancient glory had later been destroyed by Latin (Portuguese) and Oriental (Arab) colonists. German historians took a philological view of the past, privileging text sources over other kinds of information.[69] Many of the scholars who promulgated the new idea of an Indian Ocean World regularly wrote or spoke to wider audiences than to those would read purely scholarly works.[70] Most were scholars of the classical world, indicative of the academic terrain of late nineteenth century Germany. Educated Germans exalted ancient Greece as the pinnacle of human achievement prior to the nineteenth century and condemned Latin and Oriental influences that had subsequently diluted Greek culture.[71] As Germany expanded overseas, the same historical methods and models were extended to world history and geography. When surveying East Africa's history, German colonists privileged the *Periplus of the Erythraean Sea*, a first-century text, and early modern Portuguese and German written sources, which described the coast and not the interior, and naturally highlighted aspects of coastal culture recognizable to Europeans rather than connections to Africans of the interior.[72] German ideas about history and sources predisposed colonists to foreground Indian Ocean connections over connections to inland communities.

Colonial propagandists deployed Orientalist ideas about the Near East to make their argument for German control of territory. Renowned Egyptologist Heinrich Brugsch supported claims of connections between ancient Egypt and East Africa in a speech to the Berlin branch of the *Kolonialgesellschaft* on 8 February 1888. Brugsch told his audience that "Hamitic" peoples had travelled to East Africa centuries before Portugal's arrival. There, their superior knowledge of shipbuilding and construction allowed them to become "leaders and teachers of the negro races (*Negerstämme*) in the Hinterlands of the East African coast."[73] This line of argument parallels the contemporary development of the "Hamitic hypothesis," a theory that "Hamitic" peoples had travelled south from the Middle East and North Africa and were the source of all civilization in sub-Saharan Africa. The British explorer John Hanning Speke first articulated the theory to describe the Kingdom of Buganda in 1863 and it became the basis for British, Belgian, and German versions of ethnic divisions in the Great Lakes region.[74] East Africa's *hinterlands* were linked to the coast not by Zanzibari political power but by older racial and cultural connections. This idea became part of the colonial lobby's propaganda for empire.

Among the most important of these voices was explorer and naturalist Georg Schweinfurth, who became one of the leading proponents of the new narrative of Arab colonialism and representational space of East Africa. Schweinfurth had worked for Khedive Isma'il of Egypt, exploring the flora up the Nile and was serving as the head of the Geographical Institute in Cairo. He told the annual meeting of German naturalists and doctors in 1886 that Ostafrika and the Congo were the best hopes for European settlement in Africa. Europeans had finally broken through the "coastal wall," by which he presumably meant the Arab presence, that had blocked them from the vast inhabitable spaces of East Africa. Schweinfurth told the scientific audience that the "Arab settlements of the east coast lay only thin layers inland." East Africa offered great opportunities for settlement because of its variety of landscapes between the Indian Ocean and the Great Lakes. What was more, those "immeasurable territories" would offer nothing to any other Europeans so long as the German Reich claimed its place and made no more concessions to other powers.[75] Schweinfurth provided a way of thinking about the interior as linked to the coast through forms of cultivation rather than political ties.

Politicians adopted the same Indian Ocean World history and the spatial concept that what mattered for a successful colony were connections based in the movement of people and goods, not political control. National Liberal Reichstag deputy Friedrich Hammacher laid out a version of East Africa's history as connected to Europe in antiquity and as part of an Indian Ocean World at the *Kolonialverein*'s annual meeting in 1887. East Africa had traded with Greece and Rome, but with the fall of the Roman Empire, it slipped out of the European geographical imaginary and became part of the Indian Ocean World through migration of Arabs to East Africa and East African troops to Baghdad.[76] He posed long established cultural connections between East African *hinterlands* and the Middle East that had nothing to do with Zanzibar.

This history of East Africa portrayed Arab colonization as having created a "Dark Ages" for East Africa. The region had flourished until being conquered by Omani sultans, which halted all cultural progress. Only European influence could restore East Africa to its former path. According to a DOAG report, the "coast of East Africa had stood in contact with the old *Kulturvölkern* (cultured races) of Asia and the Mediterranean." The report then laid out a narrative of that contact, and offered evidence of it from sources of antiquity and the early modern period. This narrative was also a representation of space. It claimed ancient Mediterranean peoples had laid the groundwork for culture in the region, thus that East Africa had long been colonized space. Evidence came from a relief in Thebes, which showed today's Somalia, and the fact that the Sasanian Empire had created the slave trade in East Africa. Little blooming sultanates along the coast developed a relatively high level of culture from those Mediterranean roots before Zanzibar arrived on the coast, meaning the coast's culture was not created by the sultanate. For this claim, the report included an anecdote about Vasco da Gama. Da Gama was amazed by the high level of culture he found where he had expected none, a level on par with Europe. He had conquered the coast for Portugal, creating a colonial empire valued higher than Mumbai or Goa. But the Portuguese had built a plantation economy and were oriented primarily toward the "rash fleecing of the land," so they could not build lasting rule. The sultans of the coast refused subjection to Zanzibar until the Omanis moved their throne to Zanzibar, and Zanzibari claims had moved forward only through British machinations, again a European influence.[77] The signs of Arab influence in the interior thus had nothing to do with the Zanzibari sultanate.

The shared battle for European civilization in East Africa's *hinterlands* brought the DOAG and Protestant missions together after their earlier conflict. Protestant missionaries found common ground with the

DOAG and supported the representational space of East Africa as a battleground between Arabs and Africans, bridging the divide that had limited their work together over the previous years. Protestant missions worked from an idea that the "African" space west of the coast was one of heathenism, and of an Arab-directed slave trade. The East Africa Mission proposed its own version of East Africa's history. The Portuguese had never been real colonists; they only wanted to capture trade. Their presence was weak and easily brushed aside once Portuguese priorities shifted with the formation of the Iberian Union in 1580. Since then, ever more colonists had gone to the region. All the colonists, however, had focused their efforts on Zanzibar itself, meaning the "coast remained as African as from time immemorial."[78] The mission's history served to minimize Zanzibari influence and justify missionary work among the "heathens" of East Africa by claiming they existed in a state of nature untouched by the great monotheistic religions.

The representations of East Africa as space of Arab destruction discussed at the beginning of this chapter played an increasingly important role in writing those histories. The leading voice in establishing this narrative was the explorer Hermann Wissmann, who wrote about his multiple journeys across the African continent. His account of his 1886/7 expedition from the mouth of the Congo east to Quelimane on the Indian Ocean makes Wissmann's conception of the Arab presence in Africa abundantly clear: it meant only slavery. Images in the book depict East Africa's landscapes as desolate and destroyed, while its people appear merely as victims. The book's frontispiece is an inset portrait of Wissmann over a drawing of a man in Arab costume holding a chain. Before the chain holder, at the bottom of the frame, is cowering a Black African man with the chain wound around his neck.[79] A picture later in the book, shown in figure 7, depicts a slave raid in progress. While their soldiers shoot African villagers armed only with shields and spears in the background, men in Arab dress attach chains to the necks of the village's women who wear nothing but skirts. One woman flees toward the viewer in terror.[80] Another picture shows a man named Said, the leader of a party of slave raiders, using slaves chained together for pistol practice.[81] The book also includes a picture of a slave caravan. Men in Arab dress drive on slaves lashed together by the neck. The two most prominent slaves are a woman carrying an infant and walking a toddler, and a man who has fallen by the wayside, apparently dead.[82] The reader is confronted with a clear message that the story of the Arab presence in East Africa is one of slavery. Wissmann cast East Africa as a space of struggle over the fate of Black Africans being fought between slave-raiding Arabs and civilizing Europeans.

Figure 7. "The Arabs among the Benecki." From Wissmann, *My Second Journey*

Such representations found a willing audience in both Germany and other European nations. David Livingstone's example had stirred up Christian feeling and instigated broad participation in Protestant missionary efforts beginning the 1870s.[83] In 1884, Muhammad Ahmad, known in Europe as the "Mahdi," had captured Khartoum from its British governor, Charles Gordon, and killed him. In the wake of Gordon's death, the British press turned him into a hero of the British empire, a supposedly religious man defeated by Muslim fanatics bent on the destruction of European progress in Africa.[84] In Catholic parts of Europe, the Church's Primate of Africa and Cardinal of Algiers, Charles Lavigerie, had promoted a new crusade against the influence of Islam in Central Africa and had created his own mission order, the White Fathers, for the purpose. Lavigerie travelled Europe in 1887 and 1888 to win public opinion for a European movement against Islam in Africa.[85] Europeans were agreed that Africa was a space behind in time, where outsiders were struggling over which path of historical development the continent would take.

The narrative of failed Arab colonization became particularly prominent in discussions of Ostafrika, a narrative that set up East Africa as the space where questions about Africa's future were most acute. The *Deutsche Kolonialzeitung* articles mentioned at the start of this chapter provided a clear historical narrative to construct this idea in early 1888.[86] Since the fall of the Portuguese, the formerly glorious Kilwa had been overgrown by forest. Since those days of Asiatic murals and panelling, there had been only decline to late nineteenth-century bamboo and straw huts. East Africa had thus undergone a similar process of decline as Germany from its early modern heyday.[87] Africa could advance only through connections with Europe and the severance of those with Asia. These histories depicted the interior as dependent on the coast since time immemorial and as subject to the same forces of Arab colonization that the Delimitation Commission had denied. This meant a new *hinterland*, one defined by economic and cultural connections. The new definition promoted the coast as more civilized than the interior.

The new *hinterland* provided the basis for the DOAG to change its policies toward its personnel and toward Zanzibar. The company no longer saw as its primary purpose the acquisition of territory. Rather, it needed to figure out a way to make the territory it had acquired profitable. It would do so through control of coastal ports, which would link its *hinterland* to the Indian Ocean. The change in the organization's focus provided justification for its board's decision to fire Carl Peters from his leadership post.[88] The company hired Ernst Vohsen to replace Peters, demonstrating its determination to shift its focus and work with members of coastal society as fellow colonists. Vohsen arrived with a testimonial from his previous post, in Sierra Leone. A number of "influential Mahommedans" had signed, attesting to Vohsen's trustworthiness and respectfulness.[89] Vohsen would deal with Zanzibar much differently than had the always antagonistic Peters.

The DOAG's path to control of the coast was eased significantly by the death of Said Barghash on 26 March 1888, and Vohsen would make quick progress thereafter. Some members of the colonial lobby held out hope that the new sultan, Barghash's brother Khalifa bin Said, would end Britain's protectorate over his realm. Negotiations proceeded much more quickly under Khalifa. Khalifa's position on the mainland was not as strong as Barghash's had been, and he was unable to continue his brother's policy of ignoring German demands. The company would succeed in signing a lease for the coast in short order. Khalifa was thus forced into permanent direct dealings with DOAG, his lessee, a relationship his brother had tried to avoid in asserting his sovereignty and status as a ruler of equal status to the German Kaiser. The final form of the lease confirmed the new geography of East Africa; coast and *hinterland* would be inextricably linked as one territorial unit.[90] Khalifa made over "all the power which he possesses on the Mainland" (*mrima*) south of the Umba River to the Company. It was, however, only allotted that power "in His Highness [*sic*] … and subject to His Highness [*sic*] sovereign rights." Khalifa thus avoided the total cession of sovereignty that the DOAG had sought in its earlier treaties for land. The DOAG received the right to buy public land, raise taxes, appoint commissioners, pass laws and establish courts, make treaties with native chiefs, the right to acquire and regulate unoccupied land, and any other rights Khalifa possessed, except over his private property.[91] To the Sultan was reserved the right to appoint *qadis*, Islamic judges with jurisdiction over civil matters.[92] Control of the coastal strip, colonialists thought, would make possible a profitable colony through control and taxation of trade from the *hinterland*.

Conclusion

Peters declared his mission complete, that Ostafrika had reached its "natural borders."[93] He believed he had acquired greater concessions than any other European colonial company had acquired to that point. Otto Arendt, a leading figure in the colonial movement, agreed that the colony had "reached its natural borders." Germany now had a "closed-off, important territory" on the Indian Ocean with the "most consequential precondition for its prospering, unhindered access to the coast," now available. Ostafrika was ready to become the "German India."[94] The company was ready to take up the mantle of colonizing the East African *hinterlands* from Asian colonists. The historical narratives of centuries of Arab colonization in East Africa involved in reformulating *hinterland* would be central to the next few years of making space in the colony, discussed in chapters 5 and 6.

The coastal lease created the rough borders for Ostafrika as it would exist through the First World War. The evolution of representations of the region's spaces over the proceeding years and the changing meaning of *hinterland* had made this imperial project possible. A latitudinal *hinterland* based in maps had provided grounds to border Zanzibar territorially and create a sphere of interest for Germany. Some colonialists had recognized the failure of the DOAG's plans based in German settlement. A better colony, they argued, would be based in the utilization of existing spatial practices. Such a change required control of the coast. To gain control, German colonialists used historical writing to reformulate *hinterland* to mean economic connections. Economic *hinterland* had justified the completion of Peters's vision, German claims to territory stretching from the Indian Ocean to the Great Lakes. German colonialists invented longue durée narratives of economic *hinterland* to justify the territorial conquest of all of mainland Tanzania. The histories and cultural geographies they deployed in making their case made a new representational space for German audiences in which the borders of the colony appeared part of the "natural" course of history. Plans to remake East African space through the settlement of German farmers had failed, at least in the short term. Instead, colonialists would use existing spatial practice, the itineraries and networks of the long-distance trade, to create a profitable colony.

Despite creating territorial unity, the lease increased the DOAG's international problems and did not resolve Ostafrika's future. The Foreign Office did not know what to make of the newly leased territory. The GfdK's initial territory belonged, according to the company's legal argument, to the GfdK as a fully sovereign authority. The German government could therefore grant protection to the GfdK with no possible impediments on its own sovereignty. The coastal strip was different. The company had only leased the coast from the Sultan of Zanzibar, in whose name it now administered the coast. If the German government granted protection along the coast, it then would place itself under the sovereignty of the Sultan of Zanzibar, an obviously problematic relationship. It was unthinkable to late nineteenth-century Europeans that a European government could at any point be placed under the sovereignty of a non-European power, particularly one against which Germans had publicly campaigned as the enemy of European progress. But then what relationship could the Reich government have with the DOAG regarding the coastal strip?

5 The Emin Pasha Expedition, New Forms of Political Mobilization, and Reimagining East African Space, 1888–1890

In the mid- to late-1880s, the German and wider European publics cared little for the negotiations over national borders discussed in chapter 4. They fixated instead on the situation of one man, Emin Pasha. Emin's name and title, which he earned through service to the Egyptian government, belied his origins. He was born in Prussia, named Eduard Schnitzer, the son of parents who had converted from Judaism to Lutheranism. Emin had worked his way to the governorship of the province of Equatoria, in today's South Sudan and northern Uganda. When Mahdist forces captured Khartoum in 1885, they cut off Emin's communication with Europe and left him stranded in Equatoria. Beyond this basic narrative, almost nothing was known of Emin. Writers could invent their own versions of his talents and of what he could offer German colonialism in the future. Carl Peters, sidelined by the DOAG's new leadership, used Emin's situation to return to the limelight and to agitate for an expansion of Ostafrika to the west. British and German campaigns to rescue Emin raised hundreds of thousands of pounds and marks. This money funded expeditions led by Peters and Henry Morton Stanley to find Emin and bring him back to Europe. Peters's expedition would not find Emin; his failure made the Emin Pasha affair something to forget among later German colonialists.

Though the German Emin Pasha Expedition has little presence in the historiography of colonialism and mass politics, the campaign to organize Emin's rescue became a model to emulate when attempting to mobilize nationalist sentiments to achieve political goals, one most obvious in the subsequent growth of right-wing politics.[1] The growth of the Right in the early years of Wilhelm II's reign has been a central concern of the historiography of the *Kaiserreich*.[2] Its progress spanned the reigns of three Kaisers – Wilhelm I, Friedrich III, and Wilhelm II – and thus encompasses a period of transition in German politics. By the time

Emin was found, he was a figure from Germany's past under Wilhelm I as the nation moved toward Wilhelmine *Weltpolitik*. The expedition is the clearest public manifestation of the emergence of radical nationalism and the transformation of liberal imperialists such as Friedrich Ratzel into chauvinistic nationalists. The expedition's supporters pioneered the use of the media in creating public support for radical nationalism, support we can see in applications to join to the expedition. These applications also demonstrate grassroots enthusiasm for the expedition's model of German empire. The method of mobilizing existing nationalist populism became an example to emulate for Carl Peters (later chair of the Pan-German League) and other right-wing figures for new causes in the decades that followed.[3]

In tracking the campaign to create an expedition and the expedition's progress, this chapter will consider them as processes of space making. The travel accounts that pro-expedition campaigners and Peters deployed served as representations of the deep East African interior – an area to which they hoped to expand Ostafrika – and produced new representational space of the region. The debate over the expedition was primarily about what kind of space the interior of East Africa was and what kinds of things Germans could do there, more than about Emin himself. In their campaign for funding, the expedition's supporters deployed imagery of Equatoria as a dam holding back the influence of Islam on Central Africa, imagery that made Emin a symbol of European progress and civilization working in the natural landscape. On his expedition, Peters marked territory west of the coast as wilderness, available for German exploitation and empty of civilization. Peters and his supporters expanded their vision of African land as available for Germany, and maintained that all Germany needed to do to build an empire was claim territory.

Emin as a figure became the ultimate expression of the belief that a German presence alone would remake African spatial practices. Simply taking more territory and managing it well would turn the people there into loyal subjects, meaning that colonialists could continue to ignore African communities. From its outset, the public campaign argued that what mattered was control of territory, but also that there was a specifically German talent for remaking African space. Opposition to the campaign, which formed among critics of colonialism, including Chancellor Otto von Bismarck and National Liberals, also focused on space, but through a conception that African space could corrupt Europeans who spent too much time in colonies. Colonization should instead proceed through managing African people. Each side's rhetoric heightened as rumours of Peters's death arrived in Germany, and his death was used as an opportunity to argue over the empire's meaning and future. The expedition to rescue Emin Pasha enshrined a role for cultural geography and a discussion of space at the core of German colonialism. Peters's model became the radical nationalist model of colonial politics and space.

Emin Pasha

The figure at the centre of the crisis, Emin Pasha, had been born Eduard Schnitzer in Oppeln, Silesia, in 1845.[4] Oppeln was part of the Prussian territories acquired in the partition of Poland, making Schnitzer a Prussian by birth. After completing medical school, Schnitzer travelled south, where he hoped to enter the service of the Ottoman Empire.[5] His interest in the Ottomans was part of a relatively common pre-colonial German obsession with the Orient in the nineteenth century.[6] To "make communication (*Verkehr*) easier," he claimed, Schnitzer took on an identity as a Turk who had been educated in Germany.[7] He moved to Cairo and entered the service of the Egyptian government, eventually winding his way down to Khartoum in 1875. From there, he began building a reputation in Europe as a naturalist. He corresponded with other Europeans interested in Africa, collecting plants and animals, especially birds, and sending them to metropolitan museums. It was in Khartoum that Schnitzer took on the name Mehemet Emin. Emin's reputation as a naturalist and a linguist came to the attention of Charles Gordon, then governor of Equatoria.[8] Established in 1870, Equatoria was an Egyptian province, and its officials held posts under the authority of the Egyptian khedive.[9] Gordon hired Emin onto his staff as chief medical officer in 1876. When Gordon was promoted to governor of the Sudan in 1878, the khedive appointed Emin to succeed Gordon as governor of Equatoria. This promotion included the bestowal of the title, "bey" to Emin, who thenceforth went by "Emin Bey."[10]

Emin could well have disappeared from the narrative of the European colonization of Africa in 1885 with little notice, but events beyond his control intervened. He was the governor of a province, but that province was not especially important to the British Empire, and he was clearly subordinate to Gordon, the star of that empire in Africa. When the Samaniyya religious and political leader Muhammad Ahmad, better known in Europe as the Mahdi, attacked the Sudan, took Khartoum and killed Gordon in 1884, he cut Emin off from contact with Europe.[11] Gordon's death provoked memorialization verging on hagiography – poems, monuments, paintings, biographies – but Europeans paid little attention to Emin at the time. Evidence of Emin's lack of renown can be found in the German colonial lobby's first notes of his plight. His name was frequently misspelled as "Schnitzler," and little or no notice was made of his position as governor.[12] Over the next five years, Emin would become one of the most famous people in Europe; he was the object of two rescue attempts, and then eventually his star faded to near irrelevance.

The Austrian explorer Wilhelm Junker, who returned to Europe in 1886, brought Emin's case to light. Junker had also been cut off from contact with Europe, and spent several months with Emin in Wadelai, the Equatorian capital. After escaping to Zanzibar and returning to Europe, Junker lectured around Germany on his and Emin's work in late 1886.[13] He told audiences "the work of a hundred years of civilization" was at stake if Emin was not rescued.[14] Junker passed Emin's letters on to Georg Schweinfurth, a German explorer living in Cairo who had become famous in Europe for travels in Central Africa years earlier. Schweinfurth sent the letters to German and other Western European newspapers.[15] In them, Emin pleaded for assistance from Europe in fighting off forces from the north, which he claimed were part of the Mahdi's attempt to drive Europeans out of Africa. The letters make clear that Emin intended to remain in Equatoria. He wanted help fighting off the invaders; he did not ask to be rescued.

The news of Emin's survival became popular because of his connection to Charles Gordon. Gordon's death was still in the public consciousness; the news that one of his lieutenants was still alive, and asking for help, captured imaginations across the continent.[16] British and German observers claimed saving Emin was necessary to keep European civilization alive in Africa. Geographical societies made Emin an honorary member.[17] Children spoke on the streets of Germany about Emin as he became a figure of popular legend.[18] Popular support for a rescue expedition rose even as both the British and Egyptian governments lost interest in Equatoria, which was threatened with an invasion by the Kingdom of Buganda to its south. The khedive promoted Emin to Pasha in December 1886, and dismissed him from his post, leaving him free to abandon the province, but Emin remained in Equatoria.[19]

The relative lack of information about Emin was part of his appeal to Europeans, as it meant his story could be more easily moulded for political purposes. Newspapers read different meanings into the scant evidence they found in his letters. The only Germans who had had contact with Emin were naturalists to whom he had sent plant and animal samples. As the British newspaper the *Standard* put it, "So much mystery, or, if the word be preferred, so much romance surrounds the sturdy Teuton … that the thought of succouring him may well appeal to the imaginative side of the German temperament."[20] Henry Morton Stanley even thought that Emin was an Austrian.[21] Almost no one in Europe knew much about Emin, but agitation to save him increased nonetheless. Equatoria was also relatively unknown among Europeans and was one of the few places in Africa that no European state had yet claimed. This made it available for the projection of many different spatial imaginaries. Emin's story could easily be fit into the historical narratives that German colonialists had developed over the preceding years to justify conquest in East Africa.

Public enthusiasm produced action first in the United Kingdom, where Stanley organized an expedition to find Emin in 1886. Stanley, renowned as the greatest explorer of the time, put together a massive expedition to bring Emin out of Equatoria and to Britain.[22] The expedition depended on an imaginary of Equatoria much like the spaces where Stanley had made his name, a wilderness full of "savage" Africans in need of Stanley's strong hand to rescue a European who had taken a more humanitarian approach and failed. But Stanley too disappeared, which inspired even more worry about Emin's fate. Stanley did not plan for the difficulties of travelling on the Upper Congo. He did not bring sufficient transportation or supplies for his entire expedition, and he instigated attacks from many of the people he encountered along the way. No news from the Stanley expedition reached Europe after July 1887.[23] In the meantime, Gaetano Casati, Emin's right-hand man in Equatoria, got word to the Society of Commercial Exploration in Milan in December 1887 that he and Emin had still heard nothing from Stanley, and that their position was perilous.[24]

The Stanley expedition provoked concerns among German colonialists that it was an attempt to take control of territory that might become German. The DOAG wrote to Otto von Bismarck that Stanley's true goal was the acquisition of territory west of its sphere of interest, forming a "dam" against the development of Ostafrika. It threatened to send its own expeditions to the region to acquire more territory.[25] The Foreign Office took the letter seriously enough to request assurances from the British government, which the latter provided in July 1887.[26] German supporters of the Peters expedition turned rumours into a vision of British aggression. The Stanley expedition turned uncertainty about the position of one man in Equatoria into confusion about colonial rivalries and the long-term future of Central Africa. Stanley's absence provided a justification for a German Emin Pasha expedition.

The explorer Hermann Wissmann claimed that Stanley was trying to combine Emin's realm with the lands Junker had travelled through into a massive Central African state.[27] Either Emin still needed assistance, or

Germany needed to do something to prevent the United Kingdom from claiming all of Central Africa. Wissmann and Schweinfurth became active in a campaign to create a German expedition to rescue Emin. Their roles demonstrate Emin's rather unknown status. His scientific colleagues were interested in a rescue, but few Germans knew who Emin was. The small scientific community around Emin bloomed into something much larger over the next few years as others who were interested in a more active approach to empire latched on to Emin as a means to an end.

Emin Pasha as a Tool to Claim Equatorian Territory

The campaign to rescue Emin became central to the colonialist movement in the late 1880s and was formative for right-wing populist mobilization, even as the expedition itself appealed to German nationalists across the political spectrum. Germans of different classes adopted the aggressive colonialist approach pioneered by Carl Peters and the GfdK in 1884. Rhetoric around the expedition reveals the German spatial imaginary in a struggle for European and German civilization in Africa against the influence of Islam. The histories of Arab slavery developed over the preceding years were especially prominent. Discussion of Emin exposes the spatial imaginaries of German colonialists, an idea that the control of territory was what mattered, but also that a German man and German science could make a colony profitable through mere Germanness. Historical narratives of Islamic conquest and enslavement represented the caravan routes as spaces of Arab slavery in need of German intervention.

Public agitation for a German Emin Pasha expedition began in the Nuremberg branch of the *Kolonialgesellschaft* in September 1888. Organizers claimed the expedition sprang out of the grassroots of the colonial movement, that it was the product of spontaneous enthusiasm around the plight of their fellow German initiated by a "regular member" of the organization.[28] The regular member who submitted the proposal, however, was Hermann Peters, brother of Carl, belying the claims about a grassroots movement.[29] On 18 July 1888, the German Emin Pasha Commitee, as Peters and his associates named their new organization, appointed Hermann Wissmann, the anti-slavery explorer of Central Africa, as joint leader of the expedition, alongside Peters.[30] To assist Peters, the organizers selected Adolf von Tiedemann, son of a recently ennobled Prussian official and Free Conservative member of the Prussian legislature. With its leadership in place, the committee announced at a meeting on 17 September 1888 a subscription plan to raise 400,000 marks to fund the expedition.[31] Soon thereafter, the committee began advertising itself and asking for money in the German colonial press. Peters declared Emin's position the "last pillar of European civilization on the Upper Nile." Emin held the last post on a "battle line" otherwise overrun by Islam, the last hope to open Africa for Christian *Kultur*. The expedition depended on a European spatial imaginary of a "real heart of Africa," centred on the Great Lakes region.[32] Peters thereby tied his chauvinistic goals to wider European concerns about European civilization in Africa and the danger posed by Islam.

The Emin Pasha Committee emphasized national competition for African space in its arguments for the expedition. It produced a map of the region to show audiences its imagined future German East Africa.[33] The committee wrote to Bismarck in July 1888 asking for monetary support from the Reich. The letter highlighted Emin's Germanness and claimed the expedition would "open a wider *hinterland*" for German trade.[34] Emin's valuable knowledge promised to save Germany money in the colonization of Africa, as the nation could acquire a large territory in Central Africa without the expensive use of military force or administration. A *Kölnische Zeitung* article claimed that the expedition was needed to prevent the United Kingdom from its plan of the "complete displacement of Germany out of southern and eastern Africa" with Stanley's secret plan to found a Central African colony "without a coast."[35]

The expedition provided an issue that bridged the divide between the DOAG and missionaries from the preceding years with the shared representational space of the "heart of Africa." Protestant missionary organizations were among those who supported this territorial framework. Matthias Ittameier of the Leipzig Mission wrote to the Foreign Office in August 1888 expressing support for the Emin expedition to acquire territory for Germany. The German position in East Africa was "unbearable," as it was squeezed in the middle of Portuguese, Belgian, and British colonies. Without access to the Nile, Germany would have to abandon colonialism "in the grand style." Equatoria offered links to great wealth in the trade of the Nile, as well as the possibility of creating a German colony that would stretch from Cameroon in the west to Zanzibar in the east, a "belt" across Africa.[36] From there, it would block the spread of Islam south of the Sahara, a threat that missions took seriously after the rise of the Mahdi.

The representational space of the heart of Africa provided a basis for making territorial claims with no regard for Africans, imagined as passive objects. In Peters's imagination, Emin was on the frontline of the conflict between Arabs and Europeans that raged across Central Africa.

He was holding onto Equatoria, territory Europe needed for the conflict.[37] Emin also imagined himself in the middle of this conflict. In a letter written after the expedition, Emin called for a "complete separation of the land of the Blacks from the Arab territories, or those where the Arabs predominate."[38] Future Ostafrika governor Eduard von Liebert declared that the expedition would ensure the "victory of European culture over a thousand years of Semitic barbarism," a frequent pejorative about Islam's role in Africa.[39] A *Berliner Tageblatt* article claimed Emin's situation was just one result of a continent-wide Islamic movement, a movement that could be seen in Morocco and Tunis, on the Red Sea, and in Zanzibar. In this moment there had arisen a "war between crescent and cross." Islamic fatalism, dreams of paradise, polygamy, and slavery appealed more to "natural peoples" than did ascetic Christianity, so Islam was making progress in Africa.[40]

In their discussion of Islam and the slave trade, pro-expedition campaigners tapped into broader concerns about a Muslim slave trade in late-1880s Europe that shaped how Europeans imagined Africa. Both Catholics and Protestants elaborated representations of Central Africa as a battleground between Christianity and Islam. The concept of an "Arab slave trade" formed a convenient shorthand for European internationalists to organize their efforts in Africa. It combined claims that Africans needed European help with attacks on Islam as anti-modern. From the 1860s, missionary propaganda took up themes of Africa as slave-ridden and of slavery as the barrier to civilization on the continent.[41] Descriptions of slavery inevitably discussed an Arab or Muslim slave trade (often conflating racial and religious labels). They counterposed European civilization to Arab/Muslim fanaticism and savagery, claiming a competition for the future of Africa.[42] Africans appeared in their rhetoric only as objects for European subjects to save.[43] Contemporary representations evidently ignored the variety of Muslims' views on freedom and captivity. They also downplayed Europe's slave-trading past and glossed over the inconsistent implementation of anti-slavery policies.[44] Arguments about the negative influence of Islam and slavery were closely tied to historical narratives about Islam's anti-modernity. By contrast, European ideas about sovereignty, built on the work of the early modern theorists, most famously Jean Bodin, claimed that Europe was a "modern" society of subjects, and the extra-European world was made up of "feudal" societies that practised slavery.[45] The continued practice of slavery in Africa meant that the continent was behind Europe in its historical progress. Islam's presence in Africa prevented proper development and justified European intervention to assist Africans in combating the slave trade.

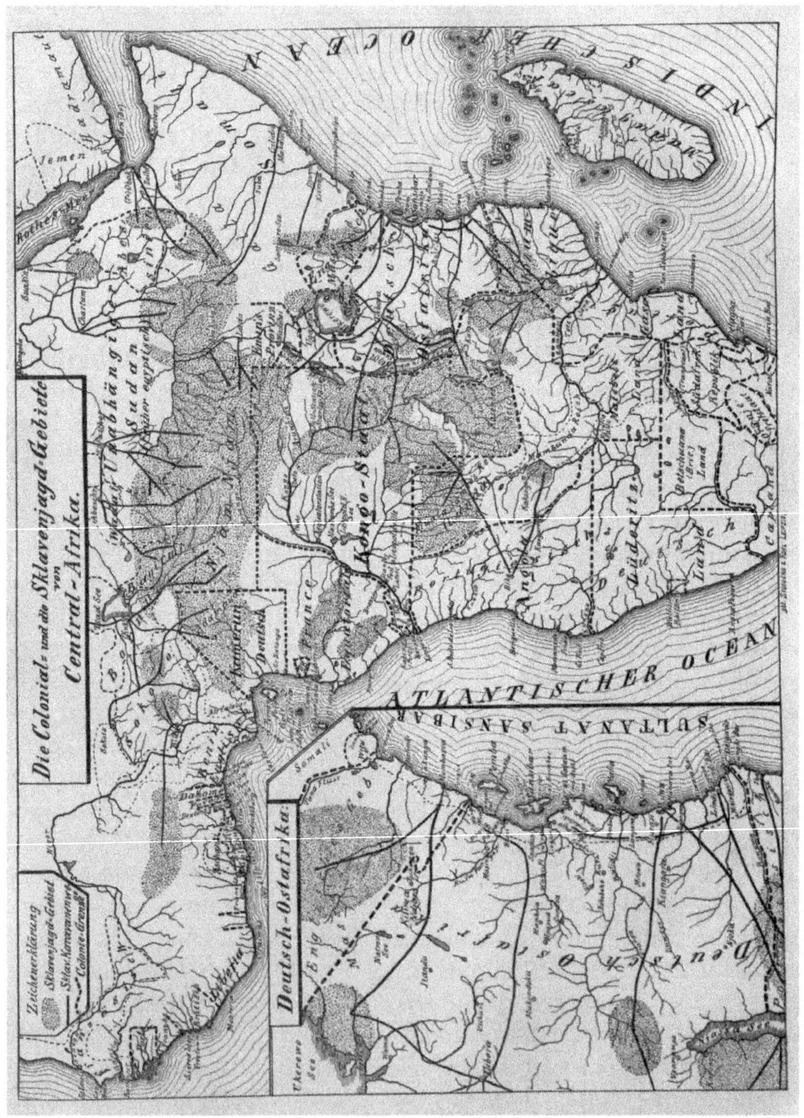

Figure 8. "Die Colonial- und die Sklavenjagd-Gebiete von Central-Afrika." From *Illustrirte Zeitung*, 24 November 1888

The spatial configuration of this conception of the Arab presence in East Africa represented the geographies of the caravan trade as geographies of slavery instead. This was represented in a map in the *Illustrirte Zeitung*, one of Germany's leading weeklies. In the map, titled "The Colonial and the Slave Raid Areas of Central Africa," the uncredited cartographer marked the primary slave trade routes from the Great Lakes region to the Indian Ocean. These routes are clearer on the inset map of Ostafrika as lines stretching between the coast and the Great Lakes. They match almost exactly the caravan routes between the coast and interior that predated German arrival. This map labels the caravan routes as slave trade routes, thereby claiming that the presence of Arabs in East Africa was driven entirely by the slave trade. Here, as well as in descriptions of the slave trade, German observers claimed that all agency in shaping the region's history lay with Arabs. The East African interior was therefore a space in need of European intervention to stave off the slave trade.[46]

For the expedition's supporters, Emin served as a tool to claim territory for Germany. The land in question was subject to the Arab slave trade and therefore stuck in the past, causing regression among the societies who lived there. These claims established Equatoria as the site of active warfare over the future of Africa. Emin's work in Equatoria had brought a degree of European progress and modernity to the region through his attempts to create *Kultur*. German assistance to him would ensure that the growth of *Kultur* continued despite the best efforts of Arab slave traders to end it. This cultural mission would become a means to define Germanness as remaking space to create progress.

German Identity in the Campaign to Rescue Emin

Participation in this struggle to assist Emin Pasha allowed Germans to formulate a national identity around empire and the idea of *Kultur*. This idea began with Emin's scientific colleagues but was adopted beyond elite scientific circles. The popular press adopted the same rhetoric to drum up support for the expedition. Several German men wrote unsolicited letters to volunteer for the expedition, hoping to participate in the German colonization of Africa. Elite and popular voices together articulated a vision of a German identity based in empire and the spread of German *Kultur* across the globe. They adopted the DOAG's spatial imaginary of empire as a national competition for scarce territory.

The celebrity explorer Georg Schweinfurth led the campaign to raise funds by convincing Germans to identify with Emin. Schweinfurth had corresponded with Emin in his role as the director of the *Institut égyptien* in Cairo while Emin was in Equatoria, but he had previously stood aloof from German colonial ventures. Schweinfurth offered his respectability as an international man of science to the nationalist agitation in 1888. He described the need to save Emin as a national duty to a fellow German. As Schweinfurth put it, Emin "is truly one of us, a German, a Prussian." An expedition to rescue Emin could have long lasting consequences for Germany in Africa, as "Germany is honoured through Emin. His work is our pride and, if it goes on, an undying glory. Every German heart must beat higher with the thought: he alone, he, the German, was able to hold onto his position in Africa."[47] Schweinfurth described Emin as a lone German maintaining *Kultur* in Central Africa through his talent and individual initiative, a veritable one-man empire.

Together with German and British scientific colleagues, Schweinfurth published a volume of Emin's correspondence meant to humanize the man and allow readers to connect with him.[48] The volume established a particular representational space of the East African interior as the standard in German discussions of the region. Ratzel's participation was based on the important role that Emin had played in the formation of his ideas about Africa. Emin had written an account of Lake Albert for Ratzel to publish, as well as two other papers, in 1886.[49] Ratzel had corresponded with Emin about the *Land* and *Leute* of Equatoria and surrounding regions before the Mahdi attacked. That correspondence formed the basis of Ratzel's descriptions of Central Africa in his *Völkerkunde* (Ethnology), a foundational text of cultural geography. Emin's work on the Kiganda and Kinyoro languages provided evidence for Ratzel's claims about their history, and Emin's work on the Mangbetu served Ratzel's claims that the Mangbetu were more evolved than other Africans in their family life.[50] Emin had thus been a formative figure in the development of ideas about *Land* and *Leute* in the cultural geography of the 1880s and would continue to serve as Ratzel's evidence for his cultural geographic theories in later writing. In the second volume of his *Anthropogeographie*, for instance, Ratzel used Emin's descriptions of Wahuma society to claim that a racial hierarchy would develop among societies at low levels of *Kultur* if their territories overlapped.[51]

Through Ratzel's and Schweinfurth's involvement with the Emin Pasha Expedition, we can see their transition from liberal imperialists to conservative expansionist nationalists, a transition that many German colonialists followed in this period.[52] Ratzel was more than just another member of the committee; he played a central role in its campaign. His involvement spanned the gamut from the edited volume mentioned above to helping schedule a room in which Wissmann could speak about the exhibition in Berlin.[53] He wrote in favour of Peters, despite Peters's reputation as anti-scientific and violent. No matter what one thought of Peters, it was time "for all politics to stop." Ratzel wrote that supporters of German work in Africa needed to support Peters, as diplomats only observed, but neither led nor fought for the advance into Central Africa.[54] Ratzel's critiques of the DOAG's aggressive expansion and his internationalist science disappeared, replaced by nationalist calls for German action.

Ratzel's role as a co-editor of the volume of Emin's writings established cultural geography as the mode for some Germans to discuss Equatoria, and the East African interior more generally. The choice of texts to publish set a representational space of the region that combined attention to people with descriptions of physical space, in contrast to Peters's writings that focused on physical geography alone. Emin's writings and the editors' comments portrayed the bey as a man who administered Equatoria successfully by making use of the racial characteristics of the region's population while introducing new flora that might prove economically valuable. While the majority of the German population shared the attitudes found in the letters volunteering for the expedition, Emin as cultural geographer appealed to educated Germans.

Stories in popular periodicals provide a window into the dreams of empire among pro-colonialist Germans and reflect identification with Emin as a fellow German. They depict Emin as a German hero, carrying on great work in the nation's name. An article in the *Berliner Tageblatt*, a newspaper owned by the Jewish scion Rudolf Mosse, claimed Emin had combined his own idealism with the practical, dogged endurance of Jews and German energy to achieve success. Though Emin's parents had converted to Lutheranism, the article focused on Emin's Jewish descent to cast Jews as members of the German community.[55] The *Kölnische Zeitung* contrasted the German expedition with Stanley's, claiming Germans were motivated out of national duty while Stanley was merely trying to claim territory.[56] The Kaiser himself took an interest in Emin's position and invited Hermann Wissmann to an audience.[57]

Grassroots supporters of the expedition formulated their own meanings for Emin and used the pasha to create an alternative version of German identity. We can see these meanings expressed in the letters of men who wrote to the Emin Pasha Committee, the Kaiser, or Bismarck to volunteer for the expedition, despite no interest in taking on volunteers on any of the recipients' parts. These letters reveal the meanings that the German public attached to the model of empire the

Emin Pasha Committee promoted. Their motivations included German nationalism, frustration with economic and political changes, and the late nineteenth-century crisis of German masculinity. The volunteers for the expedition were men from generations too young to have been active participants in unification, who sought another form of glory with none available from military service.[58] Many of the adult applicants were in precarious socioeconomic situations.[59] To strengthen his case, a tailor from Frankfurt am Main complained that he could not get a "secure position" because he was not a Social Democrat: his loyalty was only to the kaiser and to the Reich, which was why socialists would not buy from him and why he had to struggle to feed his wife and two sons.[60] Most applicants occupied skilled occupations, but they were not ones likely to have benefited from the changes in German society resulting from industrialization. These applications evoke the crisis among German men at the end of the century, as they felt destabilized by the changes of industrialization and without the traditional markers of masculine belonging.[61] In these letters, Africa appears much like it did in Peters's writings, as a wilderness space of adventure for Europeans to achieve glory not available in Europe.

Supporters of a more active approach to colonialism latched onto Emin Pasha's plight in Equatoria to make their case in 1888. We can see that these Germans attached a deeper meaning to Emin in how little time they spent discussing his scientific achievement, instead devoting their attention to ideas about German pride, or to remaking African space. To make their claims about the possibilities for a future German Central Africa, supporters represented the East African interior as a stage for competition between Christianity and Islam, its inhabitants reduced to props. That struggle would be fought in Central Africa. They argued for a German identity based in colonialism. This vision, however, was not the only one in discussions of Emin Pasha, and opponents of the expedition presented alternative understandings of world history and African space.

Chancellor Bismarck Opposes the DOAG Model of Territorial Competition

Opponents of the German Emin Pasha expedition saw it not only as frivolous but also as possibly dangerous. The vision of empire presented by the expedition's supporters threatened much of what some Germans considered the most important elements in their nation's international status. Those worries alone were unlikely to win out, so opponents developed a different vision of African space and empire. Opponents of the campaign promoted a different model of empire, one in which a focus on the people and economies of the African interior predominated, and African space could corrupt Europeans. They claimed that Emin was not what his supporters imagined him to be, undermining Emin's status as a colonial hero. But opponents of the expedition were less vocal than its supporters and, since they were already in power, did not form political movements around their opposition.

Critics claimed the expedition was not worth the investment for many reasons. These included practical ones, such as the fact that the expedition promised no clear long-term benefits to Germany, that Stanley had a long head start in looking for Emin, that it would antagonize the British government for no real reason, and that no one knew what Emin would do once relief arrived. Other reasons were fantastical, like a claim that Emin had converted to Islam and was therefore not useful for preventing the spread of Islam in the region.[62] The National Liberal *Schlesische Zeitung* claimed that Emin's status as a Muslim meant "he could no longer be counted as a German." The article's author agreed that Muslims were making the situation in East Africa dangerous, but thought the Emin expedition would only make things worse. Zanzibari sultans had traded peacefully with Europeans for centuries, and it was only now that the supposedly intractable religious conflict had broken out.[63] Opponents rejected the DOAG model of colonialism and its representational space of East Africa as requiring only territorial control for an approach that would foreground effective administrative and economic expertise, which would likely entail increased government control of the colony.

Bismarck rejected the committee's imaginary of East African space in declining its appeals for monetary or diplomatic support. His opposition was based not so much in an alternative vision of empire as in concerns about great power politics. Proposals for the expedition caused the chancellor to utter his famous line, "Your map of Africa is really quite nice. But my map of Africa lies in Europe," rebuffing colonialist plans for the primacy of colonialism.[64] He wrote in a memorandum on 14 September 1888 that "every report clearly failed" to demonstrate whether Emin could be reached from the German interest sphere, or how far his rule stretched.[65] According to Bismarck, "We know even less about the passability of the routes that lead there, and about the temper and fighting capacity of the tribes that live in the lands between the coast and Emin. The difficulties that an expedition there would have to overcome are completely unknown."[66] In his response to the committee's enquiry about government support, Bismarck raised the spectre of international opposition. He claimed the UK would see the expedition as an attack on its sphere of interest, and "the friendship of England is for us of much greater worth than anything the expedition could achieve in the best case." It would be a "criminal undertaking" for Peters to claim Equatoria. He would then truly be a "filibusterer," returning to the rhetoric of Peters's opponents discussed in chapter 3.[67] Further, any request for funds had no chance of making it through opponents in the Reichstag, so it was not even worth asking.[68]

Despite the lack of government support, the committee succeeded in raising the necessary funds for its goal of taking Equatoria for Germany. It drew on the large colonial lobby in Germany involved in interest group politics, the first time that a private organization raised a significant amount of money for a specific mission or expedition. Branches of the *Kolonialgesellschaft* and individual members from around Germany (and from Germans in Russia and the United Kingdom) sent in money, from four marks to 25,000.[69] The number of donations attests to the popularity of the Emin Pasha Committee's spatial imaginary as compared to Bismarck's more limited vision.

Bismarck worked with the British government to block Peters's landing on the African continent altogether, a decision that risked infuriating colonialists. Peters departed for Zanzibar in March 1889. Together with Sultan Khalifa, the British consul in Zanzibar prevented Peters from obtaining a ship through normal means in Zanzibar. Peters had to acquire a ship through charter with Sewa Haji to take his expedition to Lamu.[70] Peters was soon himself lost in the East African interior. The government's decision to work with the British drew the ire of the expedition's supporters. Pro-expedition campaigners came together in several Germany cities to protest British actions to block Peters in July 1889.[71] News emerged after the expedition that the German government had secretly gone even further to hinder Peters's progress. A captain in the British army said he had been sent with 1100 men, including several pieces of light artillery, to find and arrest Peters.[72] That the Bismarcks were willing to work with the British government to stop Peters, risking a conflict where British soldiers would fire on Germans, points to the primacy of the relationship with the UK over appeasing German colonial agitators.

Ultimately, attempts to prevent the German Emin Pasha Expedition failed, even though the most powerful voices in German politics, those running the government, did oppose it. Chancellor Bismarck's opposition centred around the threat it posed to improved relations with the United Kingdom and the lack of clear benefits for Germany. Such opposition entailed a rejection of the DOAG model of empire and the focus on territory as the goal of German colonialism. Acquisition of territory alone would not provide Germany with a great overseas empire. Having dedicated significant effort to stopping the expedition, the regime's failure both aggravated colonialist opposition and did not stop Peters from creating tension with the United Kingdom. West of the coast, Peters had free reign to create chaos and promote his representational space.

Peters Makes Space in East-Central Africa

Over the next year, Peters's supporters would experience a wide range of emotions as they followed the expedition's progress in the press. Through the fear, excitement and, eventually, disappointment, Peters and his supporters used the expedition to solidify their version of East African geography. Peters's and Tiedemann's post-expedition travelogues became venues in which to lay out the future vision for German colonialism. As was typical, East African space was represented primarily as text, rather than map, in order to present the entirety of Peters's construction of the region as more than coordinates. Instead of empty space, Peters had encountered what Raymond Craib calls "fugitive landscapes" – spaces in which he could not fix borders or even names of places, spaces that local communities understood through different frames of knowledge from his own. Turning landscape into textual tropes asserted Peters's ability to read and control it and present his vision of African space.[73]

Peters's and Tiedemann's travelogues of the expedition attempted to assert control over the spaces through which they passed, even as their presence had little or no effect on spatial practice. Both are replete with elements that assert the availability of territory for German conquest and their power over it, the narration that Michel de Certeau describes as making space for readers.[74] Tiedemann marked clear borders between civilization and wilderness. On approaching one town, Tiedemann felt as though he were in a fairy tale of Scheherezade, so struck was he by the "immediate transition from complete wilderness to splendid, if also barbaric, half-civilized splendour."[75] Peters was more comfortable travelling through regions he had read about. The tone of his text changes when he enters "into regions where a white man had been before us, if only for a short time."[76] He then had expectations for what he would find, a particular river or mountain he had read about before that would remove the element of danger and introduce something familiar to Peters's African adventure. Everything thereafter improved. Peters imagined the land as more fruitful, "literally flowing with milk and honey,"[77] and the vegetation "fresher and more attractive."[78]

Tiedemann's travelogue also included a sentence that made clear how tenuous the expedition's power was. He mused about his time on the "wide, still steppe, where he was unconstrained master as far as his rifle reached."[79] Though Tiedemann meant the line to claim power for himself in East Africa, it instead makes clear how dependent that power was on violence and how little he and Peters could change African spatial practices. The expedition's power stretched no farther than Peters and Tiedemann could kill people who opposed them. Through their descriptions of the spaces traversed, however, Peters and Tiedemann claimed vast territories, as landscapes bled into one another and appeared as generalized, empty, African space.

The spatial practices they observed became key features in how Peters and Tiedemann defined different parts of their route. Peters noted a difference when he left the lands of the Oromo for areas farther west, where he encountered "Wandarobbo" peoples. Whereas "among the Gallas [Oromo] there had always been a certain maintaining of relations with the coast, such relations were here entirely absent, and there was not the slightest sign that these people had even seen Suaheli or Arabs, much less white men."[80] The areas where white men had not been, which seemed in Peters's description to universally be either desert wastelands or primeval forests, landscapes absent any sign of *Kultur*, produced in him a particular form of interaction. He believed that in these places he needed to take on a teaching role, to help people "clearly to understand that lying, thieving, and cheating are not exactly the things that ought to be in this world, but that human society rests upon a certain reciprocity of responsibility and service."[81]

Though he no longer believed it necessary to disavow Zanzibari claims to the territories he was acquiring, Peters did frequently remark on the lack of knowledge of Arab culture or Islam among the peoples he encountered in Central Africa, thus representing space as empty, uncivilized, and available for German conquest. The lack of contact with Arab culture became a distinguishing characteristic in his descriptions of the people he encountered. In place of the threat from the Arab slave trade, the people of Central Africa faced a constant danger from Maasai raids. Peters imagined the Maasai as the regional power that German rule would have to overcome in the *Kultur*-less territory bordered as "Africa." When everyone Peters encountered feared Maasai rule, it meant, thereby, that they had no knowledge of Arab rule. There was so little Arab presence, in fact, that Arabs and Europeans "all have a common interest, namely, to assert ourselves against the wild natives, who, on their part, make hardly any difference between Europeans and Arabs."[82] German colonialism in the areas west of the Indian Ocean coast, therefore, faced an African enemy, not the Arab enemy that threatened the DOAG's position. They could still, therefore, be subject to German territorial claims without complication.

Peters's and Tiedemann's descriptions of the Maasai threat, as well as other elements in their descriptions, mimic the spatial imaginaries in travelogues written by coastal merchants. A poem by the Kiswahili poet Sleman bin Mwenyi Tshande describes nights spent in fear of Maasai raids.[83] Sleman's account, like Peters's, omits daily descriptions of travel and march, as well as the roles and actions of the caravan's porters, for moments in which his own action could be highlighted. Landscapes appear only where Sleman can describe them as "deserted," empty of people and houses.[84] The expedition skips from village to village, the *hongo* the expedition paid in each. He spoke dismissively of the power of chiefs in the interior, mockingly comparing them to sultans of Zanzibar.[85] Leaving the coast meant leaving civilization behind. Peters also marked a clear boundary between Arab and African space, in his terms a boundary between "civilization" and "wilderness."[86] Most of the territory through which the expedition travelled, according to Peters, was characterized by its "primeval" landscapes, with few people and entirely devoid of civilization.[87]

Peters returned to his old methods of acquiring territory during the expedition, indicating his inability to represent space differently from how he had repeatedly done. As he neared Equatoria, Peters learned of Emin's return to the Indian Ocean coast with Stanley. If Emin was no longer in Equatoria and no British governor had taken his place, it would be unclear in Europe who controlled the province. Peters saw the opportunity to make it and more of Central Africa German. He fell back on his tried-and-true method of signing treaties with local rulers. Peters signed a treaty with Mwanga II, king of Buganda, on 28 February 1890, in which Mwanga agreed to abide by the Congo Act and declare his territory a zone of free trade and passage.[88] Peters repurposed his expedition to a different method of preventing the spread of Islam and the "Arab Empire" into Central Africa. Mwanga II became Peters's proxy for Emin as the potential ruler of a German Central African province. Mwanga was not duped. He had already been deposed once with British assistance and had just again secured his position by making concessions to the British East Africa Company. Peters and Germany offered a counterbalance to a possible move by the British government to depose him again. Unlike some of Peters's previous co-signers, Mwanga understood the stakes in signing the treaty with Peters, which was written in Kiganda, Kiswahili, and French.

The Peters expedition repeated many of the tropes that had characterized his expedition earlier in the decade. His travelogue, along with Tiedemann's, created space through the narration of travel through East African landscapes. Those landscapes appeared as transplants from the primeval or medieval past into the modern present. Peters and Tiedemann used comparisons with European landscapes to make East Africa's lack of historical progress clear to German readers. Such depictions made German colonialism seem necessary to create *Kultur* where none existed and to protect people in the territory from the Arab or Maasai presence that made progress impossible. They justified greater German interventions into the territories in question in the following years.

Epilogue

While Peters was marching west, word arrived on the coast from Stanley that he had found Emin and the two were on their way to Mpwapwa. Stanley had beaten Peters to the spot and taken Emin with him back to the coast. In fact, he had already made contact with Emin on 29 April 1888, months before Peters had begun his expedition.[89] After a few weeks of negotiations, Stanley convinced Emin to leave Equatoria (a task made easier by the attrition of Stanley's expedition to the point where it could provide little military support on its own). Stanley and Emin arrived in Mpwapwa, on the route to Bagamoyo, on 11 October 1889.[90] Although there was initially much excitement at Emin's return to European society, that excitement would turn to disappointment as the real Emin was not all the expedition's supporters had claimed he was.

Military officers wrote letters about Emin's arrival in ways that supported the expedition's enthusiasts' claims about Emin's Germanness. Emin declared his thanks to Germany and the DOAG for their support. *Schutztruppe* (Germany's colonial army) officer Rochus Schmidt treated Emin as a German hero as the two celebrated a "German evening" and Emin declared his delight at seeing the German flag and looked forward to a vacation in Germany.[91] He dreamed of entering German service, first as a translator, then returning to his former province.[92] Emin broke down crying upon receiving a congratulatory telegram from the kaiser, saying he was now willing to die because he had received the highest possible honour.[93] Emin also attributed the ulterior motives to Stanley that supporters of the expedition had believed. Stanley had offered him two different posts in the service of other colonial powers, one as a British governor northeast of Lake Victoria, the other as Leopold II's governor of the Central Congo.[94]

Rochus Schmidt's letters also contained elements of the expedition's opponents' claims and one item that expressed the tension in Emin's status as the representative of white masculinity among Black Africans. Emin had an "Oriental manner," wrote Schmidt, in that he did not know how to say no to anything.[95] More troubling, however, was that Emin had a daughter, Ferida, who had been born in Equatoria. Schmidt speculated that Ferida's mother was an Oromo woman Emin had met in the province. Ferida's existence activated fears about European men "going native" while living in Africa, losing the civilization they claimed to be bringing to the continent.[96] The expedition's opponents claims about how African space would corrupt European men seemed justified. Ferida's existence undermined Emin's role as a bastion of civilization holding back the influence of Islam that had been so prominent in pre-expedition campaigning.

Emin and Stanley's return "electrified the entire civilized world," but opinions about Emin soon split along national lines.[97] The *New York Herald*, Stanley's employer, referred to Emin as an "adventurer," a characterization that the *Berliner Tageblatt* protested. If he was an adventurer, it was in the line of Columbus, Walter Raleigh, James Cook, and Francis Drake, heroes of European expansion. If only the United Kingdom had helped Emin after Gordon's death rather than leaving him to his fate, European civilization would likely still exist in Equatoria.[98] Emin had maintained a "loyal watch" on the "outermost outpost of *Kultur* in the heart of Africa," developing a "selfless dedication to the mission of *Kultur* that fell to him," showing a talent for command and organization that was rarely seen in history. He was a "rare genius."[99] The kaiser wrote to Emin that the explorer had "heroically shown true German loyalty and fulfillment of obligation."[100]

Much of the fuss over Emin Pasha seemed ridiculous when Emin did not meet German expectations. Though he had earlier called himself "half-blind," it seems colonial authorities did not really believe it until he fell out a second-story window in Bagamoyo on 4 December, having drunkenly mistaken it for a door.[101] He was laid up in the hospital for several months thereafter. Discussions with Emin revealed that he was delusional about his governorship. He was convinced that he could return to Equatoria and rule over the people there, who loved him. Reports of Emin's delusions provided further evidence against the claims the committee had made in 1888.

Peters's shift to national, rather than personal, empire during the Emin Pasha Expedition shaped his later career. After signing his treaty with Mwanga, Peters made his way back to the Indian Ocean coast, claiming success in the capture of Central Africa for German interests.

Peters was a key figure in the formation of the Pan-German League. Together with new means of ethnographic mapping, the spatial orientation of the Peters/GfdK spatial imaginary became the way that the German Right mobilized the public for national expansion. The League's members utilized historical narratives to represent space as needing German influence to progress. The acquisition of more territory would provide space for German agriculture, expansion that would relieve the pressures of modernity.[102]

By the time Peters returned, the Imperial Government, now under the direction of Chancellor Leo von Caprivi, had assumed control of Ostafrika. The DOAG would thereafter be sidelined as the primary force driving policy in the colony. Caprivi's government would attempt to head off future attempts to claim territory through an agreement with the United Kingdom on borders, the Heligoland-Zanzibar Treaty. The Peters and Emin model of empire, however, had many adherents, and the German Foreign Office would try to incorporate them into the colonial state without compromising its vision of Ostafrika's future. The resulting commissioner system and the Heligoland-Zanzibar Treaty will be part of chapter 7. But first, we turn to the expedition for which Hermann Wissmann was called away from the Emin rescue, the Bushiri War on the Indian Ocean Coast.

6 The Bushiri War and Anti-Arab Internationalism

Beginning in August 1888, residents in the coastal town of Pangani and the surrounding area took up arms to fight against the DOAG's actions upon leasing the coast from Zanzibar. The primary grievances were overly harsh methods of collecting tolls, which the company had raised to 5 per cent on all goods, and its officials' lack of respect for custom and belief, most prominently in Emil Zelewski's use of dogs to search for a suspect in a mosque.[1] German attention soon focused on Abushiri ibn Salim al-Harthi, a plantation owner near Pangani whose name Europeans shortened to "Bushiri," as the leader of what the DOAG and the government termed the "Bushiri Rebellion." The troops built up in the coastal towns quickly defeated the DOAG's small detachments of *askaris*, the title given to non-European soldiers in German service, though the DOAG did its best to resist.[2] The fighting soon spread to the entire coast as anti-German forces took control of the DOAG's stations and cities. The DOAG did not have enough forces of its own to control the situation so, after appealing to Barghash to use his *askaris* to fight the rebels through the British, it asked the German government for help.[3]

For the first time in East Africa, African actors appeared to be directly challenging German fantasies of simple colonial rule on the African continent, forcing some German colonists to reconsider those ambitions. In response, Chancellor Otto von Bismarck's government organized its first forays into formal, state-organized colonialism. Defeating anti-German forces required an eventual government outlay of 4.5 million marks, the 2019 equivalent of over 300 million US dollars, money that Bismarck had to convince the Reichstag to provide.[4] This event was a forerunner of the Kaiserreich's later state-directed imperial military campaigns in Africa, China, and the South Pacific. State-directed initiatives became central to all German colonial projects subsequent to the

Bushiri War. The response to the fighting would transform approaches to space in Ostafrika. Colonialists' belief that simply taking territory for Germany would be overtaken by a different colonial model in which Germans had to work with African communities to create *Kultur*.

The increased role for the German government depended on the invention of a new, pan-European representational space in which East Africa was decimated by the "Arab slave trade." The new space justified the expenses and provided a shared national mission for confessional conciliation. The new space was one in which the majority of the colony's development would happen through the work of African subjects. This entailed the government adopting a form of sovereignty predicated on German protection of Black Africans from Arab slave traders and the development of the colony's African population into productive subjects. To make this shift, Bismarck would elaborate historical narratives of oppression by Arab slave traders. First, Bismarck would organize a blockade of the coast in concert with other European powers to try to cut anti-German combatants off from necessary supplies. The regime's belief that a blockade could solve the DOAG's crisis reflects the victory of representations of coast and *hinterland* as inextricable and East Africa as part of the Indian Ocean World. When the blockade failed to stop the fighting, Bismarck went to the Reichstag for funding for a military operation on land. This entailed the creation of a new coalition in the Reichstag, one in which Bismarck made peace with his primary antagonists to that point, the Catholic Centre Party. Klaus Bade argued this alliance was a "manipulative instrumentalization of a colonial issue for domestic and parliamentary politics."[5] I argue that the choice by Bismarck was rather driven by a need to respond more strongly in East Africa for foreign policy purposes. The blockade as the initial response, without Reichstag approval, indicates that other concerns were supreme to winning over the Centre. Domestic politics did not trump foreign policy; Bismarck's actions in discussions out of public view reflect a real belief in racial-religious animosity against Germans.

The meaning of the Bushiri War has been the subject of great contention for subsequent Tanzanian history. In the 1960s and 1970s, it served as a touchstone in nationalist historiography, a narrative in which Bushiri was a proto-Tanzanian nationalist, fighting for the nation's independence from colonial rule.[6] Later interpretations of the anti-German actions have instead interpreted them as the continuation of older forms of protest that predated German arrival.[7] What appeared to German colonialists as a legal transfer of sovereignty from Sultan Said Khalifa of Zanzibar to the DOAG appeared to be something else to residents of coastal towns, as simply the latest change in a long cycle.[8] This chapter accepts the view of the fighting as the continuation of older forms of protest, but argues that the most important legacies of the conflict were the new visions of East African sovereignty, history, and space that it produced.

The German government adopted a representational space of East Africa that justified taking control of the colony and intrusive measures into life in the name of stopping the slave trade. Through the blockade and the Wissmann expedition, the basis of German claims to sovereignty in Ostafrika became the protection of African bodies from the Arab slave trade. The Bushiri War forced an answer to questions about the relationship between the German state and Zanzibar. The DOAG had only leased the coast, in an agreement that recognized continued Zanzibari sovereignty. But the German government's involvement in the war required different ideas about sovereignty on the coast. The GfdK/DOAG treaties had depended on a conception of East African societies as primitive and medieval in their construction of a political environment in which rulers could sign over sovereignty to Carl Peters. In the Bushiri War, German colonialists saw as their opponent not primitive Africa but "Oriental Despotism," a different kind of premodern foe for European modernity.[9] State intervention was an emergency measure to protect Black Africans from the caprices of pre-modern "Oriental" rule. The institution of the state of exception into German rule in East Africa is what set the new German colonialism apart from earlier thinking about the area. Through the Wissmann expedition, German colonialists constructed their efforts in opposition to Islam and the "Arab" population of East Africa. The emergency measures invented to assert German sovereignty over East African race and space became quotidian elements of the administration of Ostafrika.

Challenging the Gatekeeper-*Hinterland* Model

From the German side, the Bushiri War appeared to be a rebellion of an Arab element against German rule for religious and racial reasons, which served as the basis for claims that Arabs were the main impediment to development in the colony. That idea would become the standard narrative of the war among contemporary Europeans, but a closer look at anti-German motivations shows a desire to maintain existing structures of power and relationships to the world beyond East Africa. The creation of the narrative of Arab-Muslim antagonism required the silencing of alternative narratives. The spatial imaginary that had underlain the DOAG's takeover of the coast was breaking down.

An epic poem about the transfer by Makanda bin Mwenyi Mkuu reveals the roots of the antagonism in Bagamoyo in existing grievances with Khalifa rather than anger at the DOAG. The lease of the coast went into effect on 16 August and the DOAG's actions reveal that it clearly put little weight on the idea that it would be ruling in Khalifa's name thereafter. It planned to treat the coast like the interior territories it had claimed in 1884 and 1885. Ernst Vohsen, the company's director, ordered officials to cut down Khalifa's flag in Bagamoyo. When they faced resistance, German marines landed and cut down the flag by force.[10] According to Makanda, coastal people believed that Khalifa had treated them "like women" in acting alone and had not asked their opinion. It was now their duty to speak to Ambaroni (the Kiswahili nickname given to Karl von Gravenreuth, the DOAG official in Bagamoyo) and tell him to leave town, asserting their power over the town's spaces. Gravenreuth told the petitioners he thought them haughty and doubted their abilities to make war. The *liwali* warned him of the danger, but Gravenreuth went hunting for hippopotami instead of responding. On Sunday, people surrounded the city. Gravenreuth closed the gate and shooting and arson began.[11]

Though the official German line would eventually be a unified coastal resistance, differences among coastal towns reveal the importance of specific local circumstances to the outbreak of fighting. Shortly after the lease went into effect, German colonists "blundered into an explosive situation not of their own making."[12] Rebels' grievances stemmed from both patricians, who were frustrated at their growing indebtedness and political marginalization under Zanzibari rule, and recent, less wealthy arrivals in town, angry at the lack of political opportunities. Anger was directed more at Zanzibar than at Germany. Plebeian members of the community took leadership of the movement. Even as Germans blamed them for the fighting, the largest landowners (and slaveholders) largely sided with Germany to defeat the rebels.[13] In Bagamoyo, Pangani, Kilwa Kivinje, and Lindi, townspeople took up arms against the DOAG in mid- to late August. Other coastal towns were peaceful. Gustav Michahelles, the German consul general in Zanzibar, attributed the differences to whether the towns' *liwalis* supported DOAG officials, as happened in Dar es Salaam and Mikindani, or acted as though they were still in charge.[14] The differences traced back to existing grievances against the Zanzibari administrative system.

Several narratives about the causes of the fighting circulated, most of which blamed *washenzi* from the interior for bringing disorder to the more civilized coast. Khalifa blamed the DOAG for the conflict. Early German reports blamed the "chiefs in the *hinterland*."[15] Some blamed

Khalifa's officials for the fighting, saying that they were refusing to allow any Christians in their towns.[16] Khalifa was complicit in the rhetoric of civilization. He asserted that "wild people" had instigated the fighting.[17] The Berlin III Mission claimed the DOAG was to blame due to its cruelty. Scenes reminiscent of the Thirty Years' War were unfolding in East Africa, providing imagery recognizable to German readers.[18] The German consul in Zanzibar, Gustav Michahelles, wrote that what had set off the fighting was a large migration of Yao people from the south of the German sphere of interest and the north of Portuguese Mozambique.[19]

German attention increasingly fell on Pangani as the centre of all anti-German antagonism and of a unified resistance among coastal Arabs, discarding the explanations based in other populations' anti-German antagonism. Emil von Zelewski, the DOAG district official in the town, was accused of having marched soldiers and dogs into a mosque, an event that triggered the anti-German fighters to take up arms. Michahelles wrote to Bismarck that Zelewski had already been the "most hated person on the entire coast," a man who understood nothing of "how to handle the natives." He had asked Ernst Vohsen to relocate Zelewski before the mosque incident, with no luck.[20] In late September, Michahelles described Bushiri, the planter from near Pangani, as "the true leader of the rebellion."[21] Attention would focus on him thereafter.

Bushiri made his motivations clear in a conversation with Oskar Baumann and Hans Meyer, German explorers travelling back from the Kilimanjaro region. Armed men arrested the German travellers. Then Bushiri appeared, and they made the journey together to Pangani. Bushiri provided an explanation for the fighting in the actions of DOAG officials. Fighting arose when Germans challenged local control of spatial practices. In Bushiri's opinion, it would have been fine for the Germans to take over customs had they remained peaceful in the coastal towns. But "unprotected people" started raising their own flags, ordering people about, declaring they were lords of the land, and that the coastal people were their slaves. After letting the situation play out for a while, Bushiri and his followers began hunting the Germans like "cocky youth." While the British might be a powerful nation, Germans were merely "little kids" (*wadogo dogo*). He asked the explorers why Germans did not stay in Germany. They responded by asking why the Arabs had not remained in Arabia. Bushiri laughed, then said that "my forefathers had the same goal the Germans now pursue: they wanted to take possession of East Africa."[22] But they had done so with force, while Germans had merely showed up with a letter.

Bushiri's explanation provides evidence that he wished to position himself as the primary intermediary for German rule in the region and maintain the representational space of coastal communities. He told the two Germans that in East Africa there were only three "brave and important Arabs" in East Africa: Tippu Tip, Mbaruk, and himself, notably omitting Khalifa.[23] Bushiri said he hated Khalifa, that he had not been to Zanzibar in twenty years, and that he had sold Zanzibari land to foreigners. He offered his services if the Germans returned to East Africa. He would offer better terms for porters and land than did Sewa Haji. With his comparison to Tippu Tip and offer of services to future German visitors, Bushiri was offering to become the governor of the coast, in a similar role to that Tippu Tip had taken in the Stanley Falls district for the Congo Free State.[24] He was not aiming to completely eliminate the German presence, but to become its gatekeeper. German responses ignored this aspect of Bushiri's explanation and settled on the Zelewski mosque story to paint Bushiri as the leader of a religious-racial struggle against German rule.

The DOAG's position had completely deteriorated by the end of 1888 as its forces proved incapable of retaking the coastal towns. The company's leadership decided to use the narrative of Arab antagonism to "win public opinion."[25] Bushiri became the enemy Germany needed to reimagine the space of the Swahili Coast with a blockade, then an invasion. This required ignoring the evidence that the primary causes were local and that even Bushiri did not disavow working together with Germans. It disregarded Bushiri's own statements about why he had taken up arms. In creating new representations of the Swahili coast, the company had support from many other Europeans involved in East Africa. To try to hold its position, the DOAG would approach the German government for help.

Blockading the Arab Slave Trade

In response to the company's pleas, Chancellor Otto von Bismarck's government decided to take action to assist with a blockade of the East African coast in concert with other European powers. Bismarck initially rejected government intervention because he wanted a quick solution and to prevent the fighting from becoming a major political issue.[26] By late September, the German government's opinions had shifted toward intervention, based on a belief that the DOAG could no longer hold its position. The choice of a blockade reflects in part the governmental structure of the *Kaiserreich* – a blockade could be carried out with existing naval ships and personnel and would not require Reichstag budget approval. But it also reveals a spatial imaginary of how colonialism in Africa would work. The regime's belief that a blockade could solve the DOAG's crisis displays acceptance of the representational space of *hinterland* and the Indian Ocean World and a need to adopt colonial models more closely structured after other European nations to control African populations.

In the German spatial imaginary, the coastal ports were part of Zanzibar and were the gates that connected the East African interior with the Indian Ocean World. This indicated a belief that the towns of the coast depended entirely on the Indian Ocean World, as established in the previous few years, and that being cut off from that world would mean their downfall. Justus Strandes, who had run a trading house in Zanzibar since the 1870s, recommended blocking the importation of guns and powder, as they would fall into anti-German hands. The most important points on the coast needed to be occupied and the rebels strictly punished for trade to resume.[27] By cutting Bushiri's forces off from the Indian Ocean World and imports of munitions, the rebellion could be starved of necessities.

In sharp contrast to the Emin Pasha expedition, which was designed specifically to get the better of Britain and feed an imperial rivalry between Germany and the United Kingdom, the German government conducted its operations against Bushiri in cooperation with the British government, as well as the minor colonial powers of Portugal and Italy. Given the expected lack of support for bailing out the DOAG among other powers, Bismarck developed a justification for action with representations of East African space as a site of struggle for the success of European civilization. The regime based its public campaign around fighting the "more than 1000-year custom of the African slave trade" carried out by "Muslim fanaticism."[28] Government reports to the German public latched on to the fact that Bushiri claimed for himself the mantle of an Arab conqueror.[29] That narrative excused the DOAG's failure to make money with its venture, and it supplied grounds for the German state to become directly involved in the German colonization of East Africa. East Africa and the bodies of Africans became spaces of civilizational conflict that required a pan-European response. The use of historical narratives as representations of space built upon the framework developed over the preceding years to claim the link between the coast and the interior as its *hinterland* as a means to demand the coast from Zanzibar.

This idea of East Africa's history reified the model of competition for control of Africa and Africans between Europe and Asia that was so important over the preceding years. Paul Reichard, who had travelled in East Africa at the beginning of the 1880s, wrote in the *Kolonialzeitung* that government reports of Arab, Swahili, and "heathen" involvement in the rebellion should not be believed; it was only the Arabs and the Swahili who were rebelling.[30] One can see this idea circulating beyond the relatively small circle of colonialists that were part of the invention of the new *hinterland* due to the excitement generated by the war. The *Illustrirte Zeitung* laid out a vision of Africa that justified the blockade, that "wide regions" of the African interior served as "regular hunting grounds" in which Arab merchants attacked and captured Black Africans. Now that those slave traders felt threatened by the German takeover of the Zanzibari coast, they were hunting DOAG officials.[31] The DOAG in turn asked for 10 million marks for this plan to build a "lasting dam" against the slave trade.[32]

The German government adopted a cultural geographic framework to make its arguments. Cultural geography had enjoyed a reflexive relationship with the colonization of Ostafrika over the previous years. The historical geographies of the Indian Ocean World and *hinterland* established by German colonialists over the previous years were one of the cases for the model of cultural diffusion Friedrich Ratzel laid out in the third volume of his *Völkerkunde*, published the same year the war broke out. In Ratzel's telling, the Arab migration to East Africa developed as the outcome of religious and political unity in the Middle East. From Zanzibar, Arabs had spread their influence west to Lake Tanganyika. Arab influence had begun through trade, then taken over through superior *Kultur*.[33] That history was a demonstration of Ratzel's principle that most "historical activities of lower races necessitate a spatial shift."[34] Arab migration to East Africa was thus a racial struggle for control of African space. In public discussions of the war, proponents of an aggressive response leaned on Ratzel's model to make their arguments. The anti-German fighting became the manifestation of Arab-Islamic expansion into East Africa and the spread of *Kultur* from the Arabian Peninsula to Africa, with disastrous consequences for African communities. In this spatial imaginary, the European battle against Islam became a struggle over the processes of cultural diffusion.

Bismarck used the narrative of increasing Islamic attacks against European territory over the previous several years in his approach to the United Kingdom about the blockade. He asserted that "Muslim Arab" interests were "colliding" with European ones. Movements spawned by "religious fanaticism" and "business relationships" in the slave trade had led to the emir of Harare's attack on an Italian expedition in 1886, to Arab threats to the Congo State and British missionaries in Uganda and on Lake Nyasa/Malawi, to Muhammad Ahmad's attack on Khartoum, to Tippu Tip's perceived anti-European actions, and now to the Bushiri War. It was clear that a "slowly-advancing but deep-seated movement of the Muslim population in the direction of a reaction against Christian and civilizing efforts" was at work in a struggle over African space. It was especially a problem with the slave trade, which all Christian nations wished to end.[35] With or without British support, Bismarck would go to the Reichstag in the "spirit of the anti-slavery societies and the analogous efforts of Cardinal Lavigerie."[36]

Particularly important in Catholic participation in the Bushiri War was Charles Lavigerie, the archbishop of Algiers and the Church's Primate of Africa. As discussed in chapter 5, Lavigerie had promoted missionary work in Central and East Africa through a new missionary order, the White Fathers, in opposition to Islam and Arabs. Lavigerie had travelled to Germany in August to promote his new organization. He spoke in Berlin on 28 August.[37] Concern about slavery had become more strident in 1888, as members of the International Anti-Slavery Committee began plans for a conference in Brussels that was to determine Europe-wide means of combatting the slave trade. Members of the German colonial lobby were heavily involved with the international effort.[38]

In the negotiations with the UK, the German Foreign Office depicted an East Africa besieged by Arab slave traders. Diplomats claimed ending the slave trade was the "common duty" of all European nations.[39] A blockade would be the best means to defeat the "herds of the anti-Christian movement on the coast." The German government argued that a blockade of the German coast alone would prove ineffectual against the "anti-Christian and anti-*Kultur* (*kulturfeindlich*) movement," as its members would export slaves and import munitions, their "main means of support," through the British sphere of interest instead. Without British support, it would be impossible to convince Khalifa to support the blockade. He would recognize a movement of unified "Christian civilization," but otherwise his "invariable sympathies with his Arab fellow countryman and with the foundation of his house, the slave trade," would determine his conduct. A united blockade was the only way to ensure Khalifa would not support the anti-German forces.[40]

The British Foreign Office replied that its members "entirely share the vision expressed," throwing its support behind the narrative of religious and racial resentment against European colonialism. They believed in the "civilizing mission on the East Coast of Africa" and with the plan to keep Khalifa in power.[41] Lord Salisbury, the prime minister, did not doubt that "the slave-trading Arabs" were responsible for anti-European actions.[42] Mohammedan fanaticism" had kept the African slave trade alive for over 1000 years. Africans would be receptive to Christianity and European civilization were they not "repressed by the armed force, the higher intelligence, and the cohesion" of Arabs. This could be stopped by bringing superior European weapons to bear.[43] But Salisbury also suggested a different approach from German plans. He wanted to build a larger coalition of European states to participate rather than proceeding bilaterally.

Negotiations with other powers also depended on the representational space of East Africa as under threat from Arab slavers, a space that other governments participated in crafting through their conversations with German diplomats. Clear in Bismarck's rhetoric was the idea of an advancing Islamic *Kultur*, spreading to East Africa and negatively transforming people and space. To convince France, the German Foreign Office claimed the blockade was the work of all nations that "earnestly worked for the development of Christian culture and customs in European colonies in Africa," all of which needed to eliminate the "advancing Arabdom and the cruelties of slave raids."[44] The French agreement to the blockade cited the argument that "slave traders of Arab nationality oppose the suppression of the trade and legitimate commerce of Christian nations with African natives."[45] To win over Portugal, Bismarck used the same language of defeating the "advancing Arabdom and the cruelties of the slave trade" but also adopted slightly different rhetoric to appeal to Portuguese interests. He highlighted the "increasing spread of the Mohammedan movement in East Africa," the origins of which were "in part in religious fanaticism and in part a reaction of the Arabs involved in the slave trade against the penetration of European elements."[46] Henrique de Barros Gomes, the Portuguese foreign minister, agreed to participate because that the blockade was a chance to fight the "preponderance of elements contrary to the influence of European civilization."[47]

The negotiations for the blockade formed a key moment in the formation of the German spatial imaginary of East Africa. The representational space of East Africa under siege by Islam justified applying measures normally reserved for interstate conflicts to fight rebels. In trying to convince the French government to support the blockade, the German Foreign Office attempted to expand the scope of international law. The precedents were an 1827 blockade of the Greek coast by the UK, France, and Russia, and an international blockade against the Netherlands to support Belgian independence.[48] The 1827 blockade had marked a watershed in international law, in that the blockading countries never declared war on the Ottoman Empire. What would be new in East Africa was that this blockade would be directed against non-state actors, rather than a government. The agreement between Germany and the United Kingdom, issued on 8 October 1888, declared a blockade against the "fanatical and xenophobic (*fremdenfeindlich*) Arab elements" on the coast in the name of the Sultan of Zanzibar.[49] Thus, the European countries involved would apply emergency war measures against not a state, but against a group of people defined by race and economic activity.

With the blockade, which commenced on 2 December, the Bismarck government took a new step in its support for German colonialism. Negotiations with the UK and other powers threw the full support of the German government behind the colonization of East Africa. That support was based on the adoption of the representational space that had underlaid the DOAG's claims to the coast. Furthermore, they achieved international support through claims of African communities under siege by Arab slavers, in a narrative where the battlegrounds were African space and the bodies of Black Africans. Other European colonial powers adopted similar representations of space in joining the blockade. As the blockade played out, and failed to stop the fighting, the Bismarck regime would pursue that narrative even farther.

Building a Political Coalition to Win the War on Slavery

By early 1889, it was becoming clear that the blockade alone would not solve the crisis on the Swahili coast. A more extensive operation was necessary, landing a military force on the mainland to defeat the anti-German forces. Chancellor Bismarck decided to create a special force for the operation under the command of the famous explorer and anti-slavery activist Hermann Wissmann. But to pay for such a force he needed votes in the Reichstag, which controlled the imperial budget. To win funding for his emergency measure, Bismarck formed a union with the formerly antagonistic Catholic Centre Party over a conception of East Africa as the site of a world historical struggle between Christianity and Islam, inserting the version of East Africa's representational space developed over the previous years from foreign to domestic policy to make it the basis of German colonialism.

Several observers asserted that a blockade would not be sufficient to break Arab power and that military operations on land would be necessary to achieve lasting dominance over East African Arabs. Bismarck chose Hermann Wissmann to lead the expedition, a choice that provided additional support to the argument that the fighting was about holding back Arab slavers. Wissmann had made his name as an explorer of the Congo in the early 1880s; his travelogues highlighted the degradations the slave trade created in Central Africa and had been important in establishing the representational space of Africa as decimated by Arab slavers. His publications and public speaking made Wissmann probably the most well-known German witness to the horrors of slavery.[50] Wissmann had written that he was filled with "a feeling of wrath against those who had brought forth this appalling change, the Arabs."[51] In the Reichstag debate over funding for the expedition, Wissmann described the East African coast as the "most important part of the continent for fighting the slave trade." From there, Germans could establish a base for fighting the slave trade in the interior.[52]

The arguments for an expedition on land won Bismarck over, but the primary restriction placed on the kaiser and chancellor in the Imperial German political system was control over the budget, so Bismarck would need Reichstag approval of the expedition's funding. Approval for that funding would necessitate a dramatic reshuffle of parliamentary alliances and peace between the Catholic Centre Party and Bismarck. Bismarck had spent much of the 1870s and 1880s engaged in the *Kulturkampf*, an anti-Catholic campaign. He had maintained majorities in the Reichstag through an alliance of conservative and National Liberal parties, the so-called Kartell, with the Catholic Centre Party excluded from governance. In 1888, Bismarck would build a new coalition that included the large delegation of the Centre Party and bring an end to confessional conflict at the highest levels of the German government.

Catholic organizations had been involved in German expansion and the construction of ideas about geographies of the extra-European world. The Centre Party supported a German role in Africa and had shown its support through missionary work in Ostafrika and elsewhere. For Catholic missionaries, Islam was characterized by its "cruelty" and "barbarism," epitomized by the figure of the "slave raider."[53] The Bismarck government specifically latched on to Lavigerie's existing movement to win support from Germany's Catholics. Herbert Bismarck wrote to his father on 5 November 1888 that the expedition could provide the grounds for rapprochement with the Catholic Centre Party. Germany could get Catholic support to fight the "Arab element participating in the slave trade."[54] Otto von Bismarck wrote that he was going to the Reichstag "in the sense of the anti-slavery societies and the analogous efforts of Cardinal Lavigerie," aiming to win British support for operations.[55] The representational space of East Africa besieged by Arab slavers could unite Germans across confessional lines.

Signalling the new alliance, Ludwig Windthorst, leader of the Catholic Centre Party, introduced a bill to fund an expedition for "the fight against the Negro trade and slave raids in Africa," in the Reichstag. He noted the "great movement" across and outside of Europe to end the slave trade, which he claimed killed at least two million people each year and prevented the Christianization of Africa. Bismarck had raised similar concerns, highlighting the fact that many missionaries had been killed in Africa.[56] Windthorst described the bill as a "great trust vote" in Bismarck, offering millions of marks without specification for how they would be used. In normal times such a vote would be "completely unacceptable," but in this case the chancellor had all the information while the Reichstag had little and therefore needed to be trusted.[57] Emergency situations demanded emergency measures.

Protestant leaders in the Reichstag also spoke in favour of the representational space of East Africa as a battleground over African souls. In their rhetoric, Africa appears as a space of conflict, with no attention to Africans as individuals. The conservative Otto von Heldorff denied Lavigerie leadership of the anti-slavery movement in favour of William Wilberforce but announced that fighting the slave trade was now a "joint mission of the Christian confessions" for civilization.[58] Germany was now ready to take up work pursued by other great powers since the beginning of the century. The Conservative Adolf Stoecker, court chaplain to Wilhelm I, told his fellow deputies the question was "whether the Mohammedan or the Christian worldview would win in North and Central Africa."[59] These speeches signalled a desire to bridge the confessional divide in Germany in order to fight Islam.

The German mission against Arabs and Islam also attracted liberal politicians to the cause. Rudolf Virchow, renowned anthropologist and Freisinnige deputy, added his academic credentials to the narrative of anti-European agitation among Arabs. Beginning with Ahmed 'Urabi's rise to power in Egypt, Arabs had been trying to conquer all of Africa. It was impossible to imagine that a people who had "once carried out such great campaigns of conquest through the entire world and which were filled with such a glowing hatred of Christians" would simply be "driven back" by Europeans.[60] Virchow took up a line of argument popular in French demands for North African territory, where French writers had portrayed Islam as destructive of Berber culture. Their assertions cast Islam and Arabs as both medieval and as inherently violent in their entry into Africa.[61] Virchow's argument set up the struggle as one primarily over space with a representation of that space that spanned empires.

Bismarck articulated a connection between history and space during the debate, which entailed building on the base of Arab colonialism on the coast and ignoring the African wilderness of the interior. The speech marked a clear, official turn away from the DOAG's representational space of Ostafrika to a new model. The DOAG's initial territory, "beyond the Zanzibari coastal region," signed over with "a difficult to read piece of paper marked with negro crosses" could help Germany no longer. The occupation of the coast, on the other hand, was of "great significance." It made the land beyond valuable but also saddled Germany with "cultural duties" to spread Christianity inland. Arabs hated Christians as the primary disruption in their illicit trade. In Bismarck's opinion, "only from the coast can civilization cross into the interior." The future of the colony was not in the caravan trade, which was based mostly around slavery and increasingly rare elephants. Rather, it was in rubber and other plantations "in the tropical sense." Such plantations could save Germany the five hundred million marks it spent annually on importing "tropical products."[62] The coast had to be held for Germany. Bismarck's speech was based around an idea that German *Kultur*, once established on the coast, would diffuse westward over time, remaking African space as it went. This concept of cultural diffusion drew on models created by cultural geographers to explain social evolution.

Depicting Wissmann's expedition as the best means to win the struggle over East African space and bodies provided Bismarck with the opportunity to build new political alliances. Ostafrika became a representational space where Germans could imagine their union after years of interconfessional conflict. Germans united across confessional lines over the story of Ostafrika as the site of a challenge by premodern Arabs and Islam to German Christianity and glory. Confessional conflict could thus be replaced by a shared Christian conflict with Islam outside of metropolitan Germany. German Christians could direct their attentions to Ostafrika rather than struggling with one another. The Wissmann expedition departed with this powerful religious and cultural mission backing military action.

The Wissmann Expedition and the Militarization of German Rule

Wissmann set about defeating what Germans believed were Bushiri's forces from Bagamoyo on 8 May. Wissmann made short work of the anti-German forces, largely due to superior German firepower and a lack of organization among his opponents. Local leaders adapted to the shift in the balance of forces and accommodated the German presence. Such accommodation was possible due to the imagined border between the Arab space of the coast and the African space of the interior. Wissmann's actions during the campaign reflect a tension between the new representational that underlay the anti-Arab slavery campaign and the DOAG's geographies of all of East Africa as "African," an easily colonizable space.

The Wissmann expedition marked a new era in German overseas imperialism. The state was thereafter to play the leading role in shaping the empire and governing the heretofore private colonies. Bismarck told Wissmann his purpose was "to institute measures for fighting the slave trade and for the protection of German interests, to maintain peace and order in the designated areas through all designated means" on the coast.[63] Wissmann's assumption of the DOAG's authority along the coast turned a formerly civilian administration into a military one overnight. Wissmann's orders were clear: to take and fortify the cities along the coast, and to destroy whatever forces tried to stop him. That plan required almost by default a permanent German military presence in East Africa to man the fortifications that Wissmann built. This meant a long-term commitment to direct state involvement in empire, which Bismarck had tried his hardest to avoid up to that point.[64] German rule on the Swahili coast became militarized and dependent on a constant military presence. The government did not want Wissmann to leave the territory that the public campaign for the expedition had represented as Arab colonial space along the coast and enter the territory they had represented as uncivilized Africa farther inland. He was to limit his operations to the ten-mile wide strip of territory along the coast that the German, French, and British government had declared was Zanzibari territory in 1886. The government stuck to its geographical imaginary of a limited Arab presence in East Africa, one that justified the aggressive conquest of the coast and a laissez-faire approach to areas farther west.

The anti-German forces collapsed quickly in the face of the German invasion, as one might expect given their lack of overall coordination. News of Bushiri's death arrived unexpectedly and without explanation. Wissmann telegraphed on 16 December to report that Bushiri had been "sentenced to death by court martial. Sentence straightaway fulfilled."[65] Bushiri had continued his campaign, but lost allies along the way. Karl von Gravenreuth, one of Wissmann's officers, reported that almost the entire Arab population of Pangani had returned to the town, and the town's *jumbes* had promised not to undertake further hostilities.[66] With the German *Schutztruppe* in the vicinity, it made more sense to work with Germans rather than continuing the fighting. Wissmann reported that Bushiri had surrendered himself willingly "as a military leader. He did not think he would suffer a death sentence, rather he hoped I would send him to Berlin."[67] The primary goal of his expedition had seemingly been met. Bushiri's and Wissmann's interaction indicates the continued illegibility of motives between Germans and coastal elites, several years into German colonial rule. Bushiri comported himself according to a reasonable expectation of European conduct in war, surrendering and expecting to be treated as a foreign military leader. Instead, Wissmann treated him as a rebel and pursued summary justice.

Wissmann's actions after Bushiri's capture reveal the role of the narrative of racial struggle and the new representational space of Ostafrika in his thinking. After the capture, Wissmann and his soldiers dressed Bushiri in his best clothes for a photograph. They then reposed him with a bare head and in chains for another.[68] The first photo portrayed Bushiri as the figure of Arab conquest on which the whole operation was based, as a man who had grown rich from his participation in the slave trade. The second photo showed Wissmann's ability to destroy that presentation and turn Bushiri from an Arab slave trader to a Black African colonial subject. Bushiri's clothing was the outward display of the Arab connections he claimed. Wissmann asserted German power to remove such displays, write their own histories of East Africa, and classify colonial subjects as Germans desired.

Wissmann did not find the Arab slave trading threat and its space that he was looking for on the coast, so he instigated a debate over whether to project German power west or to concentrate on the coast from which German *Kultur* would diffuse westward. As Wissmann saw it, his job was to eliminate the threat to the German presence in East Africa posed by opposing forces, in the imagined Arab empire that Bushiri represented in his campaign against the DOAG. Wissmann believed that that task was not best conducted by remaining on the coast. He argued that the enemy he was sent to fight was situated in the interior, in Tabora, a town several hundred miles inland from the coast.[69] Expanding operations inland meant expanding what areas counted as part of the Arab empire of slavery that he had been sent to defeat. As he looked to expand the borders of his operations, he did so in racial terms, claiming that the Arab influence in East Africa had spread beyond the coast to Tabora. Tabora was a popular target among members of the colonial movement. Paul Reichard had declared that Zanzibar's influence in Central Africa depended on Tabora, which was "the East African Khartoum ... to whomever it belongs, belongs all of Central Africa." It could be the *hinterland* necessary to make the coast valuable.[70]

Support for expanding operations to Tabora also came from within the Foreign Office, indicating such a spatial imaginary was not only the province of aggressive colonialists. Maximilian Berchem, a Foreign Office under state secretary, noted the need to establish borders of the "Mohammedan establishment, whether from Muscat or Zanzibar." Although the coast could serve as a key point in that campaign, the "main Arab base [in East Africa] is Tabora." German interests could better be served, Berchem told his superiors, by taking and fortifying Tabora than through Wissmann's small operations along the coast. Berchem argued that "Tabora could take the role of Paris for our protected region," and "that we will reach our goals more quickly if we attack the centre rather than the periphery."[71] Berchem told the government that it would be cheaper for Germany to build a few major stations at the key points of the Arab slave trading empire than to establish a network of stations through the entire colony. This entailed a reorientation of the earlier geography of establishing a network of nodes and stations to one where Germany would police the interior as *Kultur* moved west.

Kaiser Wilhelm II threw his weight behind the new representational space Bismarck had articulated in a preview of his increasing interest in colonialism as part of his *Weltpolitik*. He determined that the coast and the interior were fundamentally different spaces, meaning the interior should be administered differently from the expensive coastal administration. The kaiser refused to go along with the idea that Tabora should be the first priority. In the margins of Berchem's letter, Wilhelm noted that "Kilwa is more important than Tabora. We will figure out the interior when we rule the coast."[72] If Wissmann could just control the entire coast of the German colony, there would be no problems in the interior. That rejection also meant a rejection of the idea that the Arab empire had moved farther into the interior than the original ten-mile strip once believed it occupied. The state's obligations could be limited to the coast and not require the same military force to rule the entire colony. Direct administration of the interior would also be unnecessary, whereas the Berchem plan meant establishing an administrative infrastructure for the interior as well as the coast. That difference was key: Berchem's plan could have saved money and work had the government imagined the entire of DOAG territory as part of the same Arab empire, and therefore in need of the same administration. If the entirety of Ostafrika was not part of that empire, however, and some of it was territory that could be administered more cheaply, or not at all, it made no sense to rule the entire colony in the same way.

Bismarck naturally took up Wilhelm's points as his own, as they fit his spatial imaginary, and he desired to keep colonial costs low. He rejected Wissmann's proposal, unwilling to devote the resources necessary to extend military rule so far west. The chancellor agreed that Wissmann should "limit himself to the pacification of the coast."[73] The Foreign Office stated that it was more important to secure the coast than to move west. Kilimanjaro and other western regions "laid outside the program, the south coast within."[74] The coast could provide enough of a basis for building an administration. Bismarck thus threw his weight behind a new approach to Ostafrika's development, working with people of the coast to eventually extend *Kultur* west.

Wissmann had successfully completed his military orders, to defeat the imagined Arab rebellion against German colonialism. His subsequent efforts to extend military operations west relied on the representational space of the previous years that had depicted coast and *hinterland* as one indivisible space. The same kind of military action would be necessary to bring the entirety of that space to heel. Bismarck and Wilhelm did accept Wissmann's emphasis on police action for the interior, and the expedition marks the beginnings of the militarization of German rule. The chancellor's decision not to allow Wissmann to travel west, however, ensured the preeminence of the new spatial imaginary of the region and models of the diffusion of *Kultur* west from the semicivilized coast.

Conclusion

Representations of Ostafrika as the site of Arab cruelty came to dominate how Germans viewed the colony. Stories of atrocities committed by Bushiri's forces were common in the German press during and after the war. Colonialists and members of the Wissmann expedition kept the narrative of a "Bushiri Rebellion" alive after Bushiri's death. Though Oskar Baumann's travelogue described mercy shown by Bushiri, Major Eduard Liebert (who had served under Wissmann) rejected Baumann's claims. Liebert wrote that Bushiri had changed and had "earned his death a hundred times over." Wissmann's forces had found hacked-off hands and feet and "slowly-killed children over a smoldering fire."[75] The narrative of the war as the uprising of Arabs was powerful. It would come to be known as the "Arab rebellion" (*Araberaufstand*) in German accounts for decades thereafter.[76]

Wissmann returned home as Germany's first war hero since unification in 1871. Colonial enthusiasts staged an elaborate celebration in Berlin at the end of June 1890, with two hundred attendees. Wissmann sat between Ludwig Windthorst, the leader of the Catholic Centre Party and the Conservative Albert von Levetzow, the president of the Reichstag, displaying the new political alliance forged through the explorer.

Levetzow hailed Wissmann as the military leader who had "triumphed to plant Christianity, humanity, and *Kultur* in Africa." Wissmann thanked Wilhelm II and the Reichstag, and reserved special thanks for Windthorst. The whole group then sang the nationalist anthem, "Stimmt an mit hellem, hohen Klang," and someone read an announcement from the new Chancellor, Leo von Caprivi, that "Ostafrika is clearly the key element (*Schwerpunkt*) in our colonial policy."[77] The celebration welcomed the show of German military might that had laid dormant since unification.

Bismarck's new representational space of Ostafrika outlived his regime. The emergency military measures introduced during the war would remain a central element in German work in East Africa for the next three decades. In the wake of the expedition, the German government took over control of the colony for the DOAG, which from that point forward became merely a private company, no longer a private sovereign. With the government takeover, administrators would pursue rationalized colonial rule based on a spatial separation between Arab and African space, with the latter governed through military measures designed to stop the influence of Arab slavery. The racial-religious emergency of 1888–90 would become part of the colonial everyday on the coast, with Germans believing their Arab enemy had been cowed and conquered.

7 Rethinking the Spread of *Kultur* West

On 5 March 1892, the Reichstag debated funding for measures to fight the slave trade in Ostafrika. The debate pivoted on a series of articles the journalist Eugen Wolf had written in the *Berliner Tageblatt* that had levelled extensive criticisms at the actions of the colony's governor, Julius von Soden. In Wolf's telling, the "Soden System" was setting back colonial progress in nearly every area. Through Soden, bureaucrats had usurped colonialism from the "*Alte Afrikaner*" (Old Africans), by whom Wolf meant the heroes of exploration and conquest such as Emin and Wissmann, who had created the colony and imposed German order through the Bushiri War. *Bwana Kartasi* ("man of paper"), the Swahili name bestowed on Soden for his bureaucratic tendencies, had lost sight of the colonial movement's original goals in his pursuit of respectability.

Beyond superficial concerns about how much paperwork Soden was generating, the debate was fundamentally about the evolving German spatial imaginaries of Ostafrika and the role of the state in colonialism. Three approaches to making *Kultur* in the interior had been circulating through colonialist circles. The first of these was based in the thinking of the aggressive colonialism of the previous years, taking *Land* to make it German. Second, some administrators promoted importing models from other colonies to create development. Railways and labourers from Southeast Asia, these administrators thought, would put Ostafrika on the successful path followed by other colonies. Finally, some colonialists more explicitly adopted models of cultural diffusion from cultural geography. Promoting perceived positive racial characteristics possessed by some of the colony's inhabitants would cause those characteristics to spread to others in the colony, creating *Kultur* over a wider area. Both the second and third schools of thought promoted a new focus on changing African spatial practices rather than just changing landscapes.

At the beginning of the 1890s, Ostafrika's future was poised between the representational space of the colonial lobby, based around an idea of the entire region as needing only German rule, and eventual German settlement, to achieve *Kultur*, and representational space based in cultural geography articulated in the Bushiri War expedition. In the latter, *Kultur* would have to be based in the colony's African population, and diffuse over time from areas in which it already existed. As the heroes of the *Land*-focused approach failed over the next few years, the new approach won out. Administrators proposed different means of changing African spatial practices through the early part of the decade. None of their plans, however, took enough account of African initiative and thus failed. What was left in 1893 was an Ostafrika severed from the Indian Ocean World and a punitive administration increasingly using theories of cultural diffusion to spread *Kultur*.

The debate took place a little over a year into a transitional period for the colony's administration. The 1890s were a rupture from past German approaches to making space in Ostafrika. As the Bushiri War was concluding, Germany and the United Kingdom reached an agreement on a treaty that settled international borders in East Africa. In exchange for the island of Heligoland in the North Sea, Germany agreed to restrict its East African territorial ambitions to areas south of British claims in today's Kenya and to buy the coastal strip the DOAG had leased from Zanzibar outright.[1] After the agreement went into effect on 1 January 1891, the coast was the exclusive domain of the German government; private empire was at its end.[2] Obsession over spatial control remained, but focus shifted from international borders to borders internal to Ostafrika, between racially defined populations, between levels of *Kultur* rather than nations. The incorporation of hierarchies of *Kultur* into economic plans meant the adoption of concepts from cultural geographers, who had used agriculture as the primary means of measuring racial groups. Ranking societies according to their level of agriculture thus entailed the explicit application of racial thinking to administrative divisions and economic development.

On the coast, Soden attempted to reconstruct and co-opt the Arab empire to which the Delimitation Commission had assigned the area in 1886. The representational space that was central to the Bushiri War expedition, of the coast as a space with some level of *Kultur*, implanted by arrivals from across the Indian Ocean, became part of administrative policy. Soden's policies on the coast replaced the emergency anti-slavery measures of the Bushiri War with bureaucratic routine based on a German version of Islamic administration and law. Those emergency measures, however, did not disappear entirely; they merely shifted to the interior. The basis of German claims to sovereignty remained the fight against the slave trade, but with new enemies.[3] The international recognition of a unified Ostafrika, freed from any confusion over Zanzibari sovereignty, thus allowed the German colonial administration to reorient its efforts to moving *Kultur* west from the coast, into space now represented as fundamentally different, a space of wilderness into which *Kultur* had never penetrated. Development in Ostafrika, then, was not divided so much along the "emigrationist" versus "economic" lines identified by Juhani Koponen, but according to the differing representational spaces of Ostafrika and the debate over how best to bring *Kultur* to the colony.[4] The division in space marked a division in the basic conception of the relationship between race and space, between an "African" interior managed in one way and an "Indian Ocean" coast managed in another.

Though the simple spatial division between coastal and interior administrations remained important in shaping development plans, the influence of cultural geographic thinking on the colony's administration was at its most apparent in the institution of programs based on the diffusion of *Kultur*. Colonialists developed several plans to create settled agriculture in Ostafrika. One entailed the immigration of workers from China and Indonesia, labelled as "coolies," who would supposedly bring their racial characteristics to bear and provide superior plantation labour to East Africans. The second element was the building of a railway, designed to run from the coast to the Great Lakes, along which *Kultur* would spread outward as modern technology was brought to bear on the existing system of trade. When neither of those programs proved successful, colonialists turned to theories of cultural diffusion created by cultural geographers like Friedrich Ratzel. Through German efforts to speed cultural diffusion along, it was believed, Germany could develop agriculture in Ostafrika.

Reconstituting and Co-opting the Coastal Arab Empire

As its responses to Wissmann's entreaties to expand operations discussed in the previous chapter had made clear, the German Foreign Office's priority was establishing complete control of the coast, from which *Kultur* could then diffuse west. To lead that effort, it appointed Julius von Soden as its first governor in Ostafrika. On the coast, Soden attempted to create rational colonial rule through "semi-cultured" intermediaries, recognizing the area as part of the Indian Ocean World. The new form of administration was based on a belief that the people of the coast were at least partially ready to administer themselves and that *Kultur* there could improve. This meant the adoption of the representational space of precolonial coastal elites into German administration. Soden built a coastal administration with significant roles for local intermediaries, one that attempted to mimic the titles and legal forms of Zanzibari rule.

Chancellor Leo von Caprivi's choice of Soden as Ostafrika's new governor, rather than someone involved in its prior administration, demonstrates a desire to bureaucratize the colony and move away from the ad hoc aggression of the previous years.[5] Soden was trained as a jurist and had been in consular service abroad, a much different background from the previous main actors in the formation of German rule in East Africa. Otto von Bismarck had appointed him the first governor of Cameroon in 1885, making Soden among Germany's first class of colonial administrators.[6] The new governor admitted that he knew little of East Africa and would depend on his "experiences and observations of analogous conditions" in West Africa.[7] His appointment meant the transfer of the Foreign Office's existing colonial bureaucracy, small as it was, to East Africa on the assumption that Soden could continue to work in the same manner he had in Cameroon. From the very outset of his tenure Soden faced opposition from members of the colonial lobby opposed to the new approach.

Soden almost immediately set about bordering the coast from the interior, a process that attempted to recreated German understandings of the Zanzibari representational space of the mainland. To draw the border between coast and interior, Soden repeated the methods of the 1886 Delimitation Commission and issued a questionnaire to the district official in each town on the coast asking how far their influence reached.[8] After reading the replies, Soden decided on the western edge of the coastal strip, ten British miles from the coast and drawn by the commission in 1886, as the "customs border" for goods from the outside world to enter the interior in August 1891, a decision that indicated a desire to restore and co-opt the form of Zanzibari power.[9] German understandings of Zanzibari power on the mainland had clearly advanced little since 1886. A focus on the coast meant a clear decision against settler colonialism, an idea that had motivated many members of the colonial lobby who promoted empire as a means of reducing the nation's emigration totals.[10] With that decision, there would be no hope for a transformation of East Africa into another Germany abroad, as Germans would not be there to implant German *Kultur* in the soil.

The adoption of the 10-mile coastal strip reified coastal geographies with their separation of the coastal world and the *shenzi* of the interior.

The desire to recreate the Arab empire is also clear in the German choice to copy Zanzibari titles for officials – *akida* and *liwali* (in a modified form as simply *wali*) – and Soden's appointments of men who had already held those positions in his own bureaucracy. Soliman bin Nasr, for example, was the *wali* for Pangani, as he had been *liwali* under Said Khalifa. Sheikh Amir remained the *wali* of Bagamoyo, as well. An 1892 report claimed there was "no choice" but to appoint local officials from the Arab population due to the "lack of knowledge of language and legal norms ... and the influence of Arabs through religious community." The report praised Soliman and Amir as "true supports of German rule."[11] Another report produced by the Colonial Office for the Reichstag made clear the intermediary role envisioned for Arabs in Ostafrika. It labelled Arabs the "most aristocratic race," which had stamped the entire coastal population through the centuries. Now that German arms had defeated Arab ones, it was "the least one could do" to hire notable Arabs as officials, given the lack of language skills or knowledge of Islam among German officials.[12] Soden admitted that he was now promoting people against whom Germans had campaigned as slave traders but thought winning over the "intelligent part of these people" was necessary.[13]

Soden adopted Swahili as the language of administration beyond the upper levels, adopting the coastal elites' belief in their superiority to people to the west. An 1892 report on the colony noted that coastal people thought themselves superior to the *washenzi*, the uncivilized people of the interior. It concurred with this line of thinking, stating that the "mixed-race people (*Mischlinge*)" of the coast had imitated Arab customs and ways, raising them to a higher level than most Africans.[14] The report took up the German readings of coastal representational space into German colonial practice. The decision to adopt Swahili reduced the authority of military officers in favour of local intermediaries. In Germany's other colonies, administrators used German to conduct day-to-day business. Paul Kayser, the director of colonial policy in the Foreign Office, noted that Swahili would be important for governance of the coast, for the reason that a non-Swahili speaker would fall in with the "foul elements" of the population out of ignorance.[15] Adopting Swahili as Ostafrika's official language was part and parcel of the new goal of spreading *Kultur* west from the coast. Swahili was the primary language of coastal towns and a lingua franca along the caravan routes between coast and interior. Using it as the language of the administration was a choice for elite coastal culture as a means of spreading the German presence rather than imposing the German language.

Prominent members of coastal communities positioned themselves as intermediaries between the German regime and the masses. Sewa Haji, the Indian merchant in Bagamoyo who had long supplied porters to European caravans, was especially clear in his positioning as an intermediary. At the beginning of 1891, Sewa paid for a celebration of Wissmann's victory.[16] Sewa donated an additional 12,400 rupees to build a hospital and school, both to serve the colony's non-European population. Both were required to bear his name in large letters visible to passers-by.[17] He doled out gifts to people with whom he wanted to win influence, to the point that Germans began referring to Sewa's gifts as first-, second-, or third-class medals, depending on what Sewa gave them.[18] Sewa continued to exercise power over porterage to extract concessions from the administration.[19] One missionary saw historical evolution in Sewa's actions. He compared Sewa's donation to the actions of "heathens (*Heiden*)" in Ancient Rome, who had first tried to imitate Christ's life before later converting.[20] Sewa used German patronage to ensure his continued economic power and to position himself as the patron of Bagamoyo and Dar es Salaam's non-European populations.

The path of Amur bin Nasur, who taught Swahili at the Oriental Seminar in Berlin, is indicative of the broader trends among coastal intermediaries. Nasur's parents had served Sultan Said Barghash in his failed expedition against Mirambo of Unyamwezi in 1880, after a storm destroyed the family's coconut and clove plantation. He thereby attached the family to Zanzibari patronage. In adulthood, Nasur entered Barghash's army and served until the Sultan's death in 1888. He then left the army and opened a tailor shop, but the shop failed, so Nasur rejoined the army, serving under Lloyd Mathews. His family had depended on Zanzibari patronage since the beginning of the 1880s and Nasur's attempt to build his own business did not succeed. By teaching Swahili in Berlin, Nasur built a similar relationship with the German government, gaining respectable status through his association with the outside power.[21] For people like Nasur, German rule was little different from Zanzibari rule; both offered patronage opportunities.

The Soden System aimed to create development on the coast of Ostafrika along lines that Soden and his allies saw as an intermediate level of *Kultur*, between the levels of Europe and Africa. He positioned the German administration as a replacement for the Zanzibari one, attempting to recreate what Germans had imagined as a Zanzibari empire. In this project, he was assisted by coastal intermediaries, who adapted to German patronage. Rebuilding the Zanzibari empire entailed adopting the precolonial geographies of the coast that separated the *shenzi* of the interior from the *staarabu* of the coast along the border drawn by the Delimitation Commission in 1886. For the interior, Soden would continue to use the approaches to economic development prevailing since the previous decade.

The Marginalization of the Heroes of Empire and Their Travelogues

West of the coast, *Bwana Kartasi* instituted far less bureaucracy, and the administration created different kinds of relationships with local intermediaries, a structure that reflected much divergent ideas about *Land* and *Leute* from those that determined governance on the coast. The Foreign Office tried to use the *Alte Afrikaner*, the "heroes of Africa" from the previous decade – Wissmann, Peters, and Emin – to control African space as commissioners, who would attempt to continue the approach to colonialism that had underlain the DOAG's earlier operations, marking off the interior as different from the coast. They continued to think in terms of earlier German representational space to interior regions even as the new model of representational space took over coastal governance. The continuance of old modes of colonialism placated colonialist critics in the short term. It did not take long, however, for commissioners' plans to fail, discrediting their colonial model.

The Foreign Office devised a plan to co-opt the heroes of empire into the administration. Colonialist Germans had pinned many of their fantasies on the *Alte Afrikaner* and continued to see them as the best hopes for German colonialism. The *Alte Afrikaner*'s popularity shows the continued provenance of the idea of a one-man African empire beyond the coastal strip. With one fell swoop, the Foreign Office would placate the colonialists who wanted a larger role for their heroes and remove those heroes as active critics of Soden's administration. Soden anxiously awaited the end of the "Wissmann circus."[22] It was "high time," he thought, to "cut off the dilettante strivers at the head of the colonial movement in Germany as soon as possible and limit them to their former realm of article-writing, money-collecting, organization-founding, and toast-holding."[23] The Foreign Office devised a plan in which Wissmann, Peters, and Emin would become commissioners, each responsible for a portion of the interior. Soden declared the plan was neither needed nor even a "useful contrivance but was rather "merely a necessary and thus expensive evil" to reduce public pressure on his administration.[24] The *Alte Afrikaner* planned to continue the methods of old in spreading German power west.

Wissmann determined his own path as commissioner through the creation of a funding campaign for a steamer on Lake Tanganyika, a plan that entailed representing the Great Lakes as spaces similar to the Indian Ocean Coast. After Bismarck's rejection of his plan to go to Tabora to continue fighting Arab slavery, Wissmann set out on his own to the area he represented as the centre of the slave trade. The steamer would capture boats used for the slave trade. In his public campaign to raise money for the steamer, Wissmann called the eastern shore of Lake Tanganyika, East Africa's "second coast," the first coast of course being the Indian Ocean coast where he had gained his fame.[25] A steamer on the Lakes could use the "natural advantages" there to end the slave trade and draw legitimate trade to the German sphere of interest.[26] Kaiser Wilhelm II wrote that he was "sympathetic" to the campaign.[27] As the organizers of the Wissmann expedition against Bushiri had imagined their mission as the defeat of an uprising by Arab colonists to win the *hinterland* for Germany, Wissmann now imagined the anti-slavery campaign on the Great Lakes as a battle against Arab slave traders for the *hinterland* of the Lakes.

Public opinion clearly stood behind the *Alte Afrikaner* rather than Soden, showing continued support for the DOAG's model of African space. All three commissioners continued to draw donations from private actors that were not forthcoming for government-directed colonialism. Although the steamer plan was half-baked, it inspired people around the world, mostly Germans but also others, to donate money to fight an imagined threat to civilization.[28] Soden thought the steamer an "absolutely misguided undertaking," as it would serve only "sport purposes" on lakes as large as Victoria and Tanganyika and the organizers had yet to study whether the steamer would be able to operate; but donations flowed in.[29] Wissmann had successfully tapped into the international humanitarian movement that had backed the Bushiri War blockade. The Anti-Slavery Lottery raised two million marks just in the second half of 1891.[30] The money did not come just from Germany; donations arrived from around Western Europe, and from places farther away: Argentina, Japan, and Baku.[31] But Wissmann failed to complete his plans, as he was struck with nervous asthma, the first of a series of illnesses.

The second commissioner, Emin Pasha, was sent to the far west, to the area where he could supposedly use his knowledge of *Land* and *Leute*, derived from his service in Equatoria, to establish a successful administration. Emin was to secure the central caravan route,[32] but he bristled at the restrictions set by the Heligoland-Zanzibar Treaty and attempted to continue the model of empire based on the acquisition of territory. He claimed that he had never surrendered his claims to Equatoria. On that basis, he asked the German government to declare the province's neutrality to ensure the "retreat of Mahdidom to the north."[33] Emin ignored his orders to go to Lake Tanganyika and instead pursued a rumour that his former charges in Equatoria were now settled west of Lake Albert, from where he hoped to take them farther west and occupy the *hinterlands* of Cameroon, creating a German belt across Africa.[34] Emin was not serving German colonial interests but instead attempting to continue the work he imagined he had been doing in Equatoria, capturing land from Europe's Muslim enemies. Arab merchants with long-standing grievances against Emin captured and hanged him on 23 October 1892.

Carl Peters, the final commissioner, continued to pursue his belief that making *Land* German was Ostafrika's future. Peters also created a campaign to fund a steamer, but his derivative campaign was less successful. The Peters campaign eventually raised 350,000 marks for a steamer on Lake Victoria, which Peters thought the most important lake for the future of trade.[35] Among the donations was twenty-five marks from Friedrich Ratzel. The sum of 350,000 marks was not enough money to build a steamer and get it to Victoria, so Soden appointed Peters as commissioner for the Kilimanjaro region.[36] Once there, Peters's plan involved building military stations to hold territory. He spent much of mid- to late 1891 marching through the area and destroying any fortifications he found, to be replaced by German ones.[37] Kayser worried as early as July 1891 that Peters's actions "undoubtedly harmed our reputation" and doubted whether the "lost positives" could be regained. Peters responded with a letter to Kayser in which he claimed the "men on the coast" did not understand the "warlike tribes in the interior. Did they really think they could sweet talk (*beschwatzen*) them with friendly words?"[38] Peters's violence eventually made his continued employment impossible. He provoked an uprising and Kayser recalled him in 1893. Peters would never again receive an administrative post.

The commissioner system envisioned the continuation of Germans' existing relationship with the East African interior; the commissioners' writing about the interior reflects their continued adherence to older ideas about empire. As was the case in the GfdK expeditions discussed in chapter 3, the *Alte Afrikaner* marked a clear border between coast and interior. Franco Moretti has argued that different genres of writing are intimately connected to particular kinds of space, a framework that helps explain German writing about the East African interior. Writing about the interior unfolded along a linear path, with only obstacles in the way of the European traveller, a model of space that made Africans appear dangerous.[39] Wissmann returned to his old style almost

immediately as he marched west of the coastal towns during the Bushiri War. His descriptions of people became ethnographic as he quit
describing battles and began describing entire societies near Moshi.[40]
Peters described the landscape where he built the Kilimanjaro station
with the language of explorers' travelogues. It made "an impression
something like the terrain of the Thuringian Forest with the Kyffhäuser
in the background."[41] Peters continued to draw comparison with European history to explain East African politics. The Chaga kingdoms in
his district were like Sparta and Athens; the recently arrived Warombo
played the role of Macedonia, a new kingdom upsetting the old duo-
poly of power.[42] Emin's scientific assistant, Franz Stuhlmann, wrote that
the expedition's ethnography was superior to that of the "newly-fash-
ionable African tourist journeys" and that studies of the "manners and
customs of the natives" were of much greater value.[43] These descriptions set up East African communities as behind Europe in time, with
polities that Germans believed exerted a positive influence describeded
as states from the European past that Germans believed had produced
historical progress.

The ethnographic style does reveal some aspects of local geographies
that might have been absent in bureaucratic reports. It represented African societies not primarily for governance purposes but for fitting them
into a scheme of human social evolution. Writers therefore included
elements that would likely have been absent from bureaucratic reports.
Peters reported that a "neutral path" existed below the "cultivated
belt" on Kilimanjaro in which the region's communities could meet and
negotiate. He saw the protection of this path as one of the most important of the duties of the German administration.[44] For Franz Stuhlmann,
Tabora's inhabitants' imitation of Arab customs and attire was a sign
that Arabs remained the "most important cultural element" for the
future of the colony.[45] Of Unyamwezi, Stuhlmann reported that there
were signs of long-lost cultivation in the middle of the forest, proof that
there had once been villages there.[46] Stuhlmann recognized local mapping practices and the ways that people measured distance through
arm and hand signals.[47]

The continuation of older representational space of Ostafrika is also
apparent in pictorial representations of space. Two examples are indicative of a wider trend toward the rationalization of space for administration under military rule in the interior and a move away from the
extravagant ideas of the colony's early years. Included with Peters's
report on the new station at Kilimanjaro were drawings by his assistant, Pechmann.[48] These drawings, such as the one in figure 9, served
no clear administrative purpose, but rather served to illustrate Peters's

Figure 9. Pechmann's drawing of Kilimanjaro station, September 1891, BArch R 1001/755, pag. 45

tales of adventure. The size of the building and its geographic location are unclear, as is the layout of the building's interior and functionality. It looks like a generic frontier outpost that could illustrate a tale of adventure anywhere. Pechmann's drawing stands in clear contrast with a sketch of Kiswani station, in the same region as Peters's station, that Kurt Johannes sent to Soden in February 1893. Johannes's sketch is practical, designed to make distance and location clear. It includes text denoting what different parts of the station were designed for and none of Peters's beautiful landscape. His space is neatly ordered with no room for flights of fancy.[49] This transition in illustration and in textual descriptions is indicative of new approaches to African space under colonial rule. The old form of writing was not useful for planning development. It was suited to precolonial travelogues, but not colonialism, and makes clear the failure of the heroes of the 1880s to adjust to the new thinking about space as administrative territory in Ostafrika in the 1890s.

The failure of all three commissioner expeditions had become clear by 1893. With the end of the *Alte Afrikaner* came the end of the representational space that had underlain German efforts in Ostafrika since the GfdK's expeditions in 1884. That mode of thinking about African geographies had proven a failure. In the Reichstag debates over the future of the colony, Ludwig Bamberger declared that the first generation of German explorers of Africa had reached its end. He was mistrustful of the "real African explorers," who with their "sport" and their "fantasies of adventure" did not realize that people back home would have to

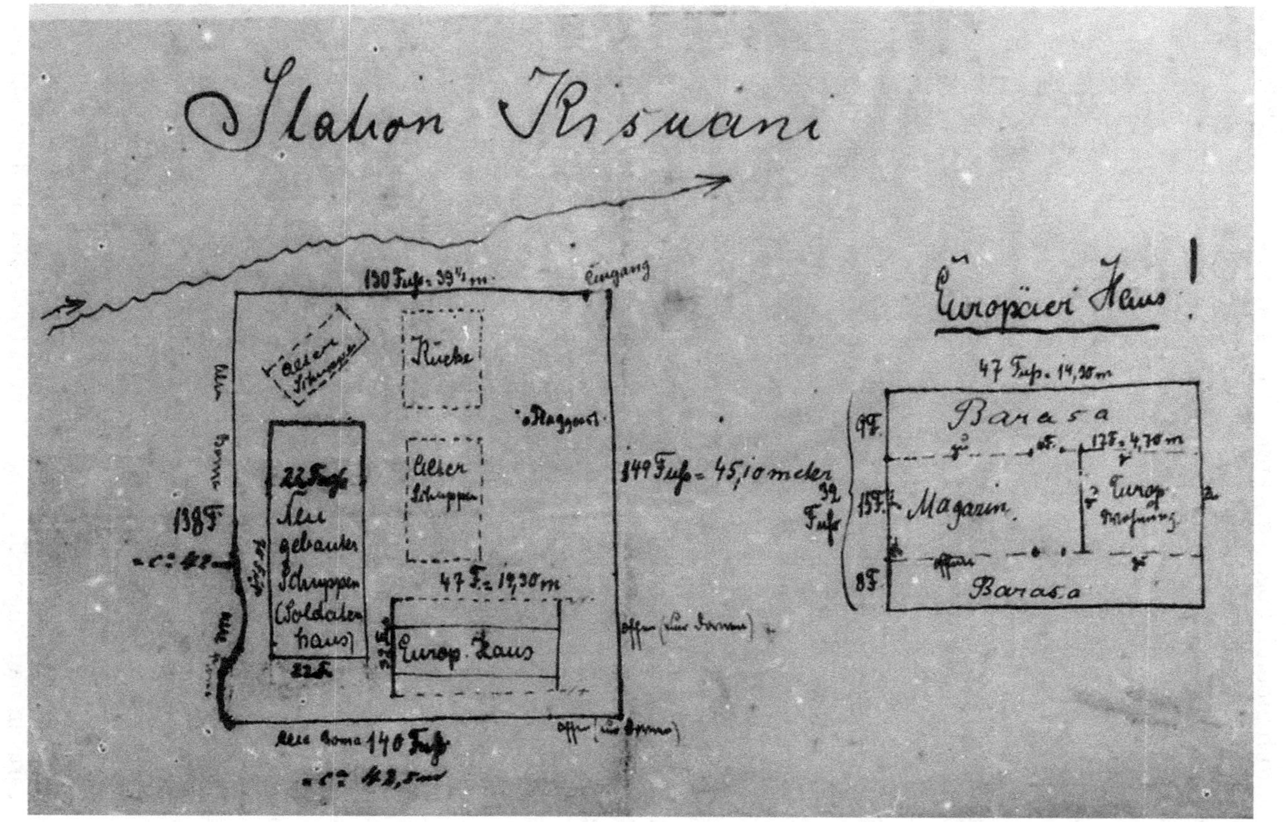

Figure 10. Schematic of Kiswani station. Kurt Johannes to Soden, 7 February 1893, TNA G 1/18

pay the bills.[50] New ways of dealing with the colony's interior would have to be found. Such new approaches had developed side-by-side with the commissioner system over the previous two years and would thereafter take centre stage. As indicated by Johannes's station sketch, the new approaches meant new representations of space, making the neat ordering of Ostafrika appear possible.

The Militarization of the Interior

While the commissioners bumbled through their attempts to link the interior to Germany, Soden's administration pursued parallel policies in the interior with longer-lasting effects. Alongside the purportedly peaceful commissioner system, Soden extended the emergency military measures of the Bushiri War to the west, to fight against the Hehe and others who fought against German measures. Ostafrika's administration did not end the state of emergency; it only moved its geographical location. But the threat to German *Kultur* in East Africa was reimagined with a new enemy. Arab communities in the interior were no longer obstacles to the progress of development. They were a positive influence, a step farther along the path to civilization than were the African societies that were truly responsible for the destruction of East African societies. The protection of Black Africans from the slave trade became the basis of German claims to sovereignty in the interior as well as on the coast; German administrators believed this was the first step to pave the path for the spread of *Kultur* west.

A new crisis for Ostafrika's administration emerged in September 1891, again triggered by responses to Emil von Zelewski's aggression. In this case, the Hehe chief Mkwawa asserted his control over spatial practice near Iringa. The campaign was the moment at which the administration of Ostafrika began to eliminate the influence of the DOAG approach to the interior, completely in favour of an administration based on punitive violence. Soden sent Zelewski to find Mkwawa after reports that Mkwawa was raiding near Kondoa.[51] Zelewski burned abandoned villages through which he passed as he made his way west.[52] On 13 September 1891, Mkwawa's forces ambushed Zelewski's column, killing its leader and most of its members. The effects were immediate: 424 of the 620 porters for Wissmann's steamer expedition to the Lakes deserted, making it clear that porters placed little faith in the administration's ability to protect the caravan routes.[53] Mkwawa's actions display an intention to preserve existing prerogatives in the caravan trade along the central route. The attack on Zelewski's column defended Hehe territory and control over caravans from German encroachment. Zelewski's approach signalled German advancement and attempts to regulate Hehe interaction with the caravan trade, so Mkwawa took up arms.

With its punitive expedition against the Hehe, the Soden regime transitioned German claims to sovereignty in fighting the slave trade away from fighting against Arab slave traders to the protection of, and control over, Black Africans. The rhetoric of the anti-slavery campaign remained central to German discussions of Ostafrika but now served to support the power of the colonial state and suppress defiance. The frontier of the slave trade, however, moved west from the Indian Ocean coast to the Great Lakes as the scope of the colonial state expanded. Grouping the Hehe with the Maasai and Mafiti, the official *Kolonialblatt* described all three as "not only enemies of German rule, but enemies of any state order" who in yearly raids "overrun the weaker tribes of the hinterland with war" and either kill or enslave them.[54] The Colonial Office's representative suggested increasing the German military presence "to assure the Indians and Arabs that … it was only the annual raid of a swarm of Wahehe, which finally shattered on the station belt." He thought it "a scandal, that we do not have the power … to conduct a sure and energetic punishment and a lasting prevention of such incidents."[55] Wissmann's operations had been designed to fight an Arab enemy. Now the representative was suggesting a military presence to protect Arabs (and Indians) from African attacks. Germany's role in East Africa was to protect the established colony it had conquered from Bushiri.

Constructing the Hehe as the enemy of German *Kultur* dovetailed with reimagining Arabs in the interior. They were no longer an impediment to progress and inextricable from the slave trade, but rather Germany's allies in bringing *Kultur* to the region. Lieutenant Tom von Prince allied German forces with the Tabora Arab community, which Wissmann had claimed was a great danger to German rule just two years earlier. As Rochus Schmidt wrote, although coastal Arabs had felt their interests threatened by anti-slavery measures, Arabs in the interior were the "main support" of the German presence. In his opinion, interior Arabs were "far" from religious fanatics, and their power had nothing to do with Islam. They had brought slavery to the region, but their practices had been copied by "powerful tribes," who regularly raided their neighbours and were now the main slave traders. What was more, Arabs, who had previously served Indian capital, "would gladly place themselves in the service of German capital."[56] Another officer reported that other than the issue of the slave trade, one "could hardly think of better harmony" between German officials and Arabs.[57]

Soden dispatched Tom von Prince to take Mkwawa's fortifications and end the slave trade in the region; Prince's actions on the march make clear the new portability of Germany's anti-slavery doctrine. Prince built fortifications along the caravan route, supposedly to stop further Hehe raids.[58] The shift to permanent militarization is evident in a new approach to Isike, the *ntemi* (ruler) of Unyanyembe. Prince undertook operations against Isike in 1893, based on claims that diplomacy with Isike was "useless" and that Isike wanted to resume the slave trade.[59] In fighting Isike, Alfred Sigl, the Tabora station chief, declared he had a "right of requisition" with regards to all civilians in the vicinity. He could therefore force members of the anti-slavery expedition to join his military expedition as ersatz officers.[60] Prince made that war and conquered Isike's capital. Defeat of Isike secured all trade routes, claimed Prince. In the wake of the defeat, people from the surrounding area "streamed" into Tabora "like hyenas and jackals to a fallen lion," ready to submit to German rule.[61] Prince had brought militarized anti-slavery space to Tabora, as Wissmann had proposed years earlier, and the new space supposedly remade African societies around it.

The different representational spaces of the militarization of the interior are evident in Prince's conduct. Prince clearly marked the borders of his military measures from those of the civilian administration. In his column to fight Mkwawa, the civilian Sigl led the way when the column was in peaceful territory. When it entered what Prince believed was Mkwawa's territory, Prince took the lead. The two men decided on an arbitrary border between the two and switched leadership as they crossed it. Sigl also led negotiations after battle against Muinimtwana, a chief who had aligned himself with the Hehe, as this was "political work" rather than military. Once Muinimtwana, the enemy of Germany, had been defeated, he became Muinimtwana, the German subject who could be handled politically.[62]

By 1893, parts of Ostafrika were in an unending state of emergency, subject to wartime police measures against enemies of the state. What was needed, according to the *Kolonialzeitung*, was to show "natives" that resistance was futile. According to the periodical, "[o]n one side stand superior races with first-rate weaponry ... filled with the proud consciousness that they are the trailblazers of a new era," while on the other there were "barbarian or half-barbarian races with poor flintlocks or other weapons" that posed no danger to Europeans. The military was a "pathfinder" for future development of the colony.[63] The Foreign Office wrote that a "long-lasting martial advance" would be necessary for development, extending the ring of stations deeper into the interior.[64] The coast was severed from the interior and the focus in administering the latter, a space of slavery and primitiveness, was breaking alternative sources of power before Germans attempted to spread *Kultur* there.

German colonialists had naively believed that the Bushiri War would create peace in Ostafrika. They were wrong, as continued arbitrary and capricious approaches to spatial practices in the interior naturally provoked anti-German fighting. The campaigns against the Hehe reoriented emergency anti-slavery measures. No longer were Arabs the primary cause of the slave trade and the foe of German *Kultur*. The threat now came from Africans. Administrators and military officers shifted thinking about Arabs and Africans to justify the separation of coast and interior and the extension of emergency military measures to fight the Hehe and recast Arabs as German allies. Military solutions would not make the East African interior economically valuable to Germany alone; *Kultur* would need to spread west alongside military force.

Technological Solutions to Development

At the outset of the 1890s, colonialists promoted technological solutions that would make East African *Land* productive. They for the most part abandoned dreams of German settlement but held on to the primacy of territorial control and a future based in German settlement. Modern transportation technology, especially railways, promised to open up new territories for the German advance. Railway promoters claimed that railways could create economic growth, primarily through plantation agriculture, in the areas along the track. That plan faced a problem, in that the labourers necessary for plantations were not clamouring for the low-paying wages that plantation work entailed. The initial solution was the importation of labourers from Southeast Asia to work. The combination of modern technology and imported labour, however, was not transformative, as expenses mounted up and workers complained of ill treatment. It became clear that plantation agriculture alone would not create immediate development. Nonetheless, the importation of foreign workers was the first project in which the German administration attempted a new approach. Rather than German settlement, the administration tried to use what it perceived as positive racial characteristics to move Ostafrika's *Kultur* to a new stage of development. These plans failed because they did not account for African spatial practice.

Economic concerns at this point focused on renewing revenue from customs and tolls on long-distance trade in the immediate wake of the Bushiri War. Many German colonialists worried that the colony would face financial ruin due to changing caravan practices, as caravans had avoided the German coast during the war and might continue their same movement after. Some believed the new British colony to the north posed the main threat. German colonialists suspected much of the trade that had gone to Tanga before European colonization had switched to Mombasa during the Bushiri War, fearing conditions on the German section of the coast. Eugen Krenzler, the district official for Tanga, reported that a Pangani *jumbe*, Mombosahsa, had moved himself and his people to Mombasa because trade was more valuable there. Another *jumbe* had moved to Wanga, north of the border, taking with him several Digo villages.[65]

Beyond recapturing trade, German emigration remained popular among supporters of the growing German agrarian movement, but most colonialists turned against it out of concerns about the growth of a European proletariat in Ostafrika. Concerns about the instability of the racial hierarchy intersected with worries about the working class that circulated in metropolitan Germany in this period, as socialism became politically powerful.[66] Carl Peters was one of the few who still supported German settlement. He claimed Germans could become accustomed to East African climates but that it would take generations for their bodies to change.[67] He still saw Kilimanjaro as a place for European settlement from which the "invigorating enterprise of the Indo-German race would increasingly permeate neighbouring territories."[68] In Peters's statement, one can read the influence of theories of cultural diffusion, the idea that German *Kultur* would reshape landscapes and, in turn, surrounding societies.

Others were less optimistic about the possibilities for German settlement, fearing it would transfer social problems from metropole to colony. Oskar Baumann described a social type that had emerged in Ostafrika, the "Kilimanjaro cobbler." This was a German who had travelled the world, experienced ups and downs living in North America and had decided to settle near Kilimanjaro. But the Kilimanjaro cobbler never made his way beyond the coastal town of Tanga.[69] Hans Hermann von Schweinitz suggested that settlement of Europeans from "lower classes" would create problems, as people in the region "decline in moral quality." The "lower classes" would be unable to resist such influence. Additionally, they would find it impossible to "tactfully handle" African workers. Used to being ordered around, they would become "half-gods" and would forget that Africans were people. African would also soon recognize the difference between European classes and see that some Europeans were subservient too. This would cause them to think they were on the same level, undermining the colony's racial hierarchy.[70] These concerns echoed similar worries about class and racial hierarchy articulated in other nations' empires.[71] They reflect a belief that the evolution from African to German *Kultur* was simply too big a leap and could not happen overnight.

With German settlement unpopular, colonialists attempted to use modern technology to capture trade from far-flung areas by changing African spatial practice. The steamers promoted by Wissmann and Peters would be one part of that plan, quickly moving modern technology all the way to the Great Lakes region without concern for the areas in between. The organizers of the Peters steamer argued it would lead the Lakes' trade to German territory and end the slave trade.[72] Steamers offered a relatively cheap solution to the perceived issue of trade from the Great Lakes travelling through the colonies of other nations. Enthusiasm for the steamers did not last, but the conviction that the railway could spur development in the spaces of the interior had greater staying power. A railway would signal German victory on the coast and project German power west.[73] The DOAG created a Railroad Company for German East Africa on 7 August 1891, hoping to connect Tanga and Korogwe.[74] The Tanga-Korogwe link would serve as a "pioneer railway" for further development in Usambara.[75] Railways could link inland areas, such as Tabora, to the coast in a way that steamers could not and end the power of non-German actors over the caravan trade. Karl von der Heydt argued that a railway could link the "zone of *Kultur*" on the Great Lakes with that on the coast, across the "wilderness lying in between." Europeans could not compete with Arab and African merchants in the existing porterage system, as non-European merchants simply purchased their porters as slaves.[76] Peters thought the railway should go to Tabora "in all cases," as Tabora was the only "*Kultur* or trade centre and the main depot for valuable trade articles." As Tabora was the "central point of trade" so far, so it must be the central point for the railway. Justus Strandes agreed with Peters; Germans had to follow the "example" set by "Arab and Swahili merchants," going to the place where there was "something to be had."[77] The railway solution was a possible workaround to the debate over German influence on *Kultur* in the interior.

Proponents of German settlement argued that they could provide the necessary impetus to the implementation of German *Kultur*. Railways, their supporters claimed, would stimulate the development of settlements alongside them, particularly the production of agricultural exports. One officer argued that a railway between Tanga and Korogwe could "economically develop the fruitful regions of Usambara" by linking them to Tanga's port. He said the railway would free seventy thousand porters for other kinds of labour. Oskar Baumann thought the "fruitful territory of Usambara," to this point "as good as worthless," could be developed with the Tanga-Korogwe route. Along with ideas about the railway creating new kinds of economies were hopes that it would transform populations who lived near it.[78] The railway expert Friedrich Lenz conducted a study for the DOAG and reported that the Tanga-Korogwe railway would make possible the settlement of Europeans and the creation of plantations in Usambara, a land "where roads and *Kultur* as well do not exist."[79] The railway was a "development rail" that would "enable the movement of the developing great production of Usambara and Kilimanjaro lands to the coast."[80] The belief in the power of railways to make *Kultur* is clear in the DOAG's 1891 annual report, which claimed the Usambara Railroad would create a new phase of development "for the entire *hinterland* landscapes of Tanga."[81]

Such claims about railways' ability to create development, particularly in agriculture, were common across Europe and North America in the second half of the nineteenth century. In Peters's opinion, railways inevitably created a "rush" for territory, the "most invigorating element" in all colonies. The example of California after the Gold Rush was the clearest example of such a rush inspiring not only the initial development but agriculture as well.[82] In the Western regions of North America, railway promoters argued that the laying of railways could remake vast regions, turning them from unproductive areas inhabited by Indigenous societies into sites of rationalized cultivation following Western models. White Americans envisioned railways as self-justifying technologies that would inherently improve the land through which they travelled. This vision left no place for Indigenous peoples, who had failed to make proper use of the land and were doomed to be replaced by white settlers if they could not adapt.[83] Similar fantasies about the power of railways inspired Sergei Witte's plans for the Trans-Siberian Railway and Ottoman-German cooperation on a railway to Baghdad.[84] In Ostafrika, German colonialists worked with similar models of what railways could accomplish, though, due to concerns about the climate in areas near the coast, the economic future was in plantations, not in small farms.

But plantations also required labour, raising concerns about slavery and the growth of a proletariat. The idea that East African labour could suffice should be "completely ruled out," according to the DOAG.[85] Peters claimed that older European colonies had an advantage in that they had been founded on slavery and "a class of labourers had been trained through slavery." It would be more difficult to create successful colonies when colonists had to accustom workers to labour for wages.[86] Peters lamented that the nineteenth century was one of "human rights" and worried about what could happen if Germany copied British examples. British attempts at the "theoretical emancipation of the coloured population" in its colonies achieved the opposite, "the barbarization (*Verwilderung*) of the native races instead of their improvement, and often even to extinction." Therefore, British colonies were often filled with "an arrogant and downright dangerous coloured rabble."[87] Europeans had arrived at their form of labour organically after many centuries. The process needed to move more quickly in Ostafrika.[88] The evangelical mission leader Alexander Merensky expressed concerns about East African labour. Slavery "had reduced the value and respect of work in the eyes of the natives."[89]

The first efforts to remake East African populations to solve the "labour question" involved moving villages to more easily supply porters. Kurt Johannes convinced the son of Muanamata, who lived near Kilimanjaro, to settle nearer the main road with villagers who would serve as porters. Service in the German caravans proved onerous. Many of the villagers began to move back home to escape such service. As his people moved away, the village's leader also wanted to return home to maintain his following. What was more, the porters Johannes could still recruit demanded hard currency instead of the cloth with which he was supplied to pay them. If they moved to be where the Germans wanted them, they expected more compensation for doing so. Johannes was convinced porters would have to be recruited on the coast.[90] The initial German plans to invent new African spatial practice failed to achieve any lasting success. Nonetheless, the plan reveals German thinking about Ostafrika's communities and their relationship to space that emerged in the early 1890s. The idea of East Africans as portable to make *Kultur* would remain influential.

In planning for plantations, German colonialists looked to models from other colonial empires. The DOAG's yearly report for 1891 declared it was of "resounding importance" to answer the question of whether "native production" could be increased, particularly through the "introduction of important varieties of cultivation from other tropical lands."[91] The Foreign Office sent Soden recommendations from a German planter in Brazil. The planter claimed Germans could thrive in higher altitudes, in which they would become accustomed to the climate, ways of life, and language of the region before striking out farther. Black people living nearby would learn to work hard and become consumers from the German example.[92] Paul Kayser looked to India, where he claimed state administrators watched over natives and made sure they planted according to the state's wishes. That system was "much more necessary in uncultivated Africa." German overseers needed to watch Africans closely.[93]

Officials decided the best solution was to import labourers who were used to colonial labour regimes rather than to try to force people from other parts of Ostafrika to move to plantation areas and work for them. Chinese workers became the most popular option, following the model of the Dutch East Indies.[94] The Foreign Office suggested the company copy Dutch methods of plantation agriculture by importing Chinese labour, as described in a forwarded translated article from the *Java Courant*, attached to a report from the German consul in Batavia.[95] The Chinese government would not allow its people to sign labour contracts to move to Ostafrika, but a German planter in Sumatra offered to supply three hundred labourers in early 1892.[96] A plan for the importation of 500 workers from Sumatra and Singapore was in place in 1892, indicating a belief that Usambara could be turned into a German East Indies. The workers would fix the "deficiency in technical strength" that existed on plantations.[97] "Coolies," it was hoped, would solve the immediate question of labour until the plantation company and plantation owners could "educate" local Africans to work.[98] The DOAG argued that without coolies, the "economic uplift of the colony" would be "impossible, however with coolies' help more or less secured."[99]

The plantation system entailed a racialized division of labour in Ostafrika that reflected the new thinking about Arabs and Indians in the colony after the Bushiri War. Kayser thought it most important to "win" Arabs and Indians to the German side as the "most intelligent and propertied" part of the population. Europeans would have to act as "teachers" for the colony's Black population so they could understand production above subsistence level.[100] One official thought that Arabs were especially suited as plantation overseers, as shown by the clove plantations in Zanzibar. They would therefore be the perfect "middlemen" for German colonization. Among the official's suggestions was a racial division of agriculture. Arabs could oversee the planting of most products, but "for the *Kultur* of the natives," whom Germans believed were incapable of difficult or complicated agriculture, he suggested oil palms and rubber, which required little care.[101] This entailed an attempt to copy the existing plantation system that predated German arrival on the coast.[102] Joachim von Pfeil articulated the attempted realignment of German interests with the "semi-cultured" members of the colony's population. He suggested Arabs could direct rice farming near government stations.[103] He thought Germany could use Indians as settlers with a method to "bring them up in our offices, make them become german [*sic*] subjects & start them with a certain credit in life."[104]

Although plantations seemed initially successful, neither they nor the railways changed spatial practice or solved the perceived development problem. By the end of 1892, the DOAG was tempering the excitement of earlier years. The "nature of our undertaking" meant that development had to be gradual.[105] The railway, tied to the plantation system, also failed to meet expectations. The rail was making progress, but was only in the "unfruitful steppe," not the areas it was supposed to link to the coast.[106] Other modern technologies failed, as well. Initial telegraph lines between Bagamoyo and Tanga were twice destroyed by giraffes.[107] It was apparently inconceivable to build the lines taller than they were in Europe. The importation of labour was also no panacea. Coolies did not prove an easier labour force to manage. Maximilian von Rode, the station chief in Pangani, reported that he had had to jail ten Chinese labourers for abandoning Lewa plantation. They had abandoned it, he wrote, because they were doing work to which they were unaccustomed and due to violent treatment by one of the plantation's German employees, who "did not understand how to deal with coolies at all."[108] Kayser followed this letter with a note that continued rough treatment of coolies would make recruitment of more impossible, "endangering the economic opening" of the colony.[109] German expectations for the work and treatment acceptable to non-white workers did not match realities.

Relatively quickly, hopes for speedy development based on plantation agriculture proved illusory. Railways were more difficult to build than initially thought, and East Asian labourers were not simply pliable material for German plantation owners. *Kultur* could not simply be created through the import of people and technology from abroad. Though such plans failed, they did mark the introduction of large-scale planning projects to Ostafrika. The importation of foreign labour was the first explicit attempt to use perceived racial characteristics to create economic development. The management of racial characteristics would remain central to all future development plans.

The Shift to Cultural Geography

Another strand of thinking emerged within colonialist circles, one based on the transformation of East African spatial practices to make productivity instead of changing territory. Such plans, based in cultural geographic models of cultural diffusion, premised development on the management of perceived positive racial traits among Ostafrika's population. Positive traits were most visible, proponents of these plans claimed, in one aspect of societies' *Kultur*: their cultivation of the soil. By the end of 1893, plans to create development through the management of racial characteristics were part of German debates over the future of Ostafrika. In these plans, the administration would use the racially defined *Kultur* of East African people to transform its *Land* to productive agricultural regions. Underlying the new thinking about development were ideas from cultural geography, particularly the concept of cultural diffusion, which became the primary means of conceptualizing race and space in Ostafrika by 1893.

This shift was only possible because African communities saw advantages in adjusting their practices to some aspects of the German presence. Makonganja, a powerful chief in Kilwa district, declared his desire to plant products for export to occupy his slaves as labourers. Many Arabs shared similar desires, reported district officials.[110] By the beginning of 1893, one official was claiming that "Negroes are easy to accustom to cultivation" and that the administration could work with Simbabwene of Mrogoro, who had expressed interest, to train people for coffee cultivation.[111] Makonganja's and Simbabwene's requests indicate a desire to tap into the relatively lucrative market for exports of agricultural products. It would be profitable for them personally to supply land and labour for coffee production. One writer argued that instead of importing European models, Germans should copy Arab and local models of recruiting labour.[112] The participation of some intermediaries in German projects increased German enthusiasm for those projects.

Settled agriculture became the primary criterion on which Germans judged East African societies and space, something most clearly observed in changing descriptions of the Maasai. John Speke's "Hamitic hypothesis" had constructed the Maasai as among the racially superior Hamitic peoples who, he claimed, were responsible for all civilization in the East and Central African interior.[113] Earlier European explorers had praised the Maasai as the best hope for the region's future. Joseph Thomson, the first European to encounter Maasai society for an extended period, described the Maasai as better than other Africans. They were more frank, less suspicious, and more intelligent, and were unsurpassed in their knowledge of geography.[114] He believed the Maasai occupied "a far higher position in the scale of humanity."[115] If cut away from their traditions, the Maasai's superior intelligence could produce economic development.[116]

German estimations of the Maasai had changed significantly by the 1890s, with possible genocidal implications. Peters wrote that the "warlike nomads of the steppes," by whom he meant primarily the Maasai, could not be good workers for some time, as they were "unreliable" and too "delicate" for humid climates.[117] The uniqueness of the Maasai, a mix of Hamitic racial characteristics with aspects of Nilotic culture, meant the Maasai would have especial difficulty learning new ways of living. One needed only look at the extermination of North American Indians to see that "the disappearance of this livestock raider and general troublemakers can only be seen as a very great felicity." Europeans could settle East African "prairies" if the Maasai were not in the way.[118] The new vision for development not only depicted the Maasai as incapable of development themselves but as a hindrance to the development of other societies. Multiple German observers declared the "mountain people" that inhabited the Pare Mountains, a "peace-loving and hardworking agricultural people," had been hindered in their development by frequent Maasai raids.[119] Their construction of walls and irrigation canals demonstrated that they were industrious. They just needed protection from Maasai raids, which Germany could offer.[120] This rhetoric hinted at a program to eliminate the Maasai population from Ostafrika, either through forced assimilation or through even more violent means.

In constructing these arguments, German colonialists depended on the cultural geographic framework established by Friedrich Ratzel. As Ratzel wrote in his *Völkerkunde*, a poor "cultured race (*Kulturvolk*)" always stood high above a nation that lacked *Kultur*, which would always live hand to mouth. Societies could be ranked by their level of agriculture.[121] For Ratzel, the border between agricultural societies and nomadic ones were the most easily observed of all borders. Simply viewing the *Land* showed clear separation between cultivated land and steppe.[122] A challenge facing colonists was that societies in East Africa fell "with a clear preference" into nomadism "with small changes in their political or economic conditions."[123] Racial and ethnological characteristics were inextricably bound up with the landscape a society inhabited; spatial practice mattered for making space. The most important sign of this link was agriculture.

The cultivation of *Land* became a stand-in for the ability of societies to contribute to Ostafrika's development and the only proof of their advancement. A Catholic missionary wrote that the Shambaa's "drive to labour" could be observed simply by looking at their territory. The whole valley that he saw "resembled a beautiful garden." They were not satisfied with self-sufficiency but increased the area under cultivation yearly. The irrigation involved attested to their "dexterity with the implementation of cultivation."[124] Even the name, "Shambaa," which translates to "farm," showed their ability to improve the *Land*.[125] Such relationships could also be negative. In his description of Ugogo, Peters declared its inhabitants possessed the same "inhospitable and abhorrent" qualities as the arid landscape.[126]

An important aspect of Ratzel's model was that levels of *Kultur* were not static and evolved over time, which is apparent in discussions of East African societies. The endpoint of *Kultur*'s evolution was always nineteenth-century German agrarianism, with African societies judged by their potential to become more like Germans. Peters compared Zulu migrations from the south to the *Völkerwanderung* of European antiquity as warlike foreigners settled among agricultural people and ruled over them.[127] Kiyao-speaking people in the South had given up fighting and adopted the agricultural of neighbouring communities, proof of what benefit could happen if societies "were forced to peaceful agriculture."[128] For Franz Stuhlmann, hunter gatherers formed the "lowest level of *Kultur*." Through agriculture, humans became sedentary and did work that would bear fruit in the future. They thereby achieved "the first stage of *Kultur*," which required fixed residence, the building of huts, and "above all" the grouping of people in a community to protect against its common enemies and to choose a leader. A society based around agriculture could reach a level of *Kultur* but could build no industry and its fixation on its livestock prevented "spiritual occupation."[129]

In Stuhlmann's travelogue from his journey with Emin, we can observe a conception that cultural diffusion was not a foolproof process, pointing toward a future in which Germans could manage it and fix what they perceived as problems. As he entered the forests of the northwestern part of Ostafrika, Stuhlmann noted that the forest made travel difficult and disorientating. The difficulty of travel meant that villages did not interact with one another. While almost everywhere else in the world there were at least a few people who were used to interacting with foreigners, in the forest "every village formed its own world … a few hours of jungle formed as great a separation as the Great Wall of China."[130] Stuhlmann devoted especial attention to the Twa, an ethnic group he referred to as "pygmy," as proof of the failures of cultural diffusion in the region. They were the remainder of the "Ur-population" of Africa from which other societies had evolved to larger physical forms through superior nutrition and the development of agriculture. They remained in the "wood age" prior the stone age and had not even developed culturally to the point of decorating themselves with jewellery and tattoos.[131] These descriptions make clear the evolutionary thinking about African societies that had begun to dominate German analyses.

Stuhlmann was one of several colonialists who fixated on the Nyamwezi as the "tribe" most qualified to spread *Kultur*. For Stuhlmann, the Nyamwezi were the best of both worlds, combining settled agriculture with herding. It was "no accident that the Great Lakes region had the most ordered states, prosperity, and industry." Strong, warlike nomads had mixed with the hard-working, conservative agriculturalists to produce stronger societies.[132] Stuhlmann suggested the "introduction of new elements" to assistant societies in "becoming somewhat more cultivated." He thought Nyamwezi farmers could be settled elsewhere.[133] Peters argued that the Nyamwezi could serve the colony in its intermediate stages of development, as they needed only one tenth of what European settlers would. This meant Nyamwezi farmers could settle areas before the railway reached them.[134] The Nyamwezi were "enterprising, peripatetic, and persevering against exertions" in a way that other Bantu societies were not. They were an "agricultural race" in the "truest sense" who "embodied a higher level of *Kultur*" than the "hunters and nomads of the steppe." In them Germany had the "material" for the "economic elevation" of areas not suitable for European settlement. The Nyamwezi could not only be workers for German but also "independent colonists" in areas decimated by war or where the inhabitants "had not risen to the economic level of agriculture." In that role, they would resemble the Dutch settlers in seventeenth- and eighteenth-century Prussia, who had brought Dutch agricultural methods and helped transform Prussia's economy.[135] Colonialists had thus adjusted their historical narratives of making space to include a more prominent role for changes in spatial practice in contrast to earlier emphases on state or civilizational action.

By 1893, cultural geography had become a primary means of conceptualizing development in Ostafrika. Technological solutions had not succeeded in transforming the region in a colonial vision modelled on the American West. In the short term, at least, development would need to proceed with the *Land* and *Leute* already in the colony, based in agriculture. This would entail instrumentalizing African spatial practices as part of development planning. Cultural geography provided a model for thinking about how agriculture could improve over time through cultural diffusion. What was new in Ostafrika in the early 1890s was the belief that the process could be managed. Germans could use the more cultivated peoples of the colony to spread *Kultur*, while the less cultivated disappeared.

Conclusion

The period 1891–3 was a rupture in German approaches to Ostafrika. Geopolitical questions had been settled by the Heligoland-Zanzibar Treaty, and the German government had assumed responsibility for development. On the coast, Soden resuscitated a version of the Zanzibari empire, meant to cheaply assert German authority and restore prosperity. Intermediaries positioned themselves in similar roles as they had with Zanzibar. Farther west, the *Alte Afrikaner* who had popularized empire in Germany attempted to continue their approaches to African space but were superseded by military measures. The biggest question for the colony's future was development. Initial efforts focused on technology as a tool to create development. When railway construction was slow, administrators shifted first to importing labour from other, "successful" colonies. When imported workers proved not to be a panacea, cultural geographic concepts became the main means of thinking about development. Cultural diffusion would create a future for Ostafrika.

Eugen Wolf's criticisms would eventually bring Soden down as governor in September 1893. The so-called Soden System angered many members of the colonialist movement and had not made Ostafrika profitable. Constant war meant expenses continued to mount. But the changes made during Soden's governorship set the course for the development debate moving forward. Administrators had adopted models from cultural geography for thinking about the relationship between *Land* and *Leute*. Spatial practice had become as important as territory in the German spatial imaginary. Over the next decade, German administrators would develop increasingly sophisticated and elaborate means of making cultural diffusion happen by managing race in the colony.

8 Creating Familiar Landscapes: *Heimatkunde* and African Spatialities, 1893–1900

Readers in 1890s Germany who had even a passing interest in the nation's East African colony would have seen images like those below. Such images did not replace existing genres for depicting Africa; they coexisted alongside others focused on the exotic or adventure. In contrast to such exciting representations of space, these images seem plain and generic. A building or a small cluster of buildings sits in a landscape, almost always containing hills or mountains, usually a German flag and often a cross, and a few stock figures depicting life at a German outpost in Ostafrika. The subjects of the images are often nearly indistinguishable. Despite their banality, these images served an important purpose for colonialists and German readers. They transformed the formerly exotic East African landscape into something familiar to German audiences. Through them, small pieces of East Africa became parts of the German *Heimat*.[1]

The study of the German *Heimat* became a field of its own in the late nineteenth century, with schoolbooks and books for adults dedicated to the field of *Heimatkunde*, a sub-discipline of cultural geography. *Heimat* was a representational space of German unity, a means to imagine Germans united across confessional and political divides through a shared experience of a home region. *Heimat* is a concept that does not translate neatly from German to other languages. It means roughly "locality," but with more depth than the English term. Celia Applegate describes *Heimat* as encompassing three elements: the natural environment, folklore and folk customs, and historical memory. For Applegate, the process of building the German nation involved the combination of regionalisms around Germany, each of which was based in the idea of its own *Heimat*. By imagining *Heimat*, Germans developed a shared pride in their local identities and found that their commonalities outweighed their differences.[2] *Heimat* images were much like those below, depicting a locality

Figure 11. Wangemannshöh. From Martha Eitner, *Berliner Mission im Njaßa-Lande*

recognizable to people who knew it well while making it generic enough to appeal to all Germans as a place they could recognize as home. They were always rural, agrarian spaces absent the troubles created by industrialization and urbanization. In *Heimat* images of Ostafrika, the German colonial project appeared to have successfully created a shared representational space of Germanness abroad.

In making the landscapes of Ostafrika familiar to German readers, *Heimat* images showed German successes in making order out of the previously uncivilized East African frontier. In this function, they served to domesticate the empire for metropolitan audiences. *Heimat* images made colonialism appear simple and beneficial for both Germans and Africans.[3] The violence of the colonial state was nowhere to be found and racial hierarchies appeared stable and unchallenged.

Figure 12. Kirche in Hohenfriedeberg. *Nachrichten aus der ostafrikanischen Mission*, August 1894

When contrasted with the images of the slave trade discussed in earlier chapters, *Heimat* images depicted the German presence as civilizing and as bringing order to the chaos of the *Land* of East Africa. Empire appears as a space safe for German family life, away from the problems of both primitive Africa and unstable urban Europe.[4]

This chapter examines *Heimat* images as part of a broader history of ideas and their dissemination. Ideas about space in Europe became increasingly entangled with ideas about space in East Africa over the course of the 1890s. By 1900, the new, state-directed cultural geography had become the primary way to conceptualize development in the colony.[5] This was not "developmentalism" of the scale and character of interwar colonial Africa, which involved European government investment in long-term economic projects, but it was an attempt to draw such investment from private sources through the deployment of expertise and centralized planning to create a colonial economy that would benefit Germany.[6] At the core of such development was the concept of *Kultur* as settled agriculture that had emerged as the central idea of progress in the previous years. The goal was to transform all irrigable *Land* into economically productive space for the benefit of Germany. Ostafrika would come to look like the images at the start of this chapter – space that, while not fully German, was recognizable as German *Heimat*.

By 1893, Ostafrika's administrators had adopted the new ideas about making space based in cultural diffusion, by changing African spatial practice. The colonial administration attempted to create a scientific and systematic approach to cultural diffusion, one that would model Ostafrika's development after that of other European colonies. Ideas flowed from Europe to Ostafrika through *Heimat* images and text representations of East African space. Systematic, scientific colonialism that followed successful models elsewhere was the best hope for a profitable colony. With cultural diffusion triumphant, a new debate emerged over how to put it into practice. One school of thought held that creating plantation agriculture was the next step in the evolution of East Africa's premodern spaces. In this line of thinking, creating productive agricultural land would in turn make Ostafrika's *Leute* into productive labourers and colonial subjects. The administration hired experts to conduct examinations of the colony to institute such scientific approaches. Another strand of planning focused instead on the instrumentalization of East African communities, advancing their evolutionary progress so that Africans would create *Kultur*. The latter approach shaped a largely unsuccessful plan to turn the caravan trade into a conduit to the interior for cash. Such interventions depended on the transformation of Ostafrika's *Leute* into market economic actors who would create further change in the *Land* around them. More successful were both military and missionary stations in the interior, though not according to German intentions. Stations became hubs around which African communities settled and cultivated the nearby landscape. Left unspoken was the basis for the success of this approach, which lay in its allowance of greater space for African initiative and for the co-optation of station personnel by African leaders.

As opposed to stories about ongoing wars, failed plantations, and evasion of caravan regulations, Germans could imagine that stations successfully planted *Kultur* in the East African interior. Ideas from Ostafrika flowed back to Europe through *Heimatkunde*, which offered a means for understanding such changes more systematically, as well as a framework for thinking about Ostafrika as undergoing historical change along the same path as Germany. The success of *Heimatkunde* marks the continued advance of cultural geography as the primary means through which Germans understood Africa. The most important work in formulating the field was by Friedrich Ratzel, who also theorized in the late 1890s that states were the primary agents in creating *Kultur*; he provided theoretical backing for increased state intervention to modify East African *Land* and *Leute*. The limited successes of existing models of development, combined with the outsized visibility of *Heimatkunde* in metropolitan discussions of Ostafrika, accelerated the adoption of models of cultural diffusion into administrative practice.

Systematic Approaches to Transform *Land*

Though Julius von Soden had lost his position as governor in part due to criticism that he did not account for Ostafrika's local circumstances, the military governors who succeeded him continued along much the same path. They imported economic models from other nations' colonies and scientific expertise from Europe. Rather than changing the spatial practices of German subjects in Ostafrika, they would import practices from elsewhere. The imports involved the rational management of natural resources as the means to create profitable economies. The reliance on models from elsewhere culminated in Ferdinand Wohltmann's 1898 expedition to study Ostafrika's agriculture. Wohltmann's negative appraisal eliminated much of the remaining optimism about approaches imported from elsewhere. The report marked the apex of several years of failed plans based on the idea that science and rational management of natural resources could produce development without a role for African communities.

Governor von Schele, who succeeded Soden in 1893, patterned his plans to increase government revenue on taxation schemes in other colonies, a further sign of trying to apply learned principles of colonialism to Ostafrika.[7] The primary tax was a hut tax, a flat tax on all buildings. In announcing the tax, Schele waved away worries that it might provoke a rebellion. A hut tax on the coast had not caused a resurgence of anti-German fighting, and other European colonies had succeeded in taxing housing. It was now time to "open this source of income," as customs duties alone had not made as much money as colonialists had claimed they would in the 1880s.[8] A hut tax would obligate all inhabitants to earn and use cash to pay the tax, which would require them to either work for wages or grow crops that Germans would buy. The need for employment would in turn force Africans to adjust their spatial practices to meet German demands.

The hut tax plan came from the British Cape Colony and Natal, which served as an important model for Germans who still dreamed of settler colonialism in East Africa. The prompt for the introduction of taxation came from one Adolf Görz, born in Mainz, who had founded one of the largest mining companies in South Africa and was well-connected in both financial and political circles. After travelling in Ostafrika, Görz wrote to the German Colonial Office with suggestions for the colony's future. The biggest difference for development between Ostafrika and South Africa, thought Görz, was that there were no taxes on "native" inhabitants of the former. Through taxes in South Africa, "[the native] was forced to earn money to pay taxes" and became "capable of consumption." To further the cash economy, Germans should establish shops to take business away from Indian- and Arab-owned shops, which earned as much as one hundred thousand marks each year. It was unreasonable that these shop owners should earn more than Germans.[9] A proper tax system, according to Görz, would ensure the resilience of racial hierarchy.

According to this line of thinking, Ostafrika's administration mimicked other colonial regimes across Africa in the same period. European administrations used the threat of punishment for failure to pay taxes to force Africans into wage labour for European employers. The British South African colonies, which served as Görz's point of comparison, was one example. There, the hut tax was introduced earlier in the 1890s to force people to work for white-owned farms and mining companies, like Görz's own. They soon became popular in European colonies across the continent for similar purposes, at times provoking armed resistance. Most prominently, the UK provoked a war in Sierra Leone with a hut tax meant to make the colony self-supporting in 1898.[10] There was nothing particularly German about the hut tax plan, as Schele based it on research in other colonial contexts.

As has been documented with other head or hut tax systems in colonial Africa, workers changed their patterns of labour as little as was necessary to meet their obligations. The plantation labour system held little appeal, as it meant the destruction of accepted labour hierarchies and systems of reciprocity.[11] The DOAG complained in 1900 that Nyamwezi labourers and porters did not serve their terms on plantations and looked to return home immediately after they had earned enough wages to pay their taxes.[12] Plantation labour had a bad reputation to the west. The station official in Mwanza, on Lake Victoria, reported that people in his district attempted to avoid it as much as possible.[13] The tax plan failed because of African initiative as labourers adjusted their practices as little as necessary to meet German demands.

Most people still paid their taxes in non-cash forms in 1898. In that sense, the system was a clear failure. This map, figure 13, from the Mpwapwa district, not far from the coast, shows tax payments by village in the district. No village paid in cash. The majority of villages paid in kind, denoted with straight underlines. Other villages paid in labour, marked with curly underlines. The continuation of tax payments in forms other than cash makes apparent the failure of German attempts to remake African practices through taxation. This map also illustrates the introduction of racial categorization to the economic system, as it breaks villages down by what "tribe" Germans determined they belonged to, denoted by the color of the underlining on the map. Even within the German system of ethnic classification, which simplified African identities, several villages were inhabited by people of different ethnicities, and ethnic borders could not be neatly drawn. An accompanying written document broke down payment by tribe.[14] At this point, however, ethnic classification did not directly bear upon tax collection.[15]

The tax plan's primary goal was mobilizing agricultural labour, and agriculture remained the most important element in economic planning. Schele authored a report in 1894 that exhibited elements of the new cultural geographic thinking. He thought many parts of the colony "worthless for the further development in the interest of Germany." For the time being, most of Ostafrika's value for Germany lay in the mountains and highlands, where a "hoard of treasure was available for the fatherland." It could make Germany independent from imports of tropical products from other nations and allow Germans to settle for agriculture and herding. But, the cultivation of "export-worthy native products through natives" was possible in most places, so production could be increased even on the steppes. This would create "greater sedentariness," the key to the development of *Kultur*, in the sense of cultivation.[16] In this statement it is possible to observe the triumph of the new representational space of Ostafrika, as a space on the first stage of an evolutionary ladder of agriculture, its ascension of the ladder

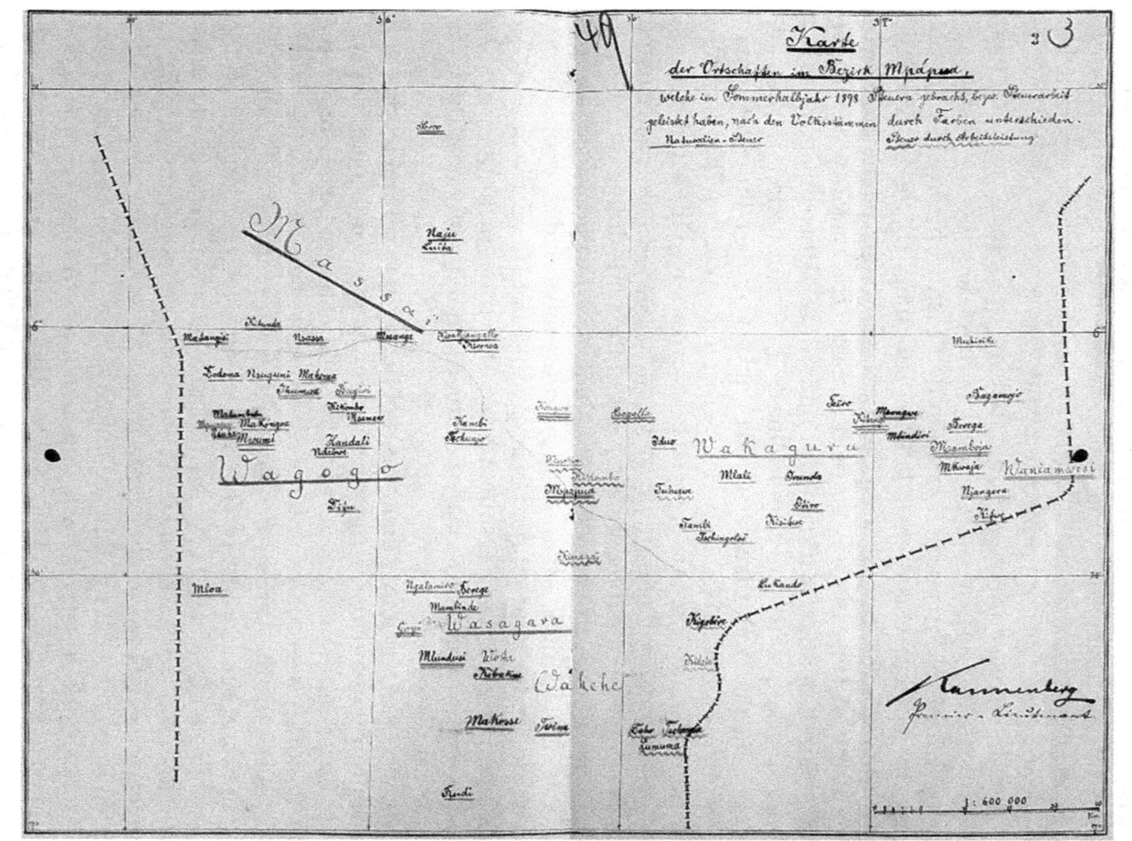

Figure 13. Mpwapwa district map. Rannenberg, "Bericht über die in Bezug auf die Häuser und Hüttensteuer gemachten Erfahrungen," 2 October 1898, TNA G 3/48

dependent on German assistance. In order to evolve that space, Germans would need to change spatial practice.

Schele continued the importation of labour from abroad to provide what he saw as the necessary labour for more advanced agriculture, but added elements of cultural diffusion to the plan. In the short term, "a race of trained labourers, Chinese or similar," was desirable. Such ideas about foreign labour could be traced to demands by planters. The director of one plantation requested more coolies from Southeast Asia in November 1893. The native population, he thought, was ill-suited for plantation work.[17] He repeated a claim, common among European administrators across colonial Africa, that African workers were lazy and unable to perform work at the necessary standard.[18] Schele's labour plan also included elements of cultural diffusion. He wrote that native workers could eventually make such foreign labourers "redundant," but only through proper training. Africans would learn how to work by observing Southeast Asians. Germans just needed to be patient and properly handle native labourers to teach them "sustained and regular labour."[19] This argument reflected the evolutionary thinking in German plans. It was not possible for Africans to learn directly from Germans, as the cultural gap was too large. An intermediary group, like Southeast Asians, would provide a more productive model.

The influence of ideas from Europe on agricultural plans for Ostafrika reached its height in 1898, when the Foreign Office hired Ferdinand Wohltmann to scientifically study agriculture in the colony. Wohltmann held a chair at the Royal Agricultural Academy Poppelsdorf, located in Bonn, where he had built a reputation as the foremost German expert on tropical agriculture. The DOAG requested that Wohltmann look at its cultivated territory and suggest some plans.[20] His tour took him mostly to the northeast of the colony, with short visits to Bagamoyo and Dar es Salaam, beginning in January 1898.[21] As vegetation reflected a region's climate, it was a good way to judge "an unknown land's value for *Kultur*," wrote Wohltmann. He conducted a detailed study that included chemical analyses of soil, studies of vegetation and the climate, and an investigation into the "ways and with what success the natives could use the land."[22]

Wohltmann wrote that German efforts had largely failed and that what success existed came not from good application of agricultural knowledge, but from following models established before German arrival. The existing plantation system must be declared a failure, Wohltmann asserted. The "clumsiness" of their creation, the "uncertainty" of the climate, the poor soil, and a bad choice of what to grow all contributed to the failure. Failure of oversight of inexperienced African labourers in the planting of coffee at Lewa, the creation of which had cost two million marks, was the clearest example. He recommended alternative arrangements. A government-owned planation near Dar es Salaam was assured of profitability, as Arab plantations in the vicinity were profitable, whereas German ones in the north were not.[23] Germans were best served by copying Arabs, who had learned through experience what crops and techniques were effective in the region.

On the other hand, Wohltmann continued the pattern of ignoring the majority of the colony's population in suggesting Germans copy elite practices. The science of plantation agriculture, with its focus on changing the natural environment, left little space for attention to East African societies. In Wohltmann's report, African workers seemed simply to be another landscape condition with which planners had to deal. He listed "worker relations" with climate, soil, and other growing conditions. For instance, one plantation had good prospects for the future because of its "excellent cheap labour relations." The workers there had built parts of the railway from Tanga and then moved to the plantation. Their railway experience meant they had "knowledge of labour and skillfulness," making them the most industrious and capable worker material" in Ostafrika.[24]

Wohltmann's report does reflect a clear belief that Ostafrika's future lay in agriculture directed by Africans rather than in plantations, a shift to changing African societies rather than landscapes. Wohltmann wrote that Ostafrika was inferior to many tropical colonies in its fruitfulness and transportation capacity, so its spaces could not bear the burden of providing for economic growth. What advantage did exist was in the population on the coast, which was relatively dense and teachable, and they had money to spend. Through Indian and Arab influence, coastal people had long ago become accustomed to industrial products and desired them more and more. Every village contained at least one skilled craftsman, indicating an ability to perform skilled labour. In general, worker relations were good and were improving. It was a "great advantage" that the population of the coast was dense, making labour plentiful. Plantations that lacked for labour were usually in that situation because Europeans had mistreated earlier workers, not because the population was low.[25] Wohltmann recommended development make use of existing agriculture and manage the *Kultur* of Africans in the colony. The administration should ensure that Africans founded settlements in "suitable places" and that they used careful, rational methods of cultivation to increase the production of the soil. The German purpose was to "lift the primitive economic methods of the natives as soon as possible and to educate them to rational economics."[26] For Wohltmann, what Germans needed to create *Kultur* was to make the interior a space more like the coast, using African labour to create a semi-civilized space.

Agricultural practices, the core of *Kultur*, had been closely identified with racial and ethnic categorization, so Wohltmann's program called for the application of racial characteristics as the foundation of development. Such thinking about race came through most explicitly in Wohltmann's discussion of the few remaining possibilities for German settlement. German settlement, wrote Wohltmann, should be by "German mountain tribes." "Flatlanders" would be "unskillful" at the work required and lack the desire for the "hard work and loneliness" of life in the mountains, whereas Germans from Alpine regions could manage life in Ostafrika rather easily. It is clear in Wohltmann's report that he was little versed in existing ethnographical and ethnological thought about ethnicity in East Africa, as he made little distinction between different ethnicities in his prescriptions for agriculture. When the focus turned to Germans, however, he could apply what he knew about ethnology and cultural geography to recommend the deployment of positive racial characteristics in colonialism.[27]

Wohltmann's report announced that the idea of the systematic, scientific transformation of *Land* in Ostafrika had reached its limits as the end of the decade neared. He had criticized the previous approach to development on and near the coast, writing that much of it could not be salvaged. *Land* could not simply be transformed with different agricultural methods, even if applied scientifically. Capital was lacking, and what capital had been spent was largely wasted. Wohltmann suggested instead that administrators attempt to manage development through East African *Leute*, by adopting aspects of local spatial practice into economic planning rather than importing practices from abroad. By utilizing the existing East African population, Germans could create a new kind of agricultural space in the East African interior.

Using *Leute* in the Caravan Trade to Transform *Land*

While an approach focused on transforming *Land* continued to dominate many plans for creating economic change in Ostafrika, more and more colonialists turned to non-technical solutions that used existing African spatial practices to achieve goals. Among these plans was the instrumentalization of the existing caravan system, which had drawn German attention to East Africa in the first place, to create cash-based economies in the interior. A cash economy, in turn, would force the colony's *Leute* to work for wages or to produce agricultural products that could be sold for cash. East African *Land* would thus be transformed into productive economic space for Germany. This plan, however, depended on a version of the caravan trade based entirely upon coastal geographies, which did not account for the dynamism of the routes in the interior or the power exercised by intermediaries. Porters and coastal merchants asserted their power over the functioning of caravans, causing the instrumentalization of the caravan routes to fail.

Successive administrations increased regulation of the caravan trade to make existing African spatial practices an instrument for German economic benefit. A regulation from September 1898 is perhaps the clearest example of the attempted utilization of the caravan trade. The cash economy that Germans desired was not developing in the interior, so all porters were to be paid in cash unless they specifically requested trade goods instead. Payment in cash was "in the interest of a more emphatic implementation of monetary exchange in the interior, such a condition much desired especially for the institution of hut taxes."[28] Administrators believed that creating a cash economy would enable more taxation, which would allow Germans to pressure more people into wage labour. In 1900, a new centralized porterage system required porters be hired in Dar es Salaam and set standard rates, some of which had to be paid in cash, for journeys of different distances, set in concentric circles from the coastal towns.[29]

Coastal elites asserted their prerogatives over the trade networks that predated German arrival. Soliman bin Nasr, discussed in previous chapters as a coastal *liwali* under Zanzibari rule and a *wali* under German rule, challenged the Dar es Salaam station's demand that he sell markets he had built in the town for a low price as well as a refusal to let him charge for grazing rights on his land. He had attached himself to the German regime early, and he cited his service in demanding continued economic power. He had rescued Meyer, Oskar Baumann, Emil Zelewski, and five other Germans from imprisonment by Bushiri. Since that war, he had directed refugees to farming and prevented countless conflicts.[30] No other inhabitant of Ostafrika had served Germans so well. Soliman compared himself to the *wali* of British Mombasa, who received a higher salary and owned more property. This was even though the Mombasa *wali* had done less than one-quarter of the service for the British that Soliman had done for the Germans. What was more, the "German regime was much greater than the English," so Soliman's work had been more extensive.[31] The result was a compromise. Soliman finally agreed to sell his markets for 25,000 marks, and the administration promised not to take his earlier profits.[32]

Beyond elite challenges, the plan of centralized porterage cash payments fundamentally failed because porters adjusted their spatial practices in the face of taxation. They spent their wages on the coast or on the first stops of the journey, where cash was in demand. No money made it to the Great Lakes. Porters arrived in the west broke and without food to feed themselves. Some were so destitute that they had to find work in order to afford to return to the coast.[33] Porters were still able to acquire the trade goods that were in demand in the interior and avoided German attempts to make them use cash for purchases. Beyond the basic necessity of cash for the hut tax, caravan participants did not want cash. Porters also adjusted their routes to go to the British colony to the north to escape German regulations. Multiple officials reported that porters were taking their wages to Uganda and buying goods there, where they were cheaper. Thus, money was flowing out of Ostafrika into British hands.[34] Another reported that conditions were so bad between Bagamoyo and Tabora that Nyamwezi and Sukuma porters were no longer travelling to the coast, meaning that coastal people had taken over the entirety of the trade. Porter wages in Tabora were now two to three times what they were on the coast.[35] Porters made the space that worked for them, where they could continue to accrue the benefits they desired, and avoided the regulations of German territory near the coast.

German administrators' representational space assumed a static caravan system following standard routes and procedures, but the reality was far more dynamic. One can observe shifting spatial practices with little control by the German administration. Changes in caravan routes make clear the ability of caravan leaders and porters to adjust to the violence instigated by the German presence and create new connections to distant places. New routes and new towns sprung up to avoid taxation and requisition. The Pangani district office reported that people along the caravan routes set up markets for Nyamwezi caravans, the basis for a rich trade in the district.[36] One official reported that Ngoni raids had caused the disappearance of the villages precolonial explorers had reported from their expeditions through the southern part of the colony. Ngindo communities had completely abandoned their old villages and moved to the protective terrain of the mountains. Ngoni raids had closed the old caravan route from Kilwa, which was now "completely overgrown and unused." They were continuing to create "uninhabited wasteland." But Rashid bin Hamid, an Arab merchant, had founded a new settlement at Mangwa, which had become a "new central point for the caravan trade." The raids had also increased the importance of Masasi, which served as the primary place to stock up on necessities before caravans entered Ngoni territory on the route to Lake Nyasa. Masasi's inhabitants paid tribute to the Ngoni in salt, which guaranteed their safety.[37] The new settlement at Mangwa and the increased importance of Masasi point to the ability of caravan participants to shift their routes to ensure continued access to Lake Nyasa. African porters adjusted spatial practice to create a new space outside German control – evidence for the inability of states to remake spaces for their own agendas in contravention of local communities' needs.

Participants in the caravan trade made greater adaptations in response to African-directed changes in the interior than they did to the German presence. German power to change practices was limited. Expeditions' destinations shifted to where better markets were available. They no longer went to Unyanyembe or Manyema, for buyers there had become too clever, and prices had fallen. Now coastal merchants primarily travelled to Nguu to purchase rubber and tobacco or took wares to Uhehe to exchange them for gold. People in both places had also become more aware of the premiums they were paying, but money had to be made somewhere. The pomp and circumstance that had formerly greeted caravans arriving on the coast no longer occurred, as the caravans did not bring wealth with them.[38] A station official reported that porters used the road from Mpwapwa to Kilimatinde in the close vicinity of stations, but they used other routes for longer journeys.[39]

In the caravan trade, German administrators attempted to co-opt existing geographies and use East African *Leute* to create changes in the functioning of economies. Cash payments, they hoped, would create cash-based economies hundreds of miles from German power on the coast. What they did not account for, however, was the dynamism of spatial practices on the caravan routes. Their version of the caravan trade was based almost entirely upon representations of space produced by Europeans and coastal elites. As had been the case before German arrival, coastal elites' ideas about the interior did not necessarily apply once there, and the plan to instrumentalize *Leute* in the caravan trade largely failed. In the interior, other approaches that allowed for more local initiative were more successful.

Transforming *Land* and *Leute* Together at Interior Stations

West of the coast, the German presence remained sparse. What effects on spatial practices Germans did have were largely limited to the immediate areas around military or mission stations.[40] Those stations, however, were more successful in creating the kinds of changes Germans hoped to see than either attempts to transform *Land* through plantation agriculture and attempts to create new economies manipulating the caravan trade. Missionaries attempted to form what they saw as more organic connections to the territory they claimed and people they evangelized. Military and mission stations did become hubs of settlement, but primarily because of famine and African action. The African intermediaries through which Germans tried to work asserted their own power over the new connections and new identities that formed around the German presence, creating new spatial practices that were not under the control of Germans.[41]

Though Germans wanted to believe in the power of the gospel or security to attract migrants to stations, the reality was that most settlement resulted from catastrophe. In the middle of 1894, swarms of locusts moved across much of the colony, destroying crops and creating a famine. Locusts arrived in "uncountable" numbers, sweeping over the landscape nearly every day and bringing "complete desertification."[42] This famine was part of a worldwide failure of systems of famine mitigation at the same time, caused in large part byEuropean colonialism.[43] As was the case in other colonies, European incursion rendered existing systems of provision in case of famine defunct. An evangelical missionary reported that even people of high social status were reduced to coming to the Kisarawe mission station to beg for work in order to buy food.[44] In May of the following year, he wrote that people continued to migrate to the coastal towns, hoping to find work and food.[45] At one Leipzig mission station, missionaries reported that nearly one thousand people had settled nearby, most of them having migrated to the area because of famine.[46]

Evangelical missionaries believed stations could serve as central points from which Christian progress would radiate, a representational space of Germanness radiating outward. One mission author wrote that it was beneficial that most of the land in Ostafrika was unsuited for both European settlement and plantations. This meant that Africans could remain in possession of their *"Heimat* lands" for smallholder farming.[47] Another missionary wrote that every mission station served as an "exit point for *Kultur* and civilization for its surroundings," turning nearby communities into close allies of Germans, closer even than the government's soldiers.[48] The Leipzig mission had begun its work in the "true African wilderness," in an area where all one saw for several days' march was "brown, miserable steppe" where not a drop of water could be found. Beyond the steppe were "wild mountains."[49] Over time, the mission station had created *Kultur* from nothing and gathered a "brightly-mixed crowd." This consisted of Swahili from the coast who had come as porters and stayed on as day labourers, skilled

Indian labourers, most of them masons, Tamil converts whom the mission had hired to come to Kilimanjaro, and some Chaga and Maasai boys who had attached themselves to the mission.[50] The missionary compared the mission buildings to other frontier scenarios – log cabins in North America or individual settlements in the Germany of 1000 years earlier.[51]

Military officials represented East African space differently. Some military officers adopted aspects of local representations of space to make sense of and dominate the territory they were tasked with controlling. Tom von Prince adopted his own version of the division between *shenzi* (wildness) and *staarabu* (Arabness) that had underlain precolonial coastal geographies as he began new operations in the South. He travelled up the Zambezi and Shire rivers to Lake Nyasa. At the north end of the lake he found the "garden of Africa" in rich agriculture. Beyond that area, however, "began the genuine *Shenzi-Land*," beyond which there was no agriculture that Prince recognized as such.[52] Prince described the Ruipa River as the "Rubicon" for the German advance, as beyond it lay conflict with the Hehe, much as caravan leaders before German arrival had delineated between Arabness and the wilderness.[53]

Surrounding communities accommodated spatial practices to the German presence, turning both mission and military stations into hubs for African settlement. Michelle Moyd has described the stations as "nodes that aided in the consolidation of colonial rule," but noted that their development as nodes depended on African soldiers turning them into public spaces for nearby communities. Stations functioned only through the Africans who came to them and made them into spaces that fit their own expectations.[54] A report from a *Schutztruppe* officer noted such changes near the Donde station. People were moving near the station and working the soil. Germans left the distribution of land up to African communities, but the officer reported that all land was under cultivation for "a circle of ten hours." The new settlers trusted the station's officials, attending the meetings at which German officials administered justice, called *shauri*, providing labour, and selling provisions, claimed Fromm.[55] Many people came to the Leipzig mission's station in Machame, on the southwest slope of Kilimanjaro. Most came for trade, others came to sell building materials, many came for medical treatments. Youth came to wonder at the *wazungu*, white people.[56] Most of the traders were youths, who soon learned they could charge missionaries high prices.[57] People had settled from many different surrounding areas at the Iringa station. They had built roads and homes. Indian, Arab, Baluchi, and Greek merchants had settled nearby to sell goods to the new settlers. The town reached an estimated population of three thousand in 1897.[58] The area around the Muhanga station had been a "humanless wilderness" five months earlier, but now appeared a "well-cultivated land" in which many natives worked their fields. The farmers had built roads between settlements without any instigation or assistance by the German officials.[59]

German observers believed that settlement near their stations indicated African acceptance of the German presence and saw in it opportunities for changing African societies. They could use such settlements to bring *Kultur* to the regions in which they were located. One evangelical missionary thought that settlement near his station would result in the creation of villages, which would enable chiefly authority.[60] Missionaries believed settlement near stations would ease surveillance and prevent relapses into heathenism, and that powerful chiefs could help maintain their moral authority. It was not possible to successfully settle freed slaves on the coast, as the density of population made it difficult to successfully surveil Africans living at missions there. Missions in the interior had plenty of space, so Africans could be watched.[61] There was some evidence that African settlers adopted identities tied to the German presence. One official met an old sultan (he gave his age as 200–400 years) who had aligned himself with Germany. The sultan reported that "formerly I was an *mshenzi* (a wild person), but now I am an *mzungu* (a white person)," making clear that he had adopted an identity based on his ties to Germany.[62]

Other people took advantage of the multicultural spaces that grew around German stations and soldiers to create new spatial practices. A Greek merchant followed a column of German soldiers, setting up a post to sell goods to surrounding communities wherever the column stopped. Another Greek merchant in Ujiji had made a fortune due to fighting in the Congo Free State cutting off trade there and forcing people to Ujiji. These merchants got their wares from Dar es Salaam and charged high prices in the interior.[63] Hans Meyer reported Indian *dukas* (stores) being built in Usambara as the German presence moved west. Indian merchants built these stores on plantations to sell goods to labourers.[64] People adjusted to the German presence and produced new means of organizing economies that accommodated the Germans without adopting German plans.

Hans Meyer documented a number of changes in spatial practices on his 1898 expedition to summit Kilimanjaro. Meyer's journals from this expedition provide some of the clearest evidence for change in the 1890s, as he had travelled through much of the same area on his 1889 expedition. Meyer noted progress in the spread of *Kultur* to the region.

Near Korogwe, wrote Meyer, a region that had previously been subject to frequent raids had been settled by farmers.[65] One could observe the security Germans had brought in the fact that villages had taken down the gates to their palisades. A little further west, "native trade blooms."[66] Meyer reported that chiefs and *akidas*, the local officials of the German state, came to the Moshi station daily to discuss livestock, land purchases, and disputes. The German *boma* (fort) there was "really the centre of Kilimanjaro."[67] Around the station, one could observe the "local expansion of *Kultur* under German protection," on both the steppe and in the forest. The population was visibly increasing.[68]

The importance of stations and the lack of a German presence in most of Ostafrika meant that many of the reported changes were limited to small areas. Meyer was struck by the "tremendous contrast" between "meticulous, successful *Kultur* " and "primitive forest nature" as he travelled through Usambara.[69] Stations along the coast shifted to paying porters in cash.[70] But porters hired in Tabora still had to be paid in cloth.[71] The station chief in Langenburg, north of Lake Nyasa, reported that he had succeeded in transitioning his porter payments to cash, but this was not a panacea. People living near the station now used cash to buy necessities, but people living farther afield continued to pay in goods.[72] In the development of cash economies, then, the German stations served as points from which changing spatial practices radiated outwards. The stations were too few in number and too distant from one another for this change to reach most of Ostafrika.

Interactions between German missionaries and African leaders make clear that it was not primarily German actions that instigated changes in practice. African figures could exercise control over the movement of weak groups of Germans, particularly missionaries. Germans without military escorts were in weak positions compared to many African leaders. Merere, a Sangu ruler in the vicinity of Lake Nyasa, attempted to control missionary movements in his territory through his promises of protection or lack thereof. He instructed the Moravian missionaries to go to town when they feared a Hehe attack.[73] He ordered them, "do not cross the borders of my territory" upon hearing a plan to go view the Ruaha River. Merere told the missionaries they would be killed by the Hehe as either spies or sorcerers.[74] He thereby secured their service to his political ends. They would not move across the river and would offer protection to his realms.

Missionaries working near Kilimanjaro explicitly adopted theories of cultural diffusion and began discussing the application of racial characteristics to changing *Land* in their publications. G. Althaus depended on existing tropes about the Maasai, the "still-feared nomads in their kraals." It was impossible to think of evangelizing there, as Maasai nomadism "made teaching and learning simply into an impossibility." The area seemed completely wild and foreign, and Althaus fell back on thoughts of the "faraway beautiful *Heimat*" to comfort himself.[75] In contrast, among the Chaga the "entire land embodied a rich and graceful appearance," as the Chaga farmed like Germans in Osnabrück. Progress had been made, as some people, "licked by *Kultur*," no longer used *"washenzi"* greetings but had adopted Swahili words.[76] Another Lutheran missionary described a true "zone of *Kultur*" in areas inhabited by the Chaga. The Chaga fit the missionaries' ideal of farmers, living in family groups recognizable to Germans. Readers of the missions' newspaper, they thought, would think the region a "true paradise" from the description. In the mission's opinion, it could be a "site of earthly happiness" if "human sin had not entered these heights." The region was beset by constant war. With time, however, the Chaga's "greater strength and courage," compared to the majority of Africans, would show through.[77]

Missionaries were among the most important conduits for changing representations of East African space to reach a non-scientific audience. Evangelical missionaries adopted the vocabulary of ethnography and cultural geography in their descriptions of Ostafrika. According to one evangelical missionary, the Kinga were a "mountain people" whose lifestyle showed the influence of the steep mountains. Both their homes and their bodies were dirty and disorderly, and they shunned connections with their neighbours.[78] The Konde of lower altitudes, in contrast, were open and friendly, as they did not even fortify their villages. Their entire way of being stood "high above many other African tribes." Even their appearance was more attractive.[79] Geographic location shaped social forms in this account.

A similar shift in the thinking of the colonial military establishment is apparent in the public engagement of Lothar von Trotha. Trotha's speeches and writings were another avenue through which changing representational spaces reached Germans, including a different audience from the readers of missionary publications. Trotha, as commander of the *Schutztruppe*, had been effectively the second-in-command in Ostafrika from 1894 through 1897. His suggestions for the interior make clear Trotha's acceptance of racialized historical narratives for both East Africa and Germany, which explain his new thinking about development in the interior. As with many other colonialists, Trotha primarily represented space through historical narratives. The interior, in Trotha's vision, was stuck in an imagined distant past. Whoever was stronger raided weaker neighbours, as Friedrich the Great had captured Saxons and Austrians for his armies, the Romans and Huns had captured wives and, in the Thirty Years' War, German *Landsknechten* had taken wives from defeated regions. Even in nineteenth century Germany, "at the highest level of *Kultur*," "gypsies" stole women from villages.[80] In writing the report, Trotha articulated an ethnography of East African societies based in a narrative of historical development that explained the entire world and which portrayed the region as behind Germany in time, as had been common in plans for the colony dating back to Carl Peters in 1884.

Trotha applied these racialized narratives of history to argue for a role for racial management in development. He connected a perceived racial problem in Ostafrika to a perceived racial problem in Germany's eastern provinces, a narrative that applied the use of theories meant to remove non-German influence in the German East to the removal of harmful racial influences in East Africa. Trotha was set against policies that tried to work through South Asians to create development. Indians, in his opinion, "must be ejected from our territory," as an increased Indian presence would destroy the remaining possibility of spreading Christianity in Central Africa. One "might as well suggest that that the thinning German populations in the Eastern provinces [of Germany] be refreshed with Polish Jews." He suggested that Germany follow the model set by the United States in its Chinese Exclusion Act of 1882 and create laws banning Indian migration to East Africa. Germany "did not need these merchant Jews," for there was plenty of merchant talent "in the children of Ham." He repeated his reference to "Indian Jews" when discussing coinage in Ostafrika and in saying that "Indian and Greek Jews deceived the poor natives in an outrageous manner." Arabs in the colony were indebted to the "African Jews, the Indians."[81]

Positive racial characteristics could be best observed in *Kultur*, thought Trotha, indicating that cultural geographic theories had penetrated the thinking of the most traditional element in German society, the Prussian Army. Trotha described East African landscapes according to the changes humans had created in them. His sign for the existence of *Kultur* was the lack of "primeval forest," as "*Kultur* is the greatest enemy of the forest." Urundi lacked forest and instead had banana and sorghum fields.[82] Another area on Lake Victoria was an "absolutely dead region," as there were no villages or cultivated fields, but rather natural vegetation.[83] The people of Lake Tanganyika, on the other hand, were hardworking and practiced agriculture, creating a superior landscape.[84] These representations depended on tropes of spatial evolution that had appeared in explorers' travelogues and become part of cultural geographic understandings of Africa.

Trotha thought the deployment of positive racial characteristics would be easy, as clear divisions could be drawn between East African "tribes," which he classified according to his own whims. In Trotha's opinion, "tribal belonging" was more developed among Africans than anywhere else in the world. Borders between tribes were drawn "exclusively according to natural obstacles." African tribes were "a purely objective geographic" matter.[85] Between every "tribe" was an "uncultivated border strip," always larger than the area under cultivation.[86] This would allow Germans to easily determine the *Kultur* of individual "tribes" to deploy elsewhere. In making these assumptions, Trotha adopted a territorial basis for understanding African ethnicity, a conception of ethnicity with little connection to reality.

Over the second half of the 1890s, both military officers and missionaries tried to instrumentalize African spatial practices to solve the challenges they faced. This instrumentalization was largely unsuccessful, as Germans continued to understand African spatial practices in similar ways as they had previously. That German stations became places to settle owed little to German intention. The African intermediaries through which Germans tried to work asserted their own power over the new connections and new identities that formed around the German presence. The interactions discussed in this section became the basis for ideas that travelled back to metropolitan Germany, to which this chapter will now turn.

Ostafrika in Metropolitan Germany

Ideas about East African geographies discussed so far in this chapter circulated in metropolitan Germany through a variety of publications. From Reichstag debates, to missionary publications, to geography textbooks, Germans encountered ideas about development based in cultural geographical models in multiple venues in the 1890s. These versions of East African societies included aspects of African geographies discussed in previous sections, meaning Germans adopted elements from African spatial imaginaries as their own. Most of the representations of East Africa presented a simplified version of East African societies for German consumption, which made it possible for Germans to apply existing models to understand them. The primary framework used to understand Ostafrika was *Heimatkunde*, developed most thoroughly in the work of Friedrich Ratzel. In turn, Ratzel applied concepts from *Heimatkunde* to develop new theories of cultural diffusion in which states played the leading role in the evolution of space. East African societies featured prominently in Ratzel's spatial evolution.

The pictures discussed at the beginning of this chapter depicted the achievement of *Kultur* through social evolution, as envisioned by missionaries. They familiarized German readers with a recognizable landscape not so different from their own, using the historical narratives that had become established as the primary means of representing East African space over the preceding years. The mission stations depicted could easily be mistaken for outposts in the German wilderness. As a Lutheran missionary wrote to his readers, such images and descriptions of East African spaces could make them "more familiar with our location and enable a spiritual ... cohabitation" between metropolitan readers and colonial missionaries. He provided a long description of the Leipzig Mamba station and the surrounding landscape for readers to form that connection.[87] With such visions in mind, readers could apply familiar skills and knowledge to make sense of Ostafrika, without considering the particularities of African societies or landscapes.

Those existing skills and knowledge had become part of the burgeoning field of *Heimatkunde*, a sub-discipline of cultural geography. Friedrich Ratzel's *Deutschland. Einführung in die Heimatkunde*, published in 1898 and the most important of *Heimatkunde* texts, makes clear the new political importance of the study of *Heimat* as geography. The best way to build closer connections to the German nation, wrote Ratzel, was to recognize that "soil and the *Volk* belong together."[88] He followed with extended descriptions of German landscapes, which read almost like descriptions of African landscapes in explorers' travelogues. Such descriptions turned African settings into places for fantasies of European accomplishment. Landscapes appeared as passive objects in which European explorers could implant the influence of civilization. For Ratzel, German landscapes appeared as passive objects for his actor, an anthropomorphized *Kultur* with its own agency and momentum. At one point, Germany had been almost entirely covered by forest, but the "progress of *Kultur*" had controlled and manipulated nature, pushing back the forest to a smaller area.[89] As Germans became more closely connected with their soil, "a new landscape came into being, a landscape of *Kultur*, full of the signs of labour," different labour from the surrounding regions, which produced different landscapes.[90]

For Ratzel, *Kultur* was the fundamental condition for progress. Nomadism was the "irreconcilable foe of every economic system," as it prevented the concentration of people. Nomads possessed an uncertain relationship to the soil.[91] Treating nomadism as a necessary step in development was "one of the worst errors of older ethnography and political geography." High culture could not connect with nomadism, as *Kultur* made use of all conditions and seasons, increasing the production of the soil and offering the opportunity for more people to live in the same space. Nomadism, in contrast, only made use of what nature offered.[92] Agriculture was not just a means to get food, but a "mode of life" around which all aspects of society were organized.[93]

Deutschland was indicative of a wider trend in Ratzel's work in the 1890s, toward a new importance for the role of states in shaping landscape. In his *Der Staat und sein Boden*, published in 1896, Ratzel laid out a systematic framework for understanding the relationship between states and the land on which they developed. Ratzel declared it important to think of states not simply as aggregates of people, but as organisms. Organisms were different in that labour was divided among organs. Simpler organisms possessed less-differentiated organs. For a state, the most important conditions for building organs were the differentiation in its soil and the spatial division of its population.[94] Space became part of the "spirit of the race" of its inhabitants.[95] The connection between a race and its soil became so strongly ingrained that two became one and could not be separated.[96] Historical development could largely be traced to the discovery of new characteristics of spaces that had previously been unknown.[97] Such was the case with the transition from medieval to modern Europe and from the Inca empire to Spanish exploitation of mineral resources in South America.[98]

Ostafrika clearly loomed large in Ratzel's new conception of the relationship between states and space. He maintained an active relationship with scientific figures involved in the colony. When Hans Meyer returned to Ostafrika in 1898, he named a valley on Kilimanjaro the "Ratzel glacial valley" (*Ratzelgletscherthal*), and wrote Ratzel with information about Ostafrika that formed the basis of what Ratzel wrote about the beginnings of state formation.[99] The idea that there were no states among people at "low levels of *Kultur*" was a fantasy of "cultured people" to justify a strict division between themselves and the "naked wild people."[100] Territoriality in Africa was less important, wrote Ratzel, as plenty of space was available. If soil quit producing after a few years, space could easily be abandoned and new fields cleared from the bush. There was therefore "no value in declaring exact borders."[101] His example of the "elementary organism" that gave rise to higher levels of *Kultur* and civilization was Unyamwezi. He included a map, showing what he claimed were the basic characteristics of a "Negro state," a chief's village at the centre, rings of villages around it, then "border wilderness." In such states, the state was as fleeting as one person's life, that of the ruler, whereas in stronger states such political goals "filled all parts of the state body."[102] The map shown in figure 14 clearly reflects a desire to make the state in question appear as amoeba-like as

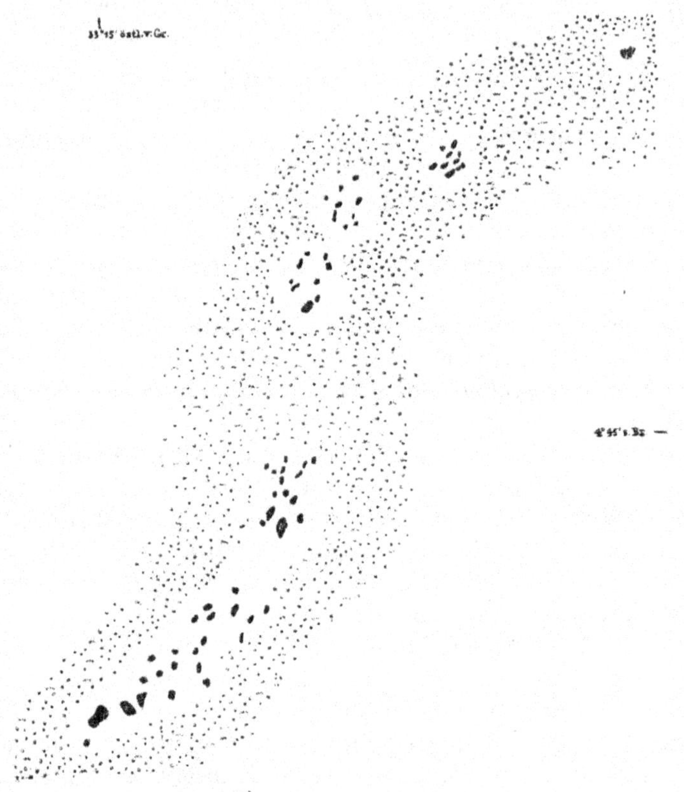

Figure 14. "Gebiet des Häuptlings Mtemi in Unjamwesi." From Ratzel, *Der Staat und Sein Boden*

possible, as an amorphous bacterial "elementary organism." Despite Ratzel's claims about the importance of soil and landscape to state formation, the map lacks any distinction of landscape. It could easily be placed anywhere in the world.

Ratzel's theories of state formation, like those of cultural diffusion, filtered down to German youth through geography textbooks. Willi Ule's textbook for more advanced students included a narrative of geographic development based on Ratzel's framework. The more a people freed itself from dependence on the natural provision of its surroundings, the greater its "spiritual development." Social conscious-ness, the first condition for the creation of states, emerged first with the raising of livestock. The next step was settled agriculture, as nomads wandered from place to place. Settlement created an "idea of property and therefore a sense of law and order."[103] Thus, what German admin-istrators were doing in Ostafrika could appear as part of the natural course of events, merely accelerated by human action.

New textbooks and theoretical works, which set a clear framework for the systematic study of *Heimat*, appeared in this period as *Heimatkunde* became a more common subject in schools. In 1908, the subject became mandatory in all *Volksschulen* in Prussia, Germany's largest state.[104] J.M. Blaut has argued that textbooks provide an "important window into a culture," as they are "semiofficial statements of what exactly the opinion-forming elite of the culture want the educated youth of that culture to believe to be true about the past and present world."[105] The main place where Germans learned about empire was geography les-sons in school, where "school geography reformers folded a broadly acceptable patriotic and colonial worldview into efforts driven chiefly by professional and pedagogical imperatives."[106] Tacit support for Ger-man colonial projects became part of regular school curricula.

Geography textbooks worked within a frame of the *Heimat* as a basis from which students were to compare other parts of the world. Willi Ule's *Lehrbuch der Erdkunde für höhere Schulen* makes this clear, begin-ning with a model for studying the students' home region, then apply-ing that model to Germany's colonies and other parts of the world. Ule wrote in the book's introduction that he wanted to move away from geography as a collection of names by grounding the book in "natu-ral science methods." Students began by learning how to create their own representations of spaces with which they were familiar. The first step was "the basics of *Heimatskunde*, which the teacher can easily illus-trate through connections with the particular conditions of the school's location."[107] Students learned to make observations of the landscape and relate different elements to one another. By taking those skills and applying them to other parts of the world, they "fulfilled the purpose of the science called geography."[108] For human societies to thrive, the first condition was to have enough food. In most cases, nature would not provide this alone, so "man must help nature" by working the soil, by creating *Kultur*.[109]

Further on in the book, students were instructed in the fundamen-tals of the representational space on which development planning in Ostafrika was founded, indicating its entry into school curricula. Hav-ing learned the appropriate vocabulary for understanding German colonial plans, they then received its ideals as facts. According to Ule, Ostafrika's population carried out agriculture and livestock raising, and that there was a racial division between farmers and the immigrant Arabs and Indians who controlled trade.[110] The next volume in the series, for higher grades, attributed lack of *Kultur* in Africa to natural conditions. Africa lacked Asia's fruitful agricultural lands and Europe's access to the sea, which had advanced *Kultur*. The continent had contributed nothing to the development of human *Kultur*.[111] Ostafrika's soil was mostly fruitful and suited for cultivation. The "settled Negroes" practised agriculture and raised livestock, though they were threatened by "unsettled raiding peoples," meaning primarily the Maasai and the Hehe.[112] Summing up the book, Ule wrote that "history, *Kultur*, the ways of a people depend in many ways on the location and the nature of the lands it inhabits." With *Kultur*, however, man's dependence on nature lessened. A part of the "purpose of *Kultur*" was the "advancement of such a position of power ... to order nature under men's free will."[113]

Ule was not alone in presenting these ideas to German youth; geography students read as about models of cultural diffusion as they did about physical or political geography. A child reading a geography book might learn that there were "different cultural levels of humanity." The lowest level consisted of people who lived by hunting or fishing, who lived "truly without property." A level above consisted of nomads, whose "property consists of domesticated animals." On the highest level were settled peoples with property who had "founded a home." They combined the activities of the lower levels with agriculture, industry, and trade.[114] Only with agriculture was the "creation of settled peoples" possible, and only settled peoples were capable of building lasting states.[115]

Geography textbooks normally included a discussion of Ostafrika specifically, in which these models were applied to thinking about German colonialism. The ways in which they were applied demonstrate the continued dominance of the representations of East African space colonialists had produced over the preceding years. According to one textbook, Ostafrika's climate "offered the conditions for a settled lifestyle based in cultivation and agriculture." For Europeans, however, the task was to "lift the black race culturally and economically and make it capable of business."[116] For the time being, the Indigenous population of the colony "stood at a lower level of *Kultur*." The easy availability of food and other necessities meant that people lacked an understanding of the importance of labour. Germans had to teach them.[117] Students in other classes could learn that the colony's population was "mostly peaceful farming tribes whose economic development has thus far been limited by Arab slave raids and attacks by raiding nomads."[118] One textbook author wrote that the mountains near the coast were suited for plantations, though most of Ostafrika was a "dry, almost uncultivable steppe," with few areas where Germans could settle. The coast was dangerous for European health and travel to the interior "very arduous." What was holding back the progress of plantation was that the colony's population was "too lazy for any lasting occupation."[119]

In addition to textbooks and missionary publications, models of the diffusion of *Kultur* became the main mode of talking about the colony in political circles, as a *Reichstag* debate on 18 March 1895 makes clear. The case that Paul Kayser, the director of colonial policy in the Foreign Office, presented for his proposed changed depended heavily on such models. He presented the spatial division of development planning as settled fact. It was necessary to "differentiate two zones" in Ostafrika, one of them suited for "grander plantation work" and the other for settlers of lesser means. Over time, *Kultur* would spread from the plantation zone elsewhere.[120] The military, "a very valuable cultural element," would help to spread *Kultur* west from the coast.[121] Kayser's speech articulated the spatial division between *Kultur* and wilderness, and plans to have *Kultur* spread west into the wilderness, as official colonial policy.

What metropolitan Germans could read about Ostafrika was a simplified version of local politics and geographies, manifested most explicitly in the station images at the beginning of the chapter. Those images identified Ostafrika's spaces as recognizable *Heimat*s and made them available for study through the field of *Heimatkunde*. In *Heimatkunde*, Germans applied cultural geographic concepts to create a more formal understanding of the spaces around them. Through such images and texts, cultural geography became the primary means of making sense of and discussing the future of Ostafrika among non-specialist audiences in Germany. While those ideas circulated in Germany, Friedrich Ratzel, Germany's leading cultural geographer, began a new line of argument in which state formation went hand-in-hand with cultural geography. Ratzel argued for a cultural geography that included states as its central actors. With Ostafrika as *Heimat*, Germans could conceptualize the challenge to colonialism in Ostafrika as one of *Kultur*. With the new state-based cultural geography, they could conceptualize solutions based on the German state manipulating the relationship between *Land* and *Leute*.

Conclusion

The period 1893–1900 was one in which German colonialists consolidated their rule over Ostafrika and began to implement the first centrally planned programs for the creation of *Kultur* in the colony. Initially, these programs deployed the most up-to-date agricultural and forestry science to solve the challenge. More successful approaches, however, involved the utilization of existing African spatial practices to make territory productive. The initiative in these changes, however, lay with African communities who adapted to the new German programs, fitting them into existing systems of knowledge and space making. What came back to metropolitan Germany was a simplified version of African geographies, which developed alongside Friedrich Ratzel's new state-based cultural geography into a model for thinking about progress in the colony. The flow of ideas, however, was largely detached from the reality of spatial practices on the ground. African communities continued to make space outside the grasp of the colonial state.

The next logical step was to combine the two to argue for a more active role for the German state in making *Kultur* through the manipulation of *Land* and *Leute*. An approach based on modifying either one alone had proven destructive rather than generative of *Kultur*. Private actors had been unable to make the desired economic productivity with their short-term focus and excessive violence. State-based cultural geography offered a clear model for the progress of *Kultur* in which the German state could play the leading role. In the twentieth century, German administrators would attempt to use state resources and power to foster *Kultur* throughout Ostafrika.

9 Biogeography and Resettlement for *Kultur*

In October 1899, Ostafrika's administration received a proposal for economic development from an unexpected source. The Aga Khan, imam of the Nizar Ismaili religion and future leader of Indian Muslims, suggested moving some of his followers from Punjab to Ostafrika, where they would settle as independent farmers. He hoped that they could form a community in the German colony and become German citizens.[1] In an article in the *Times of India*, the Aga Khan praised German work in Ostafrika. He had travelled there expecting little, but instead observed "splendid public works, the beautiful free hospitals, the excellent and numerous police force," and other signs of good governance.[2] Over the following two-plus years, administrators in Ostafrika and Berlin corresponded extensively about the Aga Khan's plan, eventually deciding settlement should happen in the Rufiji Delta, in the far south of the colony. Despite their efforts, the plans came to nothing, hindered by British opposition to Indian emigration to Ostafrika and by German colonialists opposed to increased Indian influence.

However, the idea of Indian settlement continued to intrigue German administrators. An Indian clerk for the British Uganda railway wrote Ostafrika's administration with a different settlement proposal in 1901. The clerk inquired about the cost of agricultural land in the German colony, as he was considering the "importing of a few good sturdy cultivators from the North Western Provinces of India."[3] He would then presumably serve as landlord for settled Indian farmers. While the proposal did not fit German administrators' vision of their colony's future, it did set off new discussions of South Asian settlement in Ostafrika. Most of those discussions centred on Tanga district, the location of the Usambara Mountains that had been the focus of plantation plans the previous decade. Over the next few years, Ostafrika's administration would fund the immigration of multiple groups of South Asian farmers and the infrastructure for their settlement.

Accompanying South Asian settlement were initiatives to settle Nyamwezi farmers from east of Lake Victoria in the same parts of Tanga district. The administration resettled several Nyamwezi *jumbes* in the district and provided land free of charge, and hundreds of Nyamwezi settled in the district on their own initiative. In the planners' imaginary, the two forms of settlement were closely bound together in creating economic development. They would operate separately, but the South Asian settlers' agricultural techniques would diffuse to the Nyamwezi farmers, who would eventually form the most important economic force in the colony. Unlike the plantation agriculture plans of the 1890s, freeholder African farmers were to serve as the basis for Ostafrika's profitability.

The resettlement plans indicate the triumph of cultural geographic thinking in German economic development plans after the turn of the century. German administrators now recognized the importance of African spatial practices in making economic development and attempted to manage those practices to German ends. German colonialism reached what James Scott calls "high modernity," where the state tries to completely remake societies according to a vision of the future that does not account for how people live its projects.[4] They believed there was too large a gap between the levels of *Kultur* possessed by Europeans and Africans. That gap, they thought, was the cause of earlier failures in attempts to create productive colonial subjects out of Ostafrika's inhabitants, as it was too wide for *Kultur* to diffuse from Germans to Africans. South Asian farmers, however, were on an intermediate stage of *Kultur*, which could be transmitted to African farmers in a way that German *Kultur* could not. Theoretical backing for resettlement came from Friedrich Ratzel's conceptualization of Social Darwinist claims of racial struggle as a struggle for space. In *Lebensraum*, Ratzel introduced his new concept of *biogeography*, which applied biological theories to understanding human societies and their relationship to space. Biogeography combined biological theories of evolution with geographic theories of cultural diffusion to explain the history of all life. He thus theorized both *Kultur* and colonialism as outcomes of processes of natural selection as different races competed over the limited space on the earth's surface.

To make reality out of their political vision, administrators crafted racialized resettlement policies to implant *Kultur* in Ostafrika. The plans for resettlement of South Asian and Nyamwezi farmers in the Tanga district represent the clearest manifestation of active state intervention into the racial makeup of Ostafrika. Germans would take spatial practice from elsewhere and implant it into territory that they thought should be more economically productive. The creation of *Kultur* through the management of race thus served as a form of social engineering, an attempt to create ideal societies. Administrators inserted themselves into the narrative of biogeography as the agents that would instigate the process of racial competition over space. German intervention would assure the evolution of a society made up of peasant freeholders, a form of agrarian conservatism with implications beyond Ostafrika. Administrators' ideal societies would avoid the divisive problems of modern Germany. There was more room for government action outside the German legal space, so administrators could remove troublesome African leaders and prevent the formation of a proletariat.

This system all came crashing down in 1905. Then, people in the colony's south started an uprising against German rule that has come to be known as the Maji Maji War. Settlers in Ostafrika and their metropolitan allies blamed South Asians for starting the uprising and attempted to stir up a movement for their exclusion. The German Foreign Office rejected the settler movement in favour of reconciliation with the United Kingdom, mindful of the fact that Ostafrika's South Asians were British subjects. In pursuing reconciliation, the German government began reforming Ostafrika's administration to fall more in line with accepted colonial models used by other European states. The resulting reforms curtailed the use of cultural diffusion to create *Kultur*.

Lebensraum as a Theory of Empire

Lebensraum, due to Adolf Hitler's adoption of its title as a term to provide ideological backing for his conquest of Eastern Europe, has been the subject of much scholarship that considers connections between pre-1918 German expansion and the Nazi period. Common to these works is a conception of Ratzel's work on *Lebensraum* as simply a forerunner of more aggressive imperialism later in the twentieth century. He is frequently treated as part of the ideological foundation of Nazism, a perspective that looks back from the 1930s and 1940s rather than forward from the Wilhelmine period.[5] If one reads Ratzel within his own context, however, it becomes clear that the appeal of *Lebensraum* and Ratzel's other works to contemporaries was in their all-encompassing theoretical basis for New Imperialism, something previously lacking in European expansion. The title, *Lebensraum*, denotes a new concept for thinking about natural history and its relationship to space. All living things, wrote Ratzel, required "living space," parts of the earth's surface they could inhabit and from which they could exploit natural advantages. Unlike the other currents of Social Darwinist thought used to justify the New Imperialism of the late nineteenth and early twentieth century, which posited European biological superiority as in and of itself sufficient to justify colonialism, *Lebensraum* justified territorial conquest as an element in a racial struggle for space. *Lebensraum* thus provided a theory that German administrators could use to create new policies in Ostafrika that more actively intervened in agriculture through attempts to change the colony's biological attributes.

Lebensraum was part of a wider intellectual phenomenon, a search for a science to explain European colonial expansion. The early twentieth century saw the publication of several theories of empire, each meant to explain a phenomenon that was clearly so important in contemporary politics. The most influential text of the early twentieth century was J.A. Hobson's *Imperialism*, which blamed international capitalism and finance for imposing imperialism on the rest of society, hiding its roots in capitalism. Hobson attempted to pull the wool from the eyes of European societies, which he believed were being exploited by finance capitalists who had drummed up national enthusiasm around the acquisition of overseas empire. Hobson's categorization of "finance capitalists" was decidedly anti-Semitic; he explicitly cited an international Jewish conspiracy to control the world economy.[6] Other theorists looked to biological science for an explanation. Nineteenth-century Social Darwinist thinkers such as Herbert Spencer and Francis Galton had claimed empire was a matter of the survival of the fittest, a racial struggle for the future of the world. Around the turn of the century, the famed biologist Ernst Haeckel proposed a biological framework for understanding human races as part of evolutionary processes, which became the basis for the foundation of the Monist League, an organization devoted to the promotion of social policies based on Haeckel's theory.[7]

Ratzel's biogeography turned Social Darwinists' struggle for existence into a struggle for space. There was a contradiction between the unending drive for expansion and the static space of the earth.[8] Over time, "cultured peoples" (*Kulturvölker*) had driven back "natural peoples" (*Naturvölker*) in their search for more space. Natural peoples did not die out but were restricted to smaller and smaller spaces. He cited the westward expansion of the United States and the elimination and forced migrations of Indigenous peoples as an example.[9] The geographer adopted a popular mode among German colonialists of taking the American case as the norm of colonial expansion.[10] Ratzel thus created a "logic of extermination," justifying territorial conquest and the destruction of colonized people and societies as part of the natural course of history. This kind of thinking was part of settler colonialism around the world, which depended on the elimination of Indigenous populations to clear territory for European settlement.[11]

In contrast to other popular theorists of empire from the period, Ratzel centred space. He provided an all-encompassing framework, which he termed biogeography, that fit the New Imperialism into the broad span of human history and provided a means for conceptualizing the representational space of empire. Biogeography incorporated beliefs about European racial superiority without resorting to pseudoscientific claims that Europeans were better because of genetics. *Lebensraum* unfolds as a description of biology as Ratzel posited states as life forms that followed biological rules like any other organism. All life was dependent on the conditions of the space in which it developed and lived. It would always seek increased space for nourishment, and so expansion could be seen as "the striving for the further expansion of the space of sustenance."[12] In *Die Erde und Das Leben*, published in two volumes over 1901 and 1902, Ratzel wrote that the core principle of biogeography was an understanding that all life on earth "lived as one," and that human beings belonged to the same whole as other life forms.[13] Though aspects of Ratzel's thought were adopted into National Socialist ideas of racial purity, he made an argument for a racial division of labour rather than racial purity. Racial purity was impossible and, in fact, racially mixed people such as the Anglo-Celts and Macedonian-Greeks had been the most successful.[14] When two races occupied the same space, they would come to carry out "different social functions."[15] As an example, he cited Arabs in East Africa, who had enslaved Africans and taken on different economic roles.[16]

Biogeography posited *Kultur* as a basis for claiming European superiority in the competition for space. Ratzel drew a clear hierarchy of *Kultur* and political forms with Europeans at the top. Peoples of Asia had achieved much in material things and art, but only Europeans had advanced scientifically.[17] Higher levels of *Kultur* created a movement to more valuable soil, endowing its practitioners with the means to acquire and make use of such soil, both economically and politically.[18] The root of all *Kultur* was the cultivation of the soil. "Civilization" meant simply that people had become burghers, self-sufficient farmers working for themselves.[19] Through their *Kultur*, stronger races became more difficult to move, as "*Kultur* binds humans to the soil."[20] Ratzel's vision of a strong race conveniently resembled his idealized version of European agricultural societies.

The lack of clear borders among precolonial African states became evidence for the necessity of European transformations of African space.

Without *Kultur*, there was little need for the control of much territory, as population density was low.[21] As levels of *Kultur* rose, populations increased and societies expanded to the point where there were no longer spaces between them.[22] Ratzel argued one could rank races according to their political development. One could then trace their political forms to their "cultural level," particularly the size of its territory and the number of inhabitants. States of cultured peoples were "closed, surrounded by exact borders," often maintained for centuries. In contrast, the states of the "lowest standing races" were small, weak, and without exact borders. In between stood peoples who might be politically gifted, but who could not project power over a large territory.[23]

The progress of *Kultur* naturally created expansion, meaning that European imperialism was driven by world-historical forces, not greed. What was more, such forces extended beyond humanity to all life. Ratzel suggested thinking of all movement of life forms not as migration, but as colonization. "Colonization" better captured the realities of movement, as the main challenge for relocating organisms was the holding of new territory, rather than the movement to it. Plants, animals, and human beings could all colonize.[24] If a people lacked the land for its progress, it was called upon to "improve its position" through the expansion of its borders, whether through conquest of neighbouring regions or colonization of more distant ones.[25]

The relatively small intellectual network of German colonialists and the rhetoric they used make Ratzel's influence apparent. Ratzel and Governor Gustav von Götzen, who took office in 1901, were members of the same key circles in the colonialist movement. Ratzel headed the Leipzig branch of the *Kolonialgesellschaft* and was a member of the advisory council of the Leipzig Geographical Society, while Götzen was a member and lectured at the Society's meetings. The German colonialist movement remained a small group of elites, which excluded most Germans through high membership costs in colonialist organizations. Ideas thus rapidly circulated through the movement and became received wisdom. As John Phillip Short has argued, "These networks and affiliations defined the scope of the organized colonial movement and the fundamental limits of its ambitions."[26] Although *Lebensraum* gained wider currency in the interwar period, Woodruff Smith has charted the quick acceptance of the ideas in *Lebensraum* among the wider conservative movement through similar process of intellectual circulation, demonstrating that Ratzel's work was adopted as the ideological framework for German expansion soon after publication.[27] It is inconceivable that administrators were not familiar with Ratzel's ideas. Their policies reflect their application of them.

Historians have traced the genealogy of Ratzel's thought to the American West or to German expansion in Eastern Europe and Southwest Africa,[28] but the influence of Ostafrika and the historical narratives colonialists had generated over the preceding decades is equally apparent in his thought. Ratzel attended to the specific context of Ostafrika and the historical narratives Germans had used to represent the colony's space since the previous decade. In *Die Erde und das Leben*, Ratzel described the East African coast as the clearest example of the division of labour created as races at different levels of *Kultur* mixed in the same space. The horizon of people on the sea "was always farther" than that of people living in the interior, so "the sea thus became the carrier of progress in history."[29] The Indian Ocean had been the first of the world's great oceans to "become historical," meaning it was the first to experience the cultural diffusion so important to Ratzel's narrative of historical change.[30] Coastal peoples slowly spread *Kultur* through the coastal region, binding the sea and the land ever closer until "the two crossed into one another without borders."[31] Trade over the sea "enriched *Kultur*" and improved coastal people's understanding of the world beyond their local experience. People migrated over the sea, spreading elements of their *Kultur* as they settled down. The settlement of foreigners from over the sea "not infrequently improved the race" of coastal communities, as was the case with the Swahili.[32] Colonialist writing about pre-German settlement in East Africa thus became the basis for Ratzel's understanding of the movement of *Kultur* across oceans, which would be the basis of economic planning under Götzen's administration.

Ratzel's biogeography provided a new framework for thinking about economic development in German colonialism. It created clear theoretical support for German claims to superiority and racial hierarchy in Ostafrika, while establishing colonialism as part of natural history. *Kultur* was the basis of human life, meaning it was not just a categorization German administrators forced on societies, but a fact. It could serve as theoretical backing for government policies to remake East African societies to produce more agricultural products. To do so, administrators first needed to create a comprehensive history of the colony and mould it to fit their objectives.

The Indian Ocean World as Biology

Biogeography created a need for a history in which race and colonization were the primary factors. Such a history would make administrative management of race the furtherance of a historical process rather than an imposition on local societies. Following Ratzel's lead, administrators adapted the existing representational space of the Indian Ocean World developed through the historical narratives of the 1880s and 1890s to include biology as an important element in the development of *Kultur*. They produced a new historical framework for understanding coastal spaces and their connections to the Indian Ocean World. The result of the new line of thinking was the adoption of new elements of the spatial imaginaries of coastal elites into German plans. Administrators applied concepts about the interior drawn from African actors to create economic development plans based in racial thinking.

Through a decade and a half in East Africa, German colonialists had changed little in their thinking about the interior. Much writing about journeys to the interior still depended on the tropes of precolonial travel in the region. One evangelical mission writer described poor roads, a single-file march, and the "strapping, strong forms" of Nyamwezi porters.[33] The mountains of Usagara reminded him of "hoary German villages" of the Middle Ages.[34] Georg Prittwitz und Gaffron celebrated the "old, well-known bush life," which he had always enjoyed.[35] Descriptions of territory and space still depended on rough estimations of distance and relative location instead of the rationalized planetary grid Germans claimed they were implementing. Leases or sale documents for land stated the location of the property relative to local landmarks, and then described its size. For instance, one sale in Bagamoyo described a farm solely as "on both sides of the road that leads to the leper station."[36]

Though he did not use the terms, the spatial division Ratzel used closely resembled the division between *ustaarabu* (Arabness) and *ushenzi* (wildness) in the representational space of precolonial coastal elites. The majority of Ostafrika's population appear as a nameless mass, people without a clear economic role compared to the named groups of people with *Kultur*. Ratzel broke down coastal society by race, with a clear hierarchy of economic roles. Indians were "important businessmen on the coast," Arabs led caravans and were merchants in the interior, the Swahili, "who imitate the Arabs," were porters, and the Nyamwezi were porters and merchants in the interior.[37] The rest of the people in the region remained undefined.

Governor Gustav von Götzen believed the best hope for the future of the colony were the "racial elements" that had arrived from across the Indian Ocean. Arabs were to thank for the "limited amount of civilization" that predated German arrival. They had brought writing, superior crops, cities, and "population discipline." Unfortunately, they had held back improvements because of their "cruel exploitation" of

Ostafrika for the acquisition of slaves and ivory.[38] The remaining Arabs in Ostafrika had become the best supporters of the German cause and lent important assistance, but in the process had become increasingly indebted to Indian merchants.[39] Indian merchants were the only link between large firms and the mass of the colony's population, and so played an important role.[40] Kiswahili served an important function for spreading German influence. It moved west on its own, with no work by Germans, bringing greater understanding of German administration on the coast.[41]

The new policy developed side by side with competing claims that the South Asian influence on Ostafrika was negative, based on claims similar to anti-Semitic rhetoric in Europe that South Asians were parasitic. Lothar von Trotha had written Chancellor Chlodwig von Hohenlohe-Schillingsfürst in 1895 that "Indians were a great danger for the colony because they suck out economic growth without achieving anything positive." The land that Indian merchants controlled was not productive, as the Indians in Ostafrika were "merely merchants and not farmers at all." He suggested banning Indians from buying land in the colony.[42] One plantation owner wrote that the legal position of Indians and Arabs in the colony was "completely untenable" if Ostafrika was to serve the "German national interest" and not become a colony of "the fanatical hands of Islam opposed to Occidental culture."[43] Much like claims about European Jews, these ideas about South Asians portrayed as unable to perform any kind of agriculture, the only truly productive work according to many German conservatives.

Such racial sentiments were particularly prominent among German settlers. One issue of the settler *Deutsch-ostafrikanische Zeitung* included a poem, titled "Yellow Peril" (*Gelbe Gefahr*), that focused on a belief that South Asian settlers were satisfied with lower living standards than were Europeans. Over a millennium, South Asians would come to dominate Europeans,[44] applying racist fears of an invasion of people from Asia. Kaiser Wilhelm II had been a leading figure in stoking such fears.[45] South Asians sought to "freeload" (*schmarotzen*) off the work of others. The paper claimed a clear civilizational difference existed between Europeans or westerners and "Orientals" such that even the employment of South Asians as low-level officials was dangerous. The "peril" of Indian settlement was that they would turn the African population, particularly the coastal society that had already shown it was "susceptible" to Arab influence, against Germans.[46] Another article described the "Pesa-Pesa-Indian" as a "cancer" on Ostafrika that would take all profits from agriculture and send them to India.[47] Settlers saw settlement from Asia as an economic threat to their interests.

Götzen, however, envisioned a role for migrants from Asia in Ostafrika's future. In regions where there were no "influential natives" or where economic conditions had been "complicated" by European undertaking, Götzen thought it necessary to appoint Arabs or "mixed-race people" (*Mischlinge*) from the coast as *akidas*, even though such appointees were "afflicted with the complete unreliability of Orientals."[48] Over time, pacification and increased economic activity would allow districts to transition from military to civilian rule.[49] The need for such intermediaries derived from the fact that African borders were "arbitrary, the result of a competition with other European powers, in which each attempted to secure as much territory as possible, without any consideration for the governmental establishments of the natives." Thus, Ostafrika's borders cut across existing ones. Local leaders had also been sidelined by German policies, which were too focused on direct control of territory. Germany would have been better served by working through existing power structures and strengthening "native state power."[50]

Götzen adopted such racial thinking based in cultural geography wholescale into his development planning. He wrote that environment and climate were the most basic elements to understand if success was to be achieved in Ostafrika. Götzen took a different path from what administrators did in Germany's other colonies. Southwest Africa became a space for German settlement, while Togo and Cameroon were subjected to more extensive plans for plantation agriculture based on ideas about their climates and the racial characteristics of their inhabitants.[51] He explained that his reasoning lay in geography, the "first instructor of prudent colonial policy." It had made clear that Southwest Africa was "white man's land," that Togo was "the domain of the Black man," and that East Africa would call for "both systems next to each other." In Ostafrika, administrators needed to differentiate between territories according to race. Territories at altitude needed to be reserved for whites and "spared" settlement by people of colour unless they were under labour contracts. Other areas, more densely populated, would be "opened only to trade." Thus, the extent of territories with "serious frictions between Black and white" would be minimized. The idea that the two races could live near one another with the same rights was a "fantasy image."[52] German action would attempt to move the border of *ustaarabu* west.

Most of the ideas about the East African coast in the new biogeography were not new and were part of coastal geographies predating German arrival. Though such ideas had long circulated in German administrative circles, they became more prominent after the turn of the century. German administrators largely adopted the conception of a division between coastal *ustaarabu* based in connections to the Indian Ocean World and interior *ushenzi* into their descriptions of Ostafrika. The management of race was thus wrapped up with the management of space. There was a split among colonialists over how to conceptualize the South Asian population, with some of the more chauvinistic among them calling for the exclusion of South Asians from the colony. Götzen would choose the other side of the dispute and attempt to increase South Asian settlement for the benefit of *Kultur*.

South Asian *Kultur* in East Africa

The clearest manifestation of biogeography in the administration of Ostafrika was the scheme to settle South Asian and Nyamwezi agriculturalists in the Tanga district. In biogeography, the institution of *Kultur* was the means through which races claimed territory in the struggle for space. With the new framework for understanding the historical evolution of space, administrators created a plan for government intervention into the process of spatial competition. Administrators began with a belief that South Asian farmers could simply implant their *Kultur* into East Africa soil. Nyamwezi farmers would learn from the South Asian farmers through work on the latter's farms and transfer the techniques to their own land. The scheme ultimately failed, as South Asian settlers resisted or failed to create profitable farms, and the Nyamwezi farmers succeeded with their own methods of cultivation. Administrators' insistence on pursuing it, even as setbacks mounted, makes apparent the wholesale acceptance of cultural geographic theories of diffusion and a belief that state power could speed the process along.

Ferdinand Wohltmann's 1898 report on agriculture had forced administrators to rethink their plans for plantations, so they shifted to freeholder agriculture. According to Götzen, he realized that the "lifting of the state of *Kultur* of the natives" was a policy followed by all "modern cultured states."[53] Franz Stuhlmann, Götzen's second in command, believed the only way economic progress could be made was through the "lifting of existing Negro *Kultur*." The most promising avenue for improvement was to "groom a population element that had better economic relations and stands at a higher level of *Kultur*." For this, he recommended the Nyamwezi and Indian farmers, the latter for rice and cotton cultivation.[54] Götzen described the change in concepts of development as administrators beginning to understand that Africans were more vital than Australians or Indigenous Americans and would not simply "disappear" in the face of the advancement of white settlers, ignoring the violence in the European conquest of the Americas and Australia. Götzen took a page from geographical theories of cultural diffusion to assert that any advancement in *Kultur* could not proceed too quickly. He cited the United States South as a warning of what would happen if one tried to lift a race too many "cultural levels" at once.[55] Whereas the administration of German Togo looked to the Southern United States as its model, in Ostafrika the perspective continued to look across the Indian Ocean because of their historical narratives of racial settlement in East Africa.[56]

Administrative hopes for the future focused on Nyamwezi agriculturalists, who many colonialists had believed the element of Ostafrika's population with the highest level of *Kultur* since the early 1890s. Ludwig Meyer, the district official in Tanga, thought it "very advantageous" that Nyamwezi had begun settling in the district and making better use of farmland formerly under cultivation by the "lazy" Digo population.[57] Nyamwezi settlement had begun as temporary housing for railway workers but became permanent. As progress was slow, Nyamwezi workers began to construct grass huts and plant fields where they were stationed for extended periods. Meyer had provided seeds, "Sumatran plows," knives, and axes to encourage further settlement.[58] Nyamwezi settlement, Meyer hoped, would bring an end to slavery once and for all through the creation of a low-wage workforce on the coast.[59] Meyer would become the leading proponent of attempts to cultivate positive racial characteristics in Ostafrika's population.

India, which had excited the imaginations of many German colonialist since Carl Peters had claimed in the 1880s that Ostafrika could become a "German India," served as the primary model. The Aga Khan's proposal discussed in the introduction to this chapter met with enthusiasm among German administrators. Governor Eduard von Liebert wrote the *Kolonialabteilung* of the proposal, hoping to win support for the settlement of the "first-rate rice producers" from the Punjab.[60] Victor Eschke, who had formerly worked as a planter in Ostafrika before becoming German consul in São Paulo, also recommended the plan. The "natural increase of the Negro" would not meet the need for a "numerous and industrious population," as Africans were "unusable as a cultural element." He had long hoped for Indian settlement to solve the problem. This was particularly necessary in the deep interior, where whites could not settle until more infrastructure had been built on the coast.[61] Eschke claimed the Americas and Australia "could only develop through immigration … to regions that had only populations that were weak or unusable for cultural work."[62]

Though the Aga Khan proposal came to nothing, it spurred German administrators to develop their own plans for Indian settlement. These plans constructed a racialized work force through the processes of imagining idealized workers and recruiting labour based on the idealizations.[63] South Indians, Götzen thought, would be the "best material" for Ostafrika, but it might be difficult to recruit them.[64] The Dar es Salaam district office attempted its own plan, hiring a Hindu man to operate a farm to serve as a "model" and "eventual leader" of Hindu settlers in the district. The plan failed, as the settler decided he was too old to farm without assistance.[65] Dar es Salaam requested more Indian settlers, claiming they could settle along the Rufiji river, where irrigation and herding cattle, and a settlement in "the grand style" were possible.[66] The scientific station in Amani hired Indian and Javan gardeners and overseers to supervise African workers.[67] Ludwig Meyer tried to convince Götzen that Tanga should host an Indian settlement.[68] Meyer would become the primary advocate for South Asian settlement.

Meyer made arguments for cultural diffusion based on historical comparisons between Ostafrika and Germany. Tanga's *hinterland*, West Usambara, was a region of possibly great value. Economic life there was "relatively well developed," meaning people could be taxed effectively. But more people were necessary. As things stood, work in the district was almost all done by Sukuma migrants from the northwest of the colony, who "in the way of the *Sachsengänger*" (Polish migrant laborers who travelled to Saxony) stayed for only a few months before returning home, a phrase that reflects a conception of East African development as part of geographic change worldwide.[69] Indian farmers, Meyer claimed, would show the Sukuma migrants how settled agriculture could create value and become the kernel of a large colony. The settlement of Indian farmers would demonstrate "how the possession of land really can become valuable," convincing the Sukuma migrants to take up a life of settled farming. This time, Meyer asked for money to build settlements to convince Sukuma migrants to live in the district, as well.[70]

Götzen approved a plan based on the idea that *Kultur* could diffuse from South Asian migrants to Nyamwezi farmers. The governor authorized a budget of 21,000 rupees to pay for 56 families to move from India to Ostafrika.[71] Walter Freudenberg, the German consul in Ceylon, and John Hagenbeck, a planter and animal trader there, recruited 90 Sinhalese farmers, 41 men, 21 women, and 28 children in Ceylon, and they arrived in Tanga in March 1902.[72] This was not Hagenbeck's first venture hiring Sinhalese labour for work abroad. He had organized troupes for human zoos for his half-brother, the impresario Carl Hagenbeck.[73] The district spent 4800 rupees on Sinhalese farms, for oxen, carts, agricultural implements, wells, food, and buildings.[74] At the same time, the administration plotted to move Nyamwezi farmers to Usambara, who it believed were more capable of *Kultur* than Ostafrika's other inhabitants. Götzen also wrote the *Kolonialabteilung* of hopes for cultural diffusion from the Sinhalese settlers to the Nyamwezi ones. Cultural diffusion could proceed only "when Negroes see the economic superiority of the farming methods of other coloured cultured races (*Kulturvölker*) with their own eyes."[75] Nyamwezi migrants worked as day labourers on Indian farms, while also cultivating their own lands. Meyer saw a "dual purpose" in their day labour. Not only would they earn money, they would learn "intensive agriculture" from the Indian farmers, improving their own techniques.[76] Meyer convinced a Nyamwezi *jumbe* to settle near Tanga and founded a Nyamwezi village 2.5 miles southeast of the town. What was being created, wrote Meyer, were "independent Wanyamwezi colonies." The Nyamwezi settlers were mixing with other Africans of the district, as few Nyamwezi women had moved at first, but their numbers were increasing.[77] Meyer hoped Nyamwezi settlers would "become increasingly familiar with the intensive agriculture" of Indian farmers, eventually learning to use Indian plows and draft animals and becoming settled farmers themselves to further transform the district.[78]

The Sinhalese settlement plan was a disaster, making clear that racial *Kultur* could not simply sink into the soil. Most of the Sinhalese farmers immediately wanted to leave Ostafrika, dooming the scheme. Rumours spread aboard ship that Ostafrika was "an unhealthy land full of dangerous animals," such that most of the immigrants were "so frightened that they decided they would in no case remain in Africa."[79] By 9 March, 74 of them wanted to go back to Ceylon, including all of the women and children. Sixteen unmarried men signed contracts, but Meyer believed the others would only travel to the area to be farmed with the threat of police violence and recommended returning them to Ceylon.[80] Meyer wrote that the settlement of the Sinhalese migrants who remained also had "unfortunately not developed entirely in the hoped-for manner." This could be attributed in part to poor *Land*, but more to the "character of the Indians." They were "less amenable to innovations" than were Africans and thought themselves cleverer than Europeans. Therefore, "in the way of all Orientals," the settlers, they "hid their foolish pride behind grovelling humility."[81] He estimated that they required as many African workers as there were settlers in order to operate the farms.[82] Even excitement about Sinhalese successes had to be tempered.

Indian millet grew splendidly, to "the wonder of all the natives," but two thirds of the millet was then eaten by birds, which Meyer believed was due to its quick growth, which meant that all birds concentrated on it. Aphids decimated Sinhalese sesame fields, which Meyer believed was due to the plants growing closer together than in African fields because of South Asian intensive agricultural methods.[83] Administrators' willing ignorance of local agricultural knowledge, based on the idea that African farming was primitive, thus created problems for the new program.

Meyer was determined, however, to make Ratzel's theory of cultural diffusion work in his district, even if the supposedly natural process of cultural diffusion required intervention. The administration provided additional support to the Sinhalese farms that was not available to Nyamwezi settlers. Whereas he had deemed irrigation for Nyamwezi farmers too expensive, Meyer paid for wells for Indian farmers to "remove all hindrances" to their success.[84] In order to ensure that the farmers would be "a better model" for the colony's Africans, Meyer replaced their "primitive" farm equipment with German plows, harrows, and seeding machines.[85] The provision of European farm tools and draft animals would "not only sustain the Indians as settled farming burghers, but above all … allow them to influence the local natives in the intended direction."[86] To turn his "half-cultured" South Asian farmers into a model for the "natural" African farmers, Meyer supplied tools that the Nyamwezi farmers would likely never have the means to acquire.

The Nyamwezi farms were far more successful, largely due to changing Nyamwezi spatial practices. Nyamwezi farmers did not follow the German plan; they remade it to their own ends. Meyer had been concerned that Nyamwezi settlers would return home after a short period near Tanga, as Nyamwezi labourers had returned home in earlier periods. He still worried about the "wandering nature" of the Nyamwezi, which needed to be suppressed.[87] Contrary to expectations," Nyamwezi settlements continually progressed, despite the lack of government support.[88] But such returns to Unyamwezi did not happen, at least not in significant numbers. Most of the Nyamwezi settlers remained in Tanga district and spent significant time and energy cultivating farms for the future. Meyer took it as a good sign that some Nyamwezi had begun to plant coconut palms, a type of agriculture not found in Unyamwezi, and a "sign that a large percentage of the people had the intention to remain here permanently."[89] The Nyamwezi farms indicate changes in Nyamwezi spatial practices away from the coast as a short-term destination. Settlers abandoned lifeways built around the caravan trade and yearly migrations.

Enthusiasm for further South Asian settlement faded among the central administration, and what settlement proceeded developed out of private initiatives. Götzen refused a request from a Banian merchant to settle Indian farmers six miles from town, requesting free land to do so and three years' tax exemption for the farmers, presumably seeing no benefit from offering incentives for such settlement.[90] Meyer reported that an Indian planter had bought 140 acres and imported 30 labourers from India to work it, but the enterprise had "failed completely." Only six of the 30 labourers remained and both cattle and cotton were affected by disease.[91] Meyer nonetheless agreed to further efforts by the same Indian planter, leasing 628 acres of land on the ground that he was the "only coloured landowner in the colony who carries out rational agriculture."[92] By June 1905, however, the planter was asking for a suspension of his loan payments because nearly all his cattle had died, forcing him to put some of his labourers to work with hand plows.[93]

The settlement of South Asian farmers in the Tanga district clearly did not work as intended. The farmers were meant to bring settled agricultural techniques to the region and teach Nyamwezi migrants to become productive colonial subjects through agriculture. South Asian farms largely failed, and the diffusion of *Kultur* from their proprietors to the Nyamwezi did not proceed. Settled agriculture did develop, however, but it was through the changing spatial practices of the Nyamwezi settlers in the district, who Germans believed were uncultured, rather than through the South Asian settlers. Geographic theories of cultural diffusion were at the core of the plan and seemingly failed as a method of rule.

Kultur and Politics

Around the same time Ostafrika's administration was attempting to create settled agriculture in Tanga district, it imported Boer migrants from the Transvaal and the Orange Free State in the wake of the South African War. The importance of cultural geographic theories of the diffusion of *Kultur* becomes especially clear if we compare the South Asian settlement to discussions of moving white settlement. Agricultural policies in Ostafrika reflected concerns about political stability in addition to economic development. On the surface, plans to create economic development were entirely about making Ostafrika a profitable colony for Germany. Below the surface, the plans aimed to create sweeping political changes. Some colonialists were initially excited at the possibility of attracting Germanic settlers to the German colony, who would hopefully implant their Germanic *Kultur* to make East

African soil German. Though the administration attracted several Boer settlers to East Africa, concerns about the creation of a white proletariat in Ostafrika overwhelmed hopes for the spread of positive racial characteristics. The eventual rejection of white settlement plans because of concerns about poor whites makes clear that the implementation of *Kultur* was tightly bound to the agricultural conservatism that emerged out of deeper concerns about changes in European society in the nineteenth century. Explicitly political measures that the administration took in Ostafrika reflect these concerns about political unrest. Ostafrika's administration resettled potential threats to its vision of a colony based in settled agriculture, moving them out of areas it deemed sites for the development of *Kultur*.

Many German colonialists continued to believe that white settlement was the most desirable form of settlement for Germany's colonies, but with some limitations on it that indicate political concerns about white settlers. The *Kolonialgesellschaft* published a long report on possibilities for settlement in Usambara in 1902 that included many pictures of landscapes and livestock in the region, meant to appeal to Germans considering emigration.[94] Prospectuses for settlers near Lake Nyasa and in Moshi warned that it was not possible to quickly become rich in Ostafrika, but promised that an "industrious man can establish a secure existence." Such a settler needed to possess a "strong constitution" and be able to live without luxuries, especially alcohol, and understand that he alone was responsible for his success in a new land. There would be no "romantic diversions, hunting, or war adventures," just hard work. To begin the process of creating a secure existence, the industrious settler needed at least 9000 marks, an amount far beyond the means of most Germans.[95] The immensity of such a sum is clear from an article published in the *Deutsch-ostafrikanische Zeitung*. It advertised the high pay for skilled workers, 225 to 350 marks per month, meaning settlement would cost over two years' wages for a highly paid worker.[96] Most Germans would thus be unable to move to Ostafrika. One former explorer in the region wrote that any German without extraordinary means would be better off settling in Brazil or North America rather than in Ostafrika, where they could live "near civilization" in healthier conditions.[97]

Some administrators were initially enthusiastic about the possibility of Boer settlement when the subject was raised during the South African War. The *Deutsch-ostafrikanische Zeitung* carried frequent reports from the war, nearly all of them sympathetic to the Boer cause. Suggestions of Boer migration in Ostafrika came first from Theobald Schütze, a German settler in South Africa, who wrote the Colonial Office about the possibility in 1901. Schütze imagined the Boers in the image of pioneers on the American frontier, calling on imagery popular in Germany and that Carl Peters had used to support the GfdK in the 1880s. He had become concerned after speaking to some wounded Boer soldiers in a South African hospital. Schütze lamented the Boers' condition, claiming their lives had been made more difficult "or even impossible" by the destruction of their livestock and land. The period in which one could strike out in the wilderness without hindrances to make a new life in South Africa had reached its end because of the actions of Cecil Rhodes and other British capitalists. He believed the Boers could find "asylum" in Ostafrika, which had a similar climate. Boer history, with its "adventurous treks, bloody battles against natives, and the foundation of primitive states in distant, isolated wilderness," showed that the "true mission" of Boers was "to wrest such land from their primitive conditions" to advance the "general development of the *Land*."[98] Schütze's suggestion indicates that the model of white settlers remaking *Land* still had not died among some German colonialists.

Schütze directly compared Boer history to that of Arabs in East and North Africa to promote the idea that they could improve *Kultur*. The comparison makes clear the refashioning of Arabs in East Africa as semi-cultured intermediaries rather than opponents of German colonialism had spread beyond Ostafrika. Both had carried out "cultural preparatory missions," but the Boers had done so much "more thoroughly and rationally" on a "Christian-Germanic basis." Boers could serve as the "genetic basis (*Urstamm*) of a later white population."[99] This kind of racial logic underlay the support for Boer settlement. An old member of the DOAG wrote to the Pan-German League, the core organization of right-wing German racial nationalism, promoting Boer settlement in Ostafrika. Boers would form a "splendid tribe of colonialists" in the East African interior.[100] A former *Schutztruppe* officer agreed, writing that Boer settlement could convince Germans of the feasibility of their own settlement, which would convince them to emigrate and the German government to fund a railway.[101]

Promoters of Boer settlement attempted to overcome worries about their relative poverty compared to British South Africans with claims they were superior in terms of class and ability to change territory. A "trek" leader (as the Boers termed their expeditions) wrote that he was bringing "quiet, peaceable, married men, who desire only to make new homes and settle with their wives and families."[102] He requested 200 labourers for each family for three months, free of charge.[103] The Boer settlers clearly expected labour relations of the type that they had built in South Africa. Administrators wrote to Boer settlers that they expected each family to possess at least 10,000 marks in cash when they arrived.[104] One Boer settler countered with a claim that Boers were very frugal and could succeed with far fewer means than could Europeans. Those without means were not proletarians, but people who had lost everything in sacrifice for their fatherland. Such poor Boers were superior to the "traitor class" that still had wealth.[105] It was those Boer patriots in particular who would become cause for concern among German administrators.

Boer settlement failed to create the kind of development German administrators wanted, so they ended their support for it. A follow-up letter from Schütze made apparent the reasons why administrators would eventually turn against the plan. He wrote that the Boer was "definitely not a farmer following our conception." He was primarily a pastoralist, then a hunter, then a transporter. Boers farmed only as much as they needed for their own food and were therefore unsuitable for regions with fruitful soil and fit better in open spaces.[106] In the early stages of discussions, Franz Stuhlmann wrote to the Colonial Office that Ostafrika wanted only Boer families with means as settlers.[107] Stuhlmann saw these theories confirmed by Boer settlement near Mount Meru. The Boers had found the open spaces for which they yearned, but they had not helped with economic growth more broadly. They refused to share water with Africans in the region, indicating the "tendency of Boers to seclude themselves from the outside world."[108] Fifty poor Boers arrived in 1905 and were met with concerns about the "preponderance of the Dutch-Boer over the German element" in the colony.[109]

The *Kölnische Zeitung* declared the Boer settlement a failure in July 1905, and the language in which it did so reflects the shift in thinking about the lack of historical progress among nomadic peoples that had happened over the preceding decade and a half. Such changes mirror the increasingly fierce denouncements of Sinti and Roma lifeways within metropolitan Germany at the time.[110] Africans near Boer settlements had taken to calling them the "white Dorobo" (*Wandorobbo Wasungu*), after a derogatory term for nomadic peoples in Tanganyika and Kenya. Trekboers, in the author's estimation, were differentiated from Africans only in that they wore European cloth and could read the Bible. But these "cultural achievements" meant mostly that Boers felt comfortable evading and ignoring laws. They were ignoring wildlife protection laws in Ostafrika. Trekboers would remain Trekboers, and what would be achieved through their settlement would be "only the addition to the nomadic, parasitic Negro peoples one with white skin!"[111] While early European travellers in the region had praised nomads as the best hope for the future, by 1905 opinions had so turned against nomadism that critics attacked white settlers for exhibiting aspects of nomadism in their behaviour. German sentiment had solidi-fied firmly behind settled agriculture.

The power of settled agriculture as ideology and its connections to domestic politics are also apparent in discussions of the social conse-quences of a class of "poor whites." As Ann Stoler has argued, poor white settlers threatened white elite patriarchy and were therefore a danger to the basis of colonial rule. Colonial administrators, including German ones, thus hoped to prevent the formation of a class of poor white people in their colonies.[112] This desire was clear in a response that Franz Stuhlmann sent to British men who had written in hopes of settling in Ostafrika. According to Stuhlmann, "For German East Africa only agricultural men or artizans [sic] are wanted, and as blacksmiths, carpenters, horticulturalists, engineers etc; in the offices we can only use people, who know to read and write the German language."[113] The administration also did not consider supporting an Austrian carpenter who wrote in hopes of settling. He had travelled to Ostafrika previ-ously, before trying his luck in New York. But now he wanted "to live as a German in a German colony."[114] The answer indicates clear concerns about poor whites in Ostafrika – a population of white settlers depen-dent on the government for support – as a potential source of political unrest.

In concerns about unsettled populations, we can observe how plan-ning for economic development in Ostafrika became a much larger social experiment, meant to solve fears about unrest not only in the col-onies, but in Germany as well. Attempts to create *Kultur* in Ostafrika, then, were a form of "social imperialism," not after the meaning pro-posed by Hans-Ulrich Wehler, of colonialism as an effort to manipulate popular politics, but as an experiment to solve social issues at home.[115] Conservative Germans in the Wilhelmine period became increasingly concerned about the threat posed by socialism and the proletariat as the nation entered the twentieth century. Their concerns strengthened the political power of agriculturalist conservatives who promoted imag-ined agricultural tradition as the source of Germany's strength, and policies that maintained the power of agricultural landowners. Fried-rich Ratzel himself joined the agriculturalist conservative movement.

The creation of *Kultur*, its proponents hoped, would eliminate politi-cal opposition, a vision that contrasted with the politics of most Ger-man settlers. The *Deutsch-ostafrikanische Zeitung*, the urban settlers' newspaper, declared that the colony's towns should be ruled by mer-chants and industrialists, who "most exactly knew the conditions and particularities of our colony and especially those of the city and the district." It placed its demand in a tradition of political reform in Europe with a citation of Karl vom Stein, who had created liberalizing reforms in Prussia a century before, to argue for wider diffusion of political power.[116] Such politics diverged from the agrarian conservatism at the core of the administration's settlement policy.

Interior stations put the brakes on attempts to create greater reforms. They countered suggestions of introducing tax collectors to make taxation more efficient with claims that taxation through *jumbes* ensured "peace and order," even if it was less efficient.[117] People were "very conservative-minded" and would object to the taxation system, which did manage to collect most of the taxes assessed.[118] One interior station chief reported that the poorer *Kultur* was, "the more people hung on to it and the more meager their tendency to advancement." People kept growing cassava, despite periodic crop failures, demonstrating an "absolute aversion to everything new."[119]

Attitudes toward white settlement had changed significantly in just a few short years. Many colonialists had been excited at the possibility of Boer settlement early in the decade, believing Boers could bring superior racial attributes to bear on Ostafrika's *Land*. But Boer settlement did not create the desired settled freeholder agriculture, and in therefore did not fit the larger project of creating progress through *Kultur*. Their modes of living would not diffuse to the colony's African inhabitants. Such concerns intertwined with worries about the creation of a class of "poor whites," which threatened to undermine racial hierarchy and bring socialism to Ostafrika. Agrarianism had triumphed over alternatives in planning for the future of Ostafrika's development.

The Maji Maji War

In mid-1905, an anti-German uprising spread quickly through southern Ostafrika. Most likely precipitated by July arrests of two healers, anti-German forces attacked administrative outposts, houses of Arab, Indian, and German officials, and missionary buildings over the following months. The war caught German officials by surprise. They wondered how a violent movement could spread in just a few weeks over an area of approximately 100,000 square miles inhabited by people who spoke twenty-five different languages. Their response was swift and vindictive. *Schutztruppe* officers adopted scorched earth tactics, utilizing famine to force communities to heel. Historians have estimated that the brutal German tactics killed at least 250,000 Africans between 1905 and 1907. After a period of confusion about the causes of the fighting, Germans labelled the war the "Maji Maji Rebellion," after the Swahili word for "water." German narratives of the conflict blamed African superstition for creating the war, focusing on the spirit medium Kinjikitile and his use of water as a war medicine that would supposedly stop German bullets from harming the user.[120] This narrative was in place in the German press by 1907. Its solidification marked the end of managing race for development in the way Ostafrika's administration had in the prior two decades.

The Maji Maji War has been a key landmark in the historiography of anti-colonial nationalism in Africa since the 1960s. The narrative of superstitious Africans stirred up by sorcery became the basis for the historiography of the war as the beginnings of anti-colonial nationalism in Africa, as the "first organized – quasi-national – rising" in Tanzania's history. In this story, the anti-German movement had formed out of a shared African identity and feeling of oppression under German rule.[121] The standard narrative of the war developed out of the silencing of settlers' narratives blaming Indians and aimed to resolve tensions with the United Kingdom rather than antagonize them through claims of a shared enemy in the development of Africa.

Stories in the *Deutsch-Ostafrikanische Zeitung* (*DOAZ*) about the causes of the war quickly zeroed in on the role of Indian merchants in inciting anti-German feeling and aiding anti-German forces. Multiple articles asserted that the primary reason for the start of fighting was the debts owed to Indian merchants.[122] Indian merchants' exploitation of the people of East Africa through usury was the real cause of the war, one article claimed, and their close ties to German officials made Africans think Germans were responsible for their debt.[123] Settlers asserted those merchants were agents for the United Kingdom, Germany's rival on world markets, as they sent their profits back to India.[124] The newspaper alleged that Indians had provided funding and the gunpowder and bullets for anti-German forces, a situation possible only because the colonial administration had allowed Indian merchants to control trade.[125] One colonial propagandist, a Reichstag deputy beginning the following year, wrote that Indians were British agents who had stoked rebellion against Germany.[126]

German settlers in Ostafrika heightened their rhetoric of an Indian invasion of German space in response to the conflict, relying on anti-Semitic stereotypes used in metropolitan Germany. One article asserted Germany needed to fight the "constant invasions" of Indians, as "a German colony should be German and remain German."[127] The propagandist also wrote in favour of colonists' demands for an end to Indians' rights as citizens of a foreign power in German lands. He suggested the colony's administration should require Indians to carry passes to travel around the colony, and that they conduct all business in a European language.[128] The latter idea became especially popular, and a look at the deliberations of responses to the conflict is illustrative of prevailing racial thinking. Settlers claimed that Indian merchants did not compete fairly with German ones, and the requirement that they conduct business and keep records in a European language was clearly directed only at that issue of competition, despite rhetoric about the war.[129] Indians should have to give up their Indian subjecthood, the *DOAZ* suggested, if they wanted to conduct business in German lands, and compared the situation to British actions against the Boers in South Africa.[130] Indians' dual loyalty made them a threat to Germany.

The same rhetoric also circulated in metropolitan Germany through the colonial and Pan-German press. Several metropolitan newspapers carried stories that included claims that East Africa's Indian population had supported the war. Colonialist and nationalist newspapers tried to stir up metropolitan Germans with stories of Indian hatred for Germans. The *Kolonialzeitung* included a call to protect East Africa from being "swamped" by Indians, choosing language common to racialized threats of immigration.[131] It included a claim that Indians, "as British agents," had stirred up the rebellion.[132] A story in the Pan-German League's *Deutsche Zeitung* described the Indian population as a "thorn in the eye of our impartial justice," and asked rhetorically whether East Africa would be an Indian or a German colony.[133] These newspaper reports created support for blaming Indians among the more aggressive supporters of colonialism. Metropolitan reports implied that Ostafrika was losing its German character because people of another race had entered the colony, never mind the fact that the population in question had been there before East Africa was colonized by Germans.

The most important voice in creating the narrative of Indian involvement was Hans Paasche, who served as a *Schutztruppe* officer during the war. Paasche's father was Hermann Paasche, the vice president of the Reichstag, meaning his narrative of events reached important circles in German politics. Hans Paasche's postwar account described East Africa's Indian population as "parasites" whose only *Heimat* was "money bags" full of "dirty copper coins from the hand of naked *Nyamwezi*" sent to the Chartered Bank of India.[134] This rhetoric painted a picture of an Indian exploitative class taking advantage of Black Africans in such a state of poverty or savagery that they could or would not wear clothes. The causes of the war, he thought, were high prices charged by Indian merchants at a time of shortage in August 1905. The prices had taken advantage of Africans in dire straits and produced animosity against both Germans and Indians. So long as Germany tolerated the role in Indian merchants in East African trade, the colony would never reach its potential, as those merchants aimed to take their profits out of the colony.[135] This repatriation of profits was robbing Germans of money that should be theirs. Therefore, Germany needed to exclude Indians from the colony.[136]

The push to challenge the status of Indian merchants and be more aggressive toward the British Empire did not make it into government narratives of the war, indicating a desire for moderation in the German government's thinking about race in Ostafrika. German official accounts silenced claims of Indian guilt in favour of a more conciliatory story of irrational African resistance to German development. From the start of fighting, the official *Kolonialblatt* blamed sorcery and "plentiful beer drinking" for the fighting.[137] Götzen told the Reichstag that the anti-German movement was "purely heathen," a "reaction of 'bush negroes' against the advancing *Kultur*." Territorial conquest always met such resistance, he claimed.[138] In an interview with the *Deutsche Zeitung*, he rejected the claim of Indian guilt and stated rebels were attacking Indians as well as Germans.[139] An officer who had served in East Africa the decade before wrote in several German newspapers defending the colony's Indian population.[140] The narrative of Indian involvement in the war was buried beneath one blaming African sorcerers.

Instead of blaming Indian merchants, the narrative that emerged was one of united African resistance to German colonialism, placing Germany in agreement with the United Kingdom on colonial affairs. Stories of an Indian rebellion against Germans disappeared and in their place were others about African-directed violence against all Europeans. Even in the *Kolonialzeitung*, which had published anti-Indian screeds early in the war, the war appeared the result of a movement of Africans united against the imposition of European rule on the continent.[141] Reports of an uprising in the British colony of Nyasaland were a sign that Africans were taking up arms against European colonial states, not just Germany.[142] Claims that Indian settlers had caused the rebellion had disappeared, replaced by a story of inter-imperial cooperation.

German diplomats focused on conciliation with the UK rather than raising tensions over British Indian subjects. In addition to official publications ignoring settlers' claims, German discussions with British Foreign Office officials ignored settlers' demands for actions against Indians in East Africa. Discussions between the two countries attended closely to possible conflicts over the Morocco question, and a possible conflict in East Africa emerged only over accusations by Hermann Paasche that British investors had embezzled money in the building of a railway between Uganda and Mombasa.[143] Neither the Kaiser nor the Foreign Secretary raised the fighting as an issue in a discussion with the British ambassador in August 1906.[144] Interactions between the two countries display the German Foreign Office's desire to avoid raising tensions with the United Kingdom.

Gustav von Götzen's semi-official postwar account of hostilities did not attend to the narrative of Indian guilt, either. He described anti-German fighters as motivated by a kind of nationalism against German rule. The roots of the conflict could be found in anger at the colonial administration, stirred up by sorcerers. Götzen acknowledged a role for Indians in creating difficulties for German merchants and indebting the rest of the colony's population, but insisted that they could not be discriminated against, as they were British subjects. He assigned them none of the blame for creating the war.[145] Rather, "modern humane colonial praxis" did less to eliminate the "racial opposition between Europeans and Negroes" than earlier forms of colonialism.

Like the Bushiri War, the Maji Maji War convinced Berlin that a new approach to colonialism was necessary. In the creation of a unified narrative of African resistance, one can observe the centralization of German colonial policy and limits on the influence of the most extreme elements in German racial thinking. Along with the Herero-Nama War in Southwest Africa, the Maji Maji War prompted a reorganization of Germany's colonial administration. The Reichstag split colonial issues off from the Foreign Office and created a new stand-alone Colonial Office. With Bernhard Dernburg's accession to the newly created position of State Secretary for the Colonies, more decisions would happen in Berlin, and Germany would more closely hew to models from other empires.[146] Subsequent governors in Ostafrika would have far less leeway in determining policies to create *Kultur*. Thus, the curtain was drawn on plans for managing race for development in Ostafrika, and the processes discussed in this book moderated for the remainder of the German colonial period.

Conclusion

The Maji Maji War upset Götzen's plans for *Kultur*. The production of a story that included blaming a pan-African movement against Europeans for challenges to German rule portrayed Europeans as bearing a shared burden for developing Africa despite African resistance.[147] Although recent histories of the war have shown that the narrative of a unified resistance movement does not reflect the reality of motivations for challenging German rule, the circumstances of the standard narrative's formation reveal German colonialists' desire to change the workings of German colonialism.[148] The creation of the narrative of a unified African resistance reflects a desire by the German Colonial Office to silence the aggressive racial rhetoric of German settlers in favour of conciliation with the United Kingdom. Though the government had silenced claims of Indian complicity, settler anger would forbid continued importation of agriculturalists from overseas to remake space. Settler fury and the consolidation of the pan-Africanist narrative brought an end to the processes discussed in this book, to be replaced by colonial models based less in the management of race in the colony and hewing more closely to international norms.

In the first years of the twentieth century, Ostafrika's administrators built plans for the future of the colony on the idea of cultural diffusion, believing that higher levels of *Kultur* in the vicinity of African communities would gradually create *Kultur* in those African communities. Intellectual backing for this new form of state intervention came from biogeography, the historical and geographical theory that Friedrich Ratzel developed in *Lebensraum* and *Die Erde und das Leben*. Biogeography posited that human history was a competition for space. To explain Ostafrika's racial makeup within the biogeographical framework, German administrators adopted existing coastal geographies and narratives of an Indian Ocean World for making sense of the interior and thinking about spreading the border of *Kultur* west into the interior.

But the most explicit area in which administrators applied the new biogeography was in the management of settled agriculture. Governor Gustav von Götzen and Tanga district officer Ludwig Meyer created a settlement of Sinhalese and Nyamwezi farmers in Tanga district based on a theory of cultural diffusion in which the Sinhalese farmers would implant *Kultur* in the *Land*, and Nyamwezi farmers would learn through their example. The administration's support for the settlement of non-white farmers in Tanga contrasts with growing discontent about Boer settlement in the interior. Boer settlers did not create *Kultur* and attempted to exclude African communities from their progress. The administration thus wound up supporting non-white settlement in Ostafrika over Boer settlement in its pursuit of *Kultur* above all else. *Kultur* promised to solve political problems in Ostafrika, and also in metropolitan Germany. It had become an ideology around which to base all development planning.

10 Conclusion

In the German spatial imaginary of the early twentieth century, the ability to shape space meant the ability to shape time. This book has covered significant changes in the German spatial imaginary of East Africa from 1884 to 1905. The German representational space of the region underwent numerous changes over the course of these two decades. To make those changes, German colonialists deployed varying representations of space, many of them based in historical narratives of social evolution. The attempts to create *Kultur* in Ostafrika constituted Germany's first state project to manipulate race to manage territory; *Kultur* thereby became Germany's colonial ideology. Changing the racial characteristics of the colony's population would advance societies and the spaces they inhabited through time, turning inhabitants into settled agriculturalists and transforming landscapes to be more like an idealized agrarian Germany. From an idea of East African space and people as medieval and available for German conquest, through the construction of an economic *hinterland* with its origins in the Indian Ocean World, Germans claimed territory to make a colony. From there, colonialists imagined East Africa as the main battleground in a world historical struggle with Islam for Africa and its people. After the German government took control of the colony, administrators spent the 1890s attempting to remake African societies and economies with a variety of methods. Eventually, they adopted theories of diffusion from cultural geographers as the basis for using the perceived racial characteristics of non-European populations to create an idealized agrarian society in Ostafrika. Remaking East African people became the means through which Germans would create historical progress on the desired economic growth that would make colonialism successful.

Initial German attempts to remake East Africa focused on *Land* alone. The Society for German Colonization had begun its work with a belief that successful colonialism simply required taking control of territory and making it more like Germany. It pursued policies that would establish internationally recognized German control of East African territory, from which great wealth would supposedly follow. The company signed a series of treaties claiming from African communities that depicted East African politics and landscapes as comparable to medieval Germany and in need of progress toward modern agriculture and commerce. Over the next few years, colonialists constructed evolving historical narratives to claim ever more land. These narratives located East Africa as behind Germany in time, judged by prevailing methods of agriculture. They began with histories of Zanzibar as a failed colonial power, then shifted to longue-durée descriptions of an Indian Ocean World and a new definition of *hinterland* to link coast and interior.

In 1888, German colonialists faced a pair of crises that called the DOAG's beliefs about spatial evolution into question. The first crisis, Emin Pasha's isolation in Equatoria, was largely of the colonial movement's own making. In the campaign to raise money for Carl Peters's expedition to rescue Emin, the German Right pioneered techniques for engaging with a larger public through depictions of East African space as available for German taking. The second crisis, the Bushiri War on the Indian Ocean coast, was a challenge to the German East Africa Company's model of private empire as German representational space met coastal spatial practices. In both, Germans mobilized behind a new representation of space, a narrative of conflict between European Christianity and Islam over East African space. For the first time in Germany's history, the state spent significant resources on a colonial venture: Hermann Wissmann's expedition to fight the war. German claims to sovereignty shifted from ideas about territory to assertions that the German presence was necessary to protect Black Africans from the Arab slave trade. The colonial movement split over whether to continue the focus on German settlement or to try to change Ostafrika's population to create development.

In the 1890s, cultural geography became the dominant means of representing East African space and administrators more explicitly attempted to manage race to create historical progress, judged in terms of settled agriculture, in East African space. Initially, administrators attempted to use technology to create a profitable colony. As technological plans failed, administrators turned to the instrumentalization of African spatial practices to create *Kultur*, applying geographic theories of cultural diffusion to create economic progress. Racial management became more elaborate over time as ideas flowed back and forth between metropole and colony. Representations of East African space came back to Germany through the discipline of *Heimatkunde*, which made East African people and landscapes familiar to German readers through comparisons to German communities. Administrators undertook more active operations to remake East African *Land* through the management of racial characteristics after the turn of the century as they brought in migrant groups from other colonies to implant their *Kultur* in East African soil. German projects failed to produce any kind of lasting economic development, as they either did not account for or misunderstood African practices of space. The application of cultural geography to make development came to an end in 1905, when new anti-German fighting forced Germans to rethink their colonial models.

Legacies

This history was largely lost after Germany lost its colonies at the end of the First World War. The majority of publications about Ostafrika from 1919 through the Second World War were hagiographies of General Paul von Lettow-Vorbeck, whom German nationalists held up as a hero who had never been defeated on the battlefield.[1] After 1945, colonialism was something to forget for most Germans, a relic of Germany's negative past. Even in more recent historiography, the colony features minimally in most histories of Imperial Germany and almost all of the figures involved in creating the German administration of East Africa have long been forgotten. Other than Lothar von Trotha, far more notorious for his central role in perpetrating the genocide of the Herero in Southwest Africa, and Carl Peters, who appears primarily as an example of a wider phenomenon of nationalist politics, few appear in accounts of Germany's later history. The omission of such figures that narratives produced later obscures their importance to contemporary events. Germans participated in the discourse about Ostafrika for the colony's entire existence. The German colonization of East Africa was a key period for the development of ideas of culture, sovereignty, space, and race for both Germany and Tanzania. The men involved in the conquest of the colony, like Hermann Wissman and Emin Pasha, were among the most famous people in contemporary Germany. Reports of their deeds, as well as what they themselves wrote, reached a wide audience. Colonial topics and themes permeated many other aspects of German life, meaning most Germans encountered ideas from Ostafrika. The colonial shaped German culture, especially among Germans interested in overseas empire.[2]

More consequential than Ostafrika's influence on popular culture were its roles in the formulation of *Kultur* as colonial ideology and its impact on intellectual currents through cultural geography. As this book has discussed, East African societies became the model that Friedrich Ratzel used for primitive societies and their spaces at low levels of *Kultur*. German schoolchildren learned such models of society and space through textbooks. Ratzel's evolutionary cultural geography would shape subsequent German ideas about territorial expansion in both world wars.[3] Ostafrika's administrations tested out the management of race to remake space and how governments could produce the agrarian outcomes Ratzel and his political allies desired. Ostafrika became a test case for solutions to the changes created by industrialization in Germany and possible agrarian alternatives.[4] The project in Ostafrika eventually reached a dead end, but Ratzel's ideas of cultural evolution through space, particularly as he developed them in *Lebensraum*, were enormously influential in subsequent German thinking about geopolitics and imperial governance. The experience of Ostafrika set models for agrarian politics and future attempts to remake space in imperial territory. Ratzel's models of cultural diffusion and state expansion as an expression of a race's conquest of landscape would shape subsequent German theories of empire.[5] German soldiers and administrators who invaded Eastern Europe worked within the same paradigm of *Land* and *Leute* to conceptualize Germany's imperial spaces as they had used in Ostafrika.[6]

The ideas of cultural diffusion put into practice in Ostafrika shaped thought about Africa and culture for generations to come. It impacted the subsequent development of anthropology as the means of studying the non-European world and remained an aspect of agrarian politics for generations. Ratzel's student Leo Frobenius developed the concept of *Kulturkreise*, which explained human societies as the diffusion of culture from multiple starting points across the globe. With *Kulturkreise*, geographers could explain all human history as a process of cultural diffusion.[7] Fritz Graebner created a methodological basis for work along these lines in 1911.[8] The idea of *Kulturkreise* would become the basis of the Viennese school of anthropology, the influence of which would stretch so far as to become the most important influence on anthropological practice in the United States. *Kulturkreise* was also important in the development of pan-Africanist ideologies in the 1930s, where the concept served as the basis to claim an African civilization not dependent on European influence, as cultural innovations had diffused in multiple directions rather than entirely from Europe.[9] The German colonial ideology of *Kultur* through cultural diffusion thus continued to influence understandings of Africa for decades after Germany lost its overseas empire.

Culture, social interactions, and history all became aspects of East African space in the German colonial project. Through the application of cultural geographic ideas, German administrators in Ostafrika had created a theoretical basis for the application of racial thinking to the management of territory. Racial hierarchies became spatial hierarchies, defined through histories of agricultural progress. German administrators created a series of programs to create economic development, primarily through agriculture, on the basis of a close connection between *Land* and *Leute*. By changing either East African people or landscapes, Germans thought, Ostafrika could become more like an idealized, agricultural Germany. The timeline of the development of German agriculture could be collapsed to a few short decades. By manipulating race, Germans could create *Kultur* – settled agriculture – where there had been none before. Managing race became the means of creating a different kind of historical progress on the blank slate of East Africa, one that averted attention from the problems of modern Germany. The movement of time could be readjusted into a different future by remaking space.

Notes

1 Introduction

1 For the purposes of this study, I will use "East Africa" to refer to the physical space and "Ostafrika" to refer to the colony in order to reduce confusion.

2 Though the pairing came from cultural geography, it took on a life of its own and pervaded discussions with little direct connection to geographical study. Just citing two examples, such books could be memories of military service: J. Sturz, *Land und Leute in Deutsch-Ost-Afrika. Erinnerungen aus der ersten Zeit des Aufstandes und der Blockade* (Berlin: Mittler und Sohn, 1894); or part of an anti-slavery campaign, Johannes Baumgarten, *Ostafrika, der Sudan und das Seeengebiet: Land und Leute, Naturschilderungen, charakterische Reisebilder: die Antisklavereibewegung, ihre Ziele und ihr Ausgang, kolonialpolitische Fragen der Gegenwart: nach den neuesten und besten Quellen* (Gotha: F.A. Perthes, 1890).

3 Here I draw on Anne McClintock's concepts of "panoptical time," the creation of simple visual depictions of Europe's supposed historical progress, and "anachronistic space," a trope of a static, primitive past out of place in European modernity, from *Imperial Leather: Race, Gender, and Sexuality in the Colonial Contest* (London: Routledge, 1995), 37–40.

4 I use the term "colonialist" to encompass a broader variety of Germans involved in making ideas and policies in the colony beyond Germans who were themselves colonists. In addition to colonists, I include officials in the Colonial Office, missionary organizations, colonial propagandists, and other pro-colonial Germans.

5 Matthew Fitzpatrick details the importance of colonialism to the politics of some German nationalists before unification in *Liberal Imperialism in Germany: Expansionism and Nationalism, 1848–1884* (New York: Berghahn, 2008).

6 Helmut Walser Smith, *German Nationalism and Religious Conflict: Culture, Ideology, Politics, 1870–1914* (Princeton, NJ: Princeton University Press, 1995), 22–34; Stefan Berger, *Germany* (London: Arnold, 2004), 7. Helmut Walser Smith, "Monuments, Kitsch, and the Sense of Nation in Imperial Germany," *Central European History* 49, no. 3–4 (2016): 322–40; Kara Ritzheimer, *"Trash," Censorship, and National Identity in Early Twentieth-Century Germany* (Cambridge: Cambridge University Press, 2016); Nancy Reagin, *Sweeping the German Nation: Domesticity and National Identity in Germany, 1870–1945* (Cambridge: Cambridge University Press, 2007); Alon Confino, *The Nation as a Local Metaphor: Württemberg, Imperial Germany, and National Memory, 1871–1918* (Chapel Hill: University of North Carolina Press, 1997); Celia Applegate and Pamela Potter, eds., *Music & German National Identity* (Chicago, IL: University of Chicago Press, 2002).

7 Perhaps most eloquently stated in the first line of Thomas Nipperdey, *Germany from Napoleon to Bismarck 1800–1866*, trans. Daniel Nolan (Princeton: Princeton University Press, 1996), "In the beginning was Napoleon." A similar formulation that is likely more familiar to non-Germanist readers is Frederick Jackson Turner's frontier thesis, laid out in *The Frontier in American History* (New York: Henry Holt and Company, 1921).

8 Martin M. Winkler, *Arminius the Liberator: Myth and Ideology* (New York: Oxford University Press, 2016); Ernst Baltrusch, Morten Hegewisch, Michael Meyer, Uwe Puschner, Christian Wendt, and Reinhard Wolters, eds., *2000 Jahre Varusschlacht: Geschichte, Archäologie, Legenden* (Berlin: De Gruyter, 2012). Authors in Colin G. Calloway, Gerd Gemünden, and Susanne Zantop, eds., *Germans & Indians: Fantasies, Encounters, Projections* (Lincoln, NE: University of Nebraska Press, 2002) describe similar lines of thinking at work in German identification with Indigenous Americans, who, like Germanic tribes, were resisting the incursion of an imperial power.

9 Rogers Brubaker, *Citizenship and Nationhood in France and Germany* (Cambridge, MA: Harvard University Press, 1992); Emily A. Vogt, *"Civilisation* and *Kultur*: Keywords in the History of French and German Citizenship," *Ecumene* 3, no. 2 (April 1996): 125–45; Richard L. Velkley, *Being after Rousseau: Philosophy and Culture in Question* (Chicago: University of Chicago Press, 2002). This difference was articulated most clearly in Friedrich Meinecke's *Weltbürgertum und Nationalstaat*, which posed the German nation as an organically formed community of shared *Kultur* versus the French nation created only by shared politics. Georg Schmidt, "Friedrich Meineckes Kulturnation. Zum historischen Kontext nationaler Ideen in Weimar-Jena um 1800," *Historische Zeitschrift* 284 (2007): 597–622. The concept of *Kultur* as deeper and more lasting than identity based in

political groupings is also the basis for Oswald Spengler's argument in *The Decline of the West: Form and Actuality*, trans. Charles Francis Atkinson (New York: Alfred A. Knopf, 1927).

10 Thomas Nipperdey, *Deutsche Geschichte 1866–1918*, vol. 2, Machtstaat vor der Demokratie (Munich: Beck, 1995); James Retallack, *Red Saxony: Election Battles and the Spectre of Democracy in Germany, 1860–1918* (Oxford: Oxford University Press, 2017); James Retallack, *The German Right, 1860–1920: Political Limits of the Authoritarian Imagination* (Toronto: University of Toronto Press, 2006); Geoff Eley, *Reshaping the German Right: Radical Nationalism and Political Change after Bismarck* (New Haven: Yale University Press, 1980).

11 The power of agrarian politics can be viewed quantitatively, in the membership numbers for the Agrarian League. Founded as a response to the agricultural crisis of the Long Depression, in 1893, the League's membership reached 250,000 by 1900. Hans-Jürgen Puhle, *Agrarische Interessenpolitik und preussischer Konservatismus im wilhelminischen Reich (1893–1914): Ein Beitrag zur Analyse des Nationalismus in Deutschland am Beispiel des Bundes der Landwirte und der deutsch-konservativen Partei* (Hanover: Verlag für Literatur und Zeitgeschehen, 1966) is still the standard work on agrarian politics in the 1890s. Woodruff Smith, *The Ideological Origins of Nazi Imperialism* (Oxford: Oxford University Press, 1989) traces agrarianism as part of the broader conservative movement. See also Robert M. Berdahl, *Conservative Politics and Aristocratic Landholders in Bismarckian Germany* (Chicago: University of Chicago Press, 1972); Shelly O. Baranowski, *The Sanctity of Rural Life: Nobility, Protestantism, and Nazism in Weimar Prussia* (Oxford: Oxford University Press, 1995); Kenneth Barkin, *The Controversy over German Industrialization, 1890–1902* (Chicago: University of Chicago Press, 1970); Klaus Bergmann, *Agrarromantik und Großstadtfeindschaft* (Meisenheim: Anton Hain, 1970); George Vascik, "Agrarian Conservatism in Wilhelmine Germany: Diederich Hahn and the Agrarian League," in *Between Reform, Reaction and Resistance. Studies in the History of German Conservatism from 1789 to 1945*, eds. Larry Eugene Jones and James N. Retallack (Providence: Berg, 1993), 229–60; Robert G. Moeller, *German Peasants and Agrarian Politics, 1914–1924: The Rhineland and Westphalia* (Chapel Hill: University of North Carolina Press, 2010).

12 McClintock's concepts of "panoptical time" and "anachronistic space" provide one means of understanding the role of colonial imagery to Europeans' claims to modernity and superiority. *Imperial Leather*, 37–40. Other works that demonstrate the importance of colonialism to European claims to modernity, Timothy Mitchell, "Introduction," in *Questions of Modernity*, ed. Timothy Mitchell (Minneapolis: University of Minnesota Press, 2000), xi–xiii; Timothy Mitchell, "The Stage of Modernity," in

Questions of Modernity, ed. Timothy Mitchell (Minneapolis: University of Minnesota Press, 2000), 8–9; Arjun Appadurai, *Modernity at Large: Cultural Dimensions of Globalization* (Minneapolis: University of Minnesota Press, 1996), 189–90. Daniel Headrick argues that such claims were based in technological progress in *The Tools of Empire: Technology and European Imperialism in the Nineteenth Century* (New York: Oxford University Press, 1981). Fredric Jameson argues for a belief in progress in space in *Postmodernism, or, the Cultural Logic of Late Capitalism* (Durham: Duke University Press, 1991), 410–18.

13 Johannes Fabian, *Time and the Other: How Anthropology Makes Its Object* (New York: Columbia University Press, 2014), 31. See also Kathleen Davis, *Periodization & Sovereignty: How Ideas of Feudalism & Secularization Govern the Politics of Time* (Philadelphia: University of Pennsylvania Press, 2008).

14 Walter Mignolo, *Local Histories/Global Designs: Coloniality, Subaltern Knowledges, and Border Thinking* (Princeton: Princeton University Press, 2000); Jerry H. Bentley, Introduction to "Perspectives on Global History: Cultural Encounters between the Continents over the Centuries," in *Proceedings of the Nineteenth Congress of Historical Sciences*, ed. Anders Jolstad and Marianne Lunde (Oslo: International Committee of Historical Sciences, 2000), 29–45; On German anthropology and empire, see Andrew Zimmerman, *Anthropology and Antihumanism in Imperial Germany* (Chicago: The University of Chicago Press, 2001). On the development of anthropology and other social sciences in Europe's colonies, see Emmanuelle Sibeud, *Une Science impériale pour l'Afrique? La Construction des Savoirs africanistes en France, 1878–1930* (Paris: Éditions de l'École des Hautes Études en Sciences Sociales, 2002); Helen Tilley, *Africa as a Living Laboratory: Empire, Development, and the Problem of Scientific Knowledge, 1870–1950* (Chicago: University of Chicago Press, 2011).

15 John C. Weaver, *The Great Land Rush and the Making of the Modern World, 1650–1900* (Montreal: McGill-Queen's University Press, 2003); Colin M. Coates, *The Metamorphoses of Landscape and Community in Early Quebec* (Montreal: McGill-Queen's University Press, 2000), 144–5; Graeme Wynn, "Settler Societies in Geographical Focus," *Historical Studies* 20, no. 80 (1983): 353–66.

16 Michael Adas, *Dominance by Design* (Cambridge: Harvard University Press, 2009).

17 Sarah Carter, *Lost Harvests: Prairie Indian Reserve Farmers and Government Policy* (Montreal: McGill-Queen's University Press, 1990); Adele Perry, *On the Edge of Empire: Gender, Race, and the Making of British Columbia, 1849–1871* (Toronto: University of Toronto Press, 2001).

18 Diana K. Davis, *Resurrecting the Granary of Rome: Environmental History and French Colonial Expansion in North Africa* (Athens: Ohio University Press,

2009); Yves Lacoste, *Ibn Khaldun: The Birth of History and the Past of the Third World*, trans. David Macey (New York: Verso, 1984).

19 The British, French, and Portuguese colonial empires utilized the concept of a mission to bring civilization to Africans, which justified European conquest and the forms of colonial rule. Alice Conklin, *A Mission to Civilize: The Republican Idea of Empire in France and West Africa, 1895–1930* (Stanford, CA: Stanford University Press, 1997); Gerrit W. Gong, *The Standard of "Civilization" in International Society* (Oxford: Clarendon Press, 1984); Jürgen Osterhammel, "'The Great Work of Uplifting Mankind.' Zivilisierungsmission und Moderne," in *Zivilisierungsmissionen. Imperiale Weltverbesserung seit dem 18. Jahrhundert*, ed. Jürgen Osterhammel and Boris Barth (Konstanz: UVK-Verlagsgesellschaft, 2005), 363–425; Kenneth Pomeranz, "Empire & 'Civilizing' Missions, Past and Present," *Daedalus* 134, no. 2 (2005): 34–45; Miguel Bandeira Jerónimo, *The "Civilizing Mission" of Portuguese Colonialism, 1870–1930*, trans. Stewart Lloyd-Jones (New York: Palgrave Macmillan, 2015). The United States also had something of a civilizing mission with claims about Manifest Destiny. Michael Adas, *Dominance by Design* (Cambridge: Harvard University Press, 2009); Amy Greenberg, *Manifest Manhood and the Antebellum American Empire* (Cambridge: Cambridge University Press, 2005).

20 Matthew Edney, *Mapping an Empire: The Geographical Construction of British India, 1765–1843* (Chicago: University of Chicago Press, 1997); Norman Etheringon, ed., *Mapping Colonial Conquest: Australia and Southern Africa* (Crawley: University of Western Australia Press, 2007); Giselle Byrnes, *Boundary Markers: Land Surveying and the Colonisation of New Zealand* (Wellington: Bridget Williams, 2001).

21 David Lambert, *Mastering the Niger: James MacQueen's African Geography and the Struggle over Atlantic Slavery* (Chicago: University of Chicago Press, 2013); Carsten Graebel, *Die Erforschung der Kolonien. Expeditionen und koloniale Wissenskultur deutscher Geographen, 1884–1919* (Bielefeld: Transfer, 2015). On colonial cartography, see Raymond Craib, *Cartographic Mexico: A History of State Fixations and Fugitive Landscapes* (Durham: Duke University Press, 2004); James R. Akerman, ed., *The Imperial Map* (Chicago: University of Chicago Press, 2009); Jason D. Hansen, *Mapping the Germans: Statistical Science, Cartography, and the Visualization of the German Nation, 1848–1914* (Oxford: Oxford University Press, 2015); Pascal Schillings, *Der letzte weiße Flecken. Europäische Antarktisreisen um 1900* (Göttingen: Wallstein, 2016). Spatial historians' work has been informed by that of historical geographers, who brought new attention to maps as texts containing political arguments in their claims to represent objective reality. Two of the most influential geographers in this vein were J.B. Harley and Denis Wood. Denis Wood, *Rethinking the Power of Maps* (New York: Guilford, 2010).

22 J.B. Harley, *The New Nature of Maps: Essays in the History of Cartography*, ed. Paul Laxton (Baltimore: The Johns Hopkins University Press, 2001), 144.

23 Cole Harris, *The Resettlement of British Columbia: Essays on Colonialism and Geographical Change* (Vancouver: UBC Press, 1997), 191.

24 The term "spatial history" comes from Paul Carter in *The Road to Botany Bay: An Exploration of Landscape and History* (New York: Knopf, 1988), xxiii, where he writes that "spatial history discovers and explores imperial history's lacunae."

25 Yi-Fu Tuan, *Space and Place: The Perspective of Experience* (Minneapolis: University of Minnesota Press, 2001), 6.

26 David Harvey, *Social Justice and the City* (Oxford: Blackwell, 1973), 13.

27 Edward W. Soja, *Seeking Spatial Justice* (Minneapolis: University of Minnesota Press, 2010), 4, 18.

28 Henri Lefebvre, *The Production of Space*, trans. Donald Nicholson-Smith (Oxford: Basil-Blackwell, 2001); Edward W. Soja has developed this concept further in *Seeking Spatial Justice* (Minneapolis: University of Minnesota Press, 2010).

29 Lefebvre, *Production of Space*, 57–60.

30 Fredric Jameson, *Postmodernism, or, the Cultural Logic of Late Capitalism* (Durham: Duke University Press, 1991), 410. This same issue is also at work in David Harvey, *The Condition of Postmodernity: An Enquiry into the Origins of Cultural Change* (Malden: Wiley-Blackwell, 1991); Roger Friedland and Deirdre Boden, eds., *NowHere: Space, Time, and Modernity* (Berkeley: University of California Press, 1995); Soja, *Seeking Spatial Justice*, other influential books in shaping the history of space.

31 Edward Said, *Orientalism* (New York: Penguin, 1978); Derek Gregory, "Edward Said's Imaginative Geographies," in *Thinking Space*, ed. Mike Crang and Nigel Thrift (London: Routledge, 2000), 302–48.

32 On German orientalism, see Suzanne Marchand, *German Orientalism in the Age of Empire: Religion, Race, and Scholarship* (Washington: German Historical Institute, 2009). Many scholars have argued for a conception of mainland Tanzania's history as part of an Indian Ocean World. Michael Pearson, *The Indian Ocean* (London: Routledge, 2003); *The World of the Indian Ocean, 1500–1800* (Aldershot, Hampshire: Ashgate, 2005); John Middleton, *African Merchants of the Indian Ocean: Swahili of the East African Coast* (Long Grove: Waveland Press, 2004); Sugata Bose, *A Hundred Horizons: The Indian Ocean in the Age of Global Empire* (Cambridge: Harvard University Press, 2006); K.N. Chaudhuri, *Asia before Europe: Economy and Civilization of the Indian Ocean from the Rise of Islam to 1750* (Cambridge: Cambridge University Press, 1990); K.N. Chaudhuri, *Trade and Civilization in the Indian Ocean: An Economic History from the Rise of Islam to 1750* (Cambridge: Cambridge University Press, 1985); Erik Gilbert, *Dhows and the Colonial Economy of Zanzibar, 1860–1870* (Athens: Ohio University Press,

2004); Kenneth McPherson, *The Indian Ocean: A History of People and the Sea* (Delhi: Oxford University Press, 1993); Randall L. Pouwels, "Eastern Africa and the Indian Ocean to 1800: Reviewing Relations in Historical Perspective," *The International Journal of African Historical Studies* 35, no. 2/3 (2002): 385–425; Abdul Sheriff, *Slaves, Spices and Ivory in Zanzibar* (London: James Currey, 1987); Markus P.M. Vink, "Indian Ocean Studies and the 'New Thalassology.'" *Journal of Global History* 2, no. 1 (2007): 41–62. The idea that Swahili society was not an import from the Middle East, but the creation of Africans, only began to emerge in the 1970s and 1980s. James de Vere Allen, "Swahili Culture and the Nature of East Coast Settlement," *The International Journal of African Historical Studies* 14, no. 2 (1981): 306–34; Randall L. Pouwels, *Horn and Crescent: Cultural Change and Traditional Islam on the East African Coast, 800–1900* (Cambridge: Cambridge University Press, 1987).

33 On Swahili poems as historical sources, see José Arturo Saavedra Casco, *Utenzi, War Poems, and the German Conquest of East Africa: Swahili Poetry as a Historical Source* (Trenton: Africa World Press, 2007).

34 Agrarian politics motivated the growth of cultural geography as a field. Woodruff D. Smith, "Friedrich Ratzel and the Origins of Lebensraum," *German Studies Review* 3, no. 1 (February 1980): 52; Thomas Nipperdey, *Deutsche Geschichte 1866–1918*, vol. 2, Machtstaat vor der Demokratie (Munich: Beck, 1990), 164–5. Bernhard Gissibl has written on German ideas about nature in Ostafrika, arguing that the colony came to epitomize the natural world for German audiences. Bernhard Gissibl, *The Nature of German Imperialism: Conservation and the Politics of Wildlife in Colonial East Africa* (New York: Berghahn, 2016).

35 Alexander von Humboldt and Aimé Bonpland, *Essay on the Geography of Plants*, trans. Sylvie Romanowski (Chicago: University of Chicago Press, 2009), 133.

36 Wilhelm Heinrich Riehl, *Die Naturgeschichte des Volkes als Grundlage einer deutschen Social-Politik*, vol. 1, *Land und Leute* (Stuttgart and Tübingen: J.G. Cotta, 1854); George L. Mosse, *The Crisis of German Ideology: Intellectual Origins of the Third Reich* (New York: Grosset & Dunlap, 1964), 19–20; Vejas Gabriel Liulevicius, *War Land on the Eastern Front: Culture, National Identity and German Occupation in World War I* (Cambridge: Cambridge University Press, 2004), 167.

37 Friedrich Ratzel, *Anthropo-Geographie, oder Grundzüge der Anwendung der Erdkunde auf die Geschichte*, vol. 1 (Stuttgart: Engelhorn, 1882).

38 Friedrich Ratzel, "Geschichte, Völkerkunde und historische Perspektive," *Historische Zeitschrift* 93, no. 1 (1904): 1–46.

39 John Noyes, *Colonial Space: Spatiality in the Discourse of German South West Africa 1884–1915* (Chur, Switzerland: Harwood Academic Publishers, 1992); Mary Louise Pratt, *Imperial Eyes: Travel Writing and Transculturation*

(London: Routledge, 1992); Edney, *Mapping an Empire*; Simon Ryan, *The Cartographic Eye: How Explorers Saw Australia* (Cambridge: Cambridge University Press, 1996); Carter, *Road to Botany Bay*.

40 The importance of Indigenous intermediaries in shaping European territorial knowledge has been an aspect of histories of European expansion. See Allen F. Roberts, *A Dance of Assassins: Performing Early Colonial Hegemony in the Congo* (Bloomington: Indiana University Press, 2013); Edney, *Mapping an Empire*; Pratt, *Imperial Eyes*; Dane Kennedy, *The Last Blank Spaces: Exploring Africa and Australia* (Cambridge, MA: Harvard University Press, 2013); Carter, *Road to Botany Bay*, among others.

41 Michel de Certeau, *The Practice of Everyday Life*, trans. Steven F. Rendall (Berkeley: University of California Press, 1984), 115.

42 The most common topic of studies is the Herero/Nama genocide in Southwest Africa. Isabel Hull, *Absolute Destruction: Military Culture and the Practices of War in Imperial Germany* (Ithaca, NY: Cornell University Press, 2005); David Olusoga and Casper Erichsen, *The Kaiser's Holocaust: Germany's Forgotten Genocide and the Colonial Roots of Nazism* (London: Faber and Faber, 2011); Jürgen Zimmerer and Joachim Zeller, eds., *Genocide in German South-West Africa: The Colonial War (1904–1908) in Namibia and Its Aftermath* (Monmouth: Merlin, 2008); Jeremy Sarkin-Hughes, *Germany's Genocide of the Herero: Why Kaiser Wilhelm II Gave the Order* (Oxford: James Currey, 2011); Susanne Kuss, *German Colonial Wars and the Context of Military Violence*, trans. Andrew Smith (Cambridge: Harvard University Press, 2017) looks at the Herero-Nama War together with other colonial wars; Albert Wirz, Andreas Eckert, and Katrin Bromber, eds., *Alles unter Kontrolle: Disziplinierungsprozesse im kolonialen Tansania (1850–1960)* (Cologne: Rüdiger Köppe, 2003) focuses on Ostafrika.

43 Vejas Gabriel Liulevicius, *War Land on the Eastern Front: Culture, National Identity, and German Occupation in World War I* (Cambridge: Cambridge University Press, 2000); Kristin Kopp, *Germany's Wild East: Constructing Poland as Colonial Space* (Ann Arbor: University of Michigan Press, 2012); Liulevicius, *The German Myth of the East: 1800 to Present* (Oxford: Oxford University Press, 2009); David Blackbourn, *The Conquest of Nature: Water, Landscape, and the Making of Modern Germany* (London: Pimlico, 2007).

44 Bradley Naranch, "Introduction: German Colonialism Made Simple," in *German Colonialism in a Global Age*, ed. Bradley Naranch and Geoff Eley (Durham: Duke University Press, 2014), 5; Jeff Bowersox, *Raising Germans in the Age of Empire: Youth and Colonial Culture, 1871–1914* (Oxford: Oxford University Press, 2013); Sebastian Conrad, *German Colonialism: A Short History*, trans. Sorcha O'Hagan (Cambridge: Cambridge University Press, 2012); John Phillip Short, *Magic Lantern Empire: Colonialism and Society in Germany* (Ithaca, NY: Cornell University Press, 2012); Volker M. Langbehn,

German Colonialism, Visual Culture, and Modern Memory (New York: Routledge, 2010); Sara Friedrichsmeyer, Sara Lennox, and Susanne Zantop, *The Imperialist Imagination: German Colonialism and Its Legacy* (Ann Arbor, MI: University of Michigan Press, 1998); David Ciarlo, *Advertising Empire: Race and Visual Culture in Imperial Germany* (Cambridge, MA: Harvard University Press, 2011); Fitzpatrick, *Liberal Imperialism in Germany*.

45 An older generation of historians claimed colonialism was unimportant to Germany or completely subordinate to domestic issues. See Horst Gründer, *Geschichte der deutschen Kolonien* (Paderborn, Germany: Ferdinand Schöningh, 1985) and Hans-Ulrich Wehler, "Bismarck's Imperialism 1862–1890," *Past & Present* 48 (August 1970): 119–55.

46 Sebastian Conrad, *German Colonialism: A Short History*, trans. Sorcha O'Hagan (Cambridge: Cambridge University Press, 2012), 51; Jan-Georg Deutsch, *Emancipation without Abolition in German East Africa c. 1884–1914* (Oxford: James Currey, 2006), 99.

47 Daniel Joseph Walther, *Creating Germans Abroad: Cultural Policies and National Identity in Namibia* (Athens: Ohio University Press, 2002); Jürgen Zimmerer, *Deutsche Herrschaft über Afrikaner: Staatlicher Machtanspruch und Wirklichkeit im kolonialen Namibia* (Münster, Lit, 2004).

48 Despite German claims that *Kultur* made the German nation fundamentally different from France, this transition mirrored French readings of Ibn Khaldun's histories to claim Arabs were a malignant influence in North Africa. Lacoste, *Ibn Khaldun*; Abdelmajid Hannoum, "Translation and the Colonial Imaginary: Ibn Khaldûn Orientalist," *History and Theory* 42 (February 2003): 61–81.

49 Julie MacArthur, *Cartography and the Political Imagination: Mapping Community in Colonial Kenya* (Athens: Ohio University Press, 2016); Timothy Parsons, "Being Kikuyu in Meru: Challenging the Tribal Geography of Colonial Kenya," *Journal of African History* 53 (2012): 65–86; Timothy Parsons, "Local Responses to the Ethnic Geography of Colonialism in the Gusii Highlands of British-Ruled Kenya," *Ethnohistory* 58 (2011): 491–523.

50 The scholarship on ethnicity in Africa is too vast to recount fully here, but there has been a transition away from first primordial arguments for ethnicity and then claims that ethnicity was merely an invention as scholars have attended to African agency in shaping ethnic belonging. John Iliffe made the argument for colonial invention in mainland Tanzania in *A Modern History of Tanganyika* (Cambridge: Cambridge University Press, 1979). For elsewhere on the continent, see Leroy Vail, ed. *The Creation of Tribalism in Southern Africa* (London: James Currey, 1989) and Eric Hobsbawm and Terence Ranger, eds., *The Invention of Tradition* (Cambridge: Cambridge University Press, 1983), which argue for a genesis

of ethnic divisions in colonial categories of rule. Thomas Spear, "Neo-Traditionalism and the Limits of Invention in British Colonial Africa," *Journal of African History* 44 (2003): 8–13; Paul Nugent, "Putting the History Back into Ethnicity: Enslavement, Religion and Cultural Brokerage in the Construction of Mandinka/Jola and Ewe/Agotime Identities in West Africa c. 1650–1930," *Comparative Studies in Society and History* 50, no. 4 (2008): 920–48 argue for an earlier genesis of ethnic divisions and limits to the ability of colonial states to remake them. Gabrielle Lynch, *I Say to You: Ethnic Politics and the Kalenjin in Kenya*, (Chicago: University of Chicago Press, 2011); Alexander Keese, *Ethnicity and the Colonial State* (Leiden, The Netherlands: Brill, 2016); Derek Peterson, *Ethnic Patriotism and East African Revival: A History of Dissent, c. 1935–1972* (Cambridge: Cambridge University Press, 2012) argue that ethnic belonging has been repeatedly been reinvented and reimagined.

51 Jonathon Glassman, *War of Words, War of Stones: Racial Though and Violence in Colonial Zanzibar* (Bloomington: Indiana University Press, 2011); Mahmood Mamdani, *When Victims Become Killers: Colonialism, Nativism, and the Genocide in Rwanda* (Princeton, NJ: Princeton University Press, 2001); Catherine Newbury, *The Cohesion of Oppression: Clientship and Ethnicity in Rwanda, 1860–1960* (New York: Columbia University Press, 1988); Jan Vansina, *Antecedents of Modern Rwanda: The Nyiginya Kingdom* (Madison: University of Wisconsin Press, 2004); Gérard Prunier, *The Rwanda Crisis: History of a Genocide* (New York: Columbia University Press, 1995); David Newbury, *The Land beyond the Mists: Essays on Identity and Authority in Precolonial Congo and Rwanda* (Athens: Ohio University Press, 2009). An exception is Alison Liebhafsky Des Forges, *Defeat Is the Only Bad News: Rwanda under Musinga, 1896–1931*, ed. David Newbury (Madison: University of Wisconsin Press, 2011). The debate over continuities through the colonial period dates back to the 1960s, when nationalist historians began to argue that colonialism did not create a clear break in African history. See J.F.A. Ajayi, "The Continuity of African Institutions under Colonialism," in *Emerging Themes in African History*, ed. Terence Ranger (Nairobi: East African, 1968), 189–200. This view stood in contrast to that of Marxist historians like Walter Rodney in *How Europe Underdeveloped Africa* (London: Bogle-L'Ouverture, 1972).

52 Michael Pesek, *Koloniale Herrschaft in Deutsch-Ostafrika: Expeditionen, Militär und Verwaltung seit 1880* (Frankfurt am Main: Campus, 2005); Michelle R. Moyd, *Violent Intermediaries: African Soldiers, Conquest, and Everyday Colonialism in German East Africa* (Athens: Ohio University Press, 2014); Albert Wirz, Andreas Eckert, and Katrin Bromber, eds., *Alles unter Kontrolle: Disziplinierungsprozesse im kolonialen Tansania (1850–1960)* (Cologne: Rüdiger Köppe, 2003).

53 Michael Pesek, *Koloniale Herrschaft in Deutsch-Ostafrika: Expeditionen, Militär und Verwaltung seit 1880* (Frankfurt am Main: Campus, 2005), 77–9.

54 Steven Press, *Rogue Empires: Contracts and Conmen in Europe's Scramble for Africa* (Cambridge: Harvard University Press, 2017); Weaver, *Great Land Rush*, 4–5.

55 For layers of European imperial sovereignty and the creation of different kinds of rule across bounded colonial spaces, see Lauren Benton, *A Search for Sovereignty: Law and Geography in European Empires, 1400–1900* (Cambridge: Cambridge University Press, 2010). Radhika V. Mongia, "Historicizing State Sovereignty: Inequality and the Form of Equivalence," *Comparative Studies in Society and History* 49, no. 2 (2007): 384–411; and Antony Anghie, "Finding the Peripheries: Sovereignty and Colonialism in Nineteenth-Century International Law," *Harvard International Law Journal* 40, no. 1 (1999): 1–81 argue that modern state sovereignty was produced through colonial encounter.

56 On attempts to create and challenge such narratives in the period, see Birgit Schäbler, *Moderne Muslime. Ernest Renan und die Geschichte der ersten Islamdebatte 1883* (Paderborn: Ferdinand Schöningh, 2016).

57 Juhani Koponen, *Development for Exploitation: German Colonial Policies in Mainland Tanzania, 1884–1914* (Helsinki: Tiedekirja, 1995)

58 Frederick Cooper, *Decolonization and African Society: The Labour Question in French and British Africa* (Cambridge: Cambridge University Press, 1996); James Ferguson, *The Anti-Politics Machine: "Development," Depoliticization, and Bureaucratic Power in Lesotho* (Minneapolis: University of Minnesota Press, 1994); Arturo Escobar, *Encountering Development: The Making and Unmaking of the Third World* (Princeton: Princeton University Press, 1995); Leander Schneider, *Government of Development: Peasants and Politicians in Postcolonial Tanzania* (Bloomington: Indiana University Press, 2014); Colin Leys, *The Rise and Fall of Development Theory* (London: James Currey, 1996). Monica van Beusekom dates developmentalism to the 1920s and 1930s, but still the interwar period. Monica van Beusekom, *Negotiating Development: African Farmers and Colonial Experts at the Office du Niger, 1920–1960* (Portsmouth, NH: Heinemann, 2002). Gregory Mann has also dated the emergence of developmentalism in French Africa to this period. Gregory Mann, *From Empires to NGOs in the West African Sahel: The Road to Nongovernmentality* (New York: Cambridge University Press, 2014)

59 For development defined in this way, see Tilley, *Africa as a Living Laboratory*, and James C. Scott, *Seeing Like a State: How Certain Schemes to Improve the Human Condition Have Failed* (New Haven: Yale University Press, 1998).

60 Albert Wirz, Andreas Eckert, and Katrin Bromber, eds., *Alles unter Kontrolle: Disziplinierungsprozesse im kolonialen Tansania (1850–1960)* (Cologne: Rüdiger Köppe, 2003), 12.

2 Geographies of East Africa in the Second Half
of the Nineteenth Century

1 Verney Lovett Cameron, *Across Africa* (New York: Harper & Brothers, 1877), 76–7.
2 S. Tristram Pruen, *The Arab and the African: Experiences in Eastern Equatorial Africa during a Residence of Three Years* (London: Seeley and Co., 1891), 184.
3 Cameron, *Across Africa*, 76.
4 Jane I. Guyer and Samuel M. Eno Belinga, "Wealth in People as Wealth in Knowledge: Accumulation and Composition in Equatorial Africa," *Journal of African History* 36, no. 1 (1995): 91–120.
5 In mainland Tanzania specifically, Steven Feierman, "Political Culture and Political Economy in Early East Africa," in *African History: From Earliest Times to Independence*, 2nd ed., ed. Philip Curtin, Steven Feierman, Leonard Thompson, and Jan Vansina (London: Longman, 1995), 144; generally or elsewhere on the continent, Paul Landau, *Popular Politics in the History of South Africa, 1400–1948* (Cambridge: Cambridge University Press, 2010); David Woodward and Malcolm Lewis, *History of Cartography*. Vol. 2, book 3, *Cartography in the Traditional African, American, Arctic, Australian, and Pacific Societies* (Chicago: University of Chicago Press, 1998); Allen Howard, "Re-Marking the Past: Spatial Structures and Dynamics in the Sierra Leone-Guinea Plain, 1860–1920s," in *The Spatial Factor in African History: The Relationship of the Social, Material, and Perceptual*, ed. Howard and Richard M. Shain (Leiden: Brill, 2005), 322; Donald Wright, " 'What Do You Mean There Were No Tribes in Africa?': Thoughts on Boundaries – and Related Matters – in Precolonial Africa," *History in Africa* 26 (1999): 414–15; Jan Vansina, *Antecedents of Modern Rwanda: The Nyiginya Kingdom* (Madison: University of Wisconsin Press, 2004), 62, 67; Barrie Sharpe, "Ethnography and a Regional System: Mental Maps and the Myth of States and Tribes in North-Central Nigeria," *Critique of Anthropology* 6 (1986): 38.
6 Mahmood Mamdani, *Citizen and Subject: Contemporary Africa and the Legacy of Late Colonialism* (Princeton: Princeton University Press, 1996); Mamdani, *Saviors and Survivors: Darfur, Politics, and the War on Terror* (New York: Doubleday, 2009); Mamdani, *Define and Rule: Native as Political Identity* (Cambridge: Harvard University Press, 2012).
7 Camille Lefebvre, *Frontières de sable, frontières de papier: Histoire de territoires et de frontières, du jihad de Sokoto à la colonisation française du Niger, XIXᵉ-XXᵉ siècles* (Paris: Sorbonne, 2015) argues for a similar process in the western Sudan.
8 Steven Feierman, *The Shambaa Kingdom: A History* (Madison: University of Wisconsin Press, 1974), 120, 188; Aylward Shorter, "Nyungu-Ya-Mawe and the 'Empire of the Ruga-Ruga,'" *Journal of African History* 9, no. 2

(1968): 247; Jutta Bückendorf, *"Schwarz-weiß-rot über Ostafrika!" Deutsche Kolonialpläne und afrikanische Realität* (Münster: Lit, 1997), 51–2.

9 James Sheehan, "The Problem of Sovereignty in European History," *American Historical Review* 111 (2006): 7–8. Some scholars have argued that European ideas of territorial sovereignty only really developed through the conquest of colonial empires and that sovereignties remained layered, overlapping, and contingent on local participation. Radhika Mongia, "Historicizing State Sovereignty: Inequality and the Form of Equivalence," *Comparative Studies in Society and History* 49 (2007): 384–411; Antony Anghie, "Finding the Peripheries: Sovereignty and Colonialism in Nineteenth-Century International Law," *Harvard International Law Journal* 40 (1999): 1–81; Jane Burbank and Frederick Cooper, *Empires in World History: Power and the Politics of Difference* (Princeton: Princeton University Press, 2010); Douglas Howland and Luise White, "Introduction: Sovereignty and the Study of States," in *The State of Sovereignty: Territories, Laws, Populations*, ed. Douglas Howard and Luise White (Bloomington: Indiana University Press, 2009), 1–18.

10 Carl Schmitt, *The Nomos of the Earth*, trans. G.L. Ulmen (New York: Telos, 2003), 45–6.

11 Friedrich Ratzel, *Die Erde und das Leben. Eine vergleichende Erdkunde*, vol. 2 (Leipzig: Bibliographisches Institut, 1902), 669.

12 John Iliffe, *A Modern History of Tanganyika* (Cambridge: Cambridge University Press, 1979), 8–12.

13 Stephen J. Rockel, *Carriers of Culture: Labor on the Road in Nineteenth-Century East Africa* (Portsmouth: Heinemann, 2006), 36–44.

14 On the growth of global capitalism and trade in this period see Sven Beckert, *Empire of Cotton: A Global History* (New York: Penguin, 2014).

15 Shorter, "Nyungu-Ya-Mawe," 258; Steven Feierman, "A Century of Ironies in East Africa (c. 1780–1890)," in *African History*, 2nd ed., 364.

16 Karl Weule, "Zur Kartographie der Naturvölker," *Petermanns geographische Mitteilungen* 61 (1915): 18–21, 59–62. Pesa Mbili told Weule that he would have drawn the networks in a straight line had the paper been larger, a line that holds together everywhere represented on the map. His route was not in actuality a straight line. It followed the coast north to Dar es Salaam, then turned westward.

17 Allen Howard, "Nodes, Networks, Landscapes, and Regions: Reading the Social History of Tropical Africa, 1700s–1920," in *The Spatial Factor in African History*, 66, 78.

18 Weule, "Zur Kartographie der Naturvölker," 18–21, 59–62.

19 Weule, "Zur Kartographie der Naturvölker," 18–21, 59–62.

20 Richard White, *Railroaded: The Transcontinentals and the Making of Modern America* (New York: W.W. Norton, 2011), 146.

21 Rockel, *Carriers of Culture*, 60; Feierman, *The Shambaa Kingdom*, 174–6.

22 Feierman, *The Shambaa Kingdom*, 178–9; "A Century of Ironies in East Africa," 365; Shorter, "Nyungu-Ya-Mawe," 240–1, 247. Big Men's authority mirrored that of chiefs in many respects, though they maintained authority through giving and receiving gifts rather than requiring tribute as a right. Chiefs could impose judgment on subject lineages, whereas Big Men had to use persuasion. Their positions were not heritable, in contrast to chiefs'. Chiefly authority often coexisted with that of Big Men.

23 Feierman, *The Shambaa Kingdom*; Ralph Austen, *Northwest Tanzania under German and British Rule: Colonial Policy and Tribal Politics, 1889–1939* (New Haven: Yale University Press, 1968), 17.

24 Shorter, "Nyungu-Ya-Mawe," 250.

25 "Lindi und die Handelsverhältnisse im Süden von Deutsch-Ostafrika," *Deutsches Kolonialblatt*, November 15, 1892, 578–83.

26 Johann Ludwig Krapf, *Travels, Researches, and Missionary Labours, during an Eighteen Years' Residence in Eastern Africa* (London: Trübner & Co., 1860), 260–2.

27 Joseph Thomson, *Through Masai-Land: A Journey of Exploration among the Snowclad Volcanic Mountains and Strange Tribes of Eastern Equatorial Africa* (Boston: Houghton, Mifflin, and Company, 1885), 271.

28 Thomson, *Through Masai Land*, 56. Jeremy Prestholdt has noted regional and temporal changes in fashion in the region in "On the Global Repercussions of East Africa Consumerism," *American Historical Review* 109, no. 3 (June 2004): 761–2.

29 Thomson, *Through Masai Land*, 70.

30 Thomson, *Through Masai Land*, 338.

31 Ernst Nigmann, *Die Wahehe. Ihre Geschichte, Kult-, Rechts- Kriegs- und Jagd-Gebräuche* (Berlin: Mittler und Sohn, 1908), 10.

32 Nigmann, *Die Wahehe*, 11–12.

33 Alison Redmayne, "Mkwawa and the Hehe Wars," *Journal of African History* 9, no. 3 (1968): 425–6.

34 Nigmann, *Die Wahehe*, 84.

35 Georg von Prittwitz und Gaffron journal, January 25, 1898, IfL, Prittwitz Nachlass, Box 245, 1/245, vol. 2.

36 Tom von Prince, Geschichte der Magwangwara nach Erzählung des Arabers Raschid bin Masaiid und des Fussi, Bruders des vor 3 Jahren verstorbenen "Sultans" der Magwangwara Mharuli, April 25, 1894, BArch R 1001/698, pag. 155–64.

37 "Die Volksstämme am Tana in Ostafrika (Mitgetheilt von Msgr. Le Roy)," *Die katholischen Missionen*, June 1895, 133.

38 "Die Volksstämme am Tana in Ostafrika (Mitgetheilt von Msgr. Le Roy)," *Die katholischen Missionen*, June 1895, 133–4.

39 Thomson, *Through Masai Land*, 310.

40 Bückendorf, *"Schwarz-weiß-rot über Ostafrika,"* 23–7.

41 Oskar Baumann, *Usambara und seine Nachbargebiete* (Berlin: Dietrich Reimer, 1891), 176.

42 Norman Robert Bennett, *Arab versus European: Diplomacy and War in Nineteenth-Century East Central Africa* (New York: Africana Publishing Company, 1986), 35–6.

43 Feierman, *The Shambaa Kingdom*, 18, 188.

44 Tom von Prince, Geschichte der Magwangwara nach Erzählung des Arabers Raschid bin Masaiid und des Fussi, Bruders des vor 3 Jahren verstorbenen "Sultans" der Magwangwara Mharuli, April 25, 1894, BArch R 1001/698, pag. 155–64.

45 "Kilossa in Usagara," *Deutsches Kolonialblatt*, March 1, 1892, 146.

46 Bericht des Kaiserlichen Gouverneurs für Deutschostafrika. Generalmajor Liebert über seine Reise nach Usagara und Uluguru, August 26, 1898, BArch R 1001/7806, pag. 93–109.

47 Mwalim Mbaraka bin Shomari, "Shairi la bwana mkubwa," in *Suaheli-Gedichte*, ed. Carl Velten (Berlin: Reichsdruckerei, 1918), 68–86.

48 For overviews of the coast's history, see Randall Pouwels, *Horn and Crescent: Cultural Change and Traditional Islam on the East African Coast, 800–1900* (Cambridge: Cambridge University Press, 1987); Mark Horton and John Middleton, *The Swahili: The Social Landscape of a Mercantile Society* (Oxford: Blackwell, 2000); Derek Nurse and Thomas Spear, *The Swahili: Reconstructing the History and Language of an African Society, 800–1500* (Philadelphia: University of Pennsylvania Press, 1985).

49 Adria LaViolette and Stephanie Wynne-Jones, "The Swahili World," in *The Swahili World*, eds. Wynne-Jones and LaViolette (London: Routledge, 2018), 4.

50 Iliffe, *Modern History of Tanganyika*, 36–41.

51 James de Vere Allen, "Swahili Culture and the Nature of East Coast Settlement," *The International Journal of African Historical Studies* 14, no. 2 (1981): 317–18; Felicitas Becker, *Becoming Muslim in Mainland Tanzania 1890–2000* (Oxford: Oxford University Press, 2008), 208; Iliffe, *Modern History of Tanganyika*, 47. *Uungwana* was the common concept of cosmopolitanism prior to the late nineteenth century, when freed slaves co-opted it to mark themselves off from enslaved people. Elites then developed *ustaarabu* to further differentiate themselves. Jeremy Prestholdt, *Domesticating the World: African Consumerism and the Genealogies of Globalization* (Berkeley: University of California Press, 2008), 102.

52 Allen, "Swahili Culture and the Nature of East Coast Settlement," 315–17.

53 Jonathon Glassman, *Feasts and Riot: Revelry, Rebellion, and Popular Consciousness on the Swahili Coast, 1856–1888* (Portsmouth, NH: Heinemann, 1995), 92–3.

54 TNA G 32/8, page 25–6, Abschrift from Etienne Baur and Alidina Jaffer to Georg von Rechenberg, March 11, 1897; Walter Thaddeus Brown, *A Precolonial History of Bagamoyo: Aspects of the Growth of an East African Coastal Town*. PhD diss., Boston University, 1971, iii–iv, 186, 188.

55 John Middleton, *African Merchants of the Indian Ocean: Swahili of the East African Coast* (Long Grove: Waveland Press, 2004), 4; Allen, "Swahili Culture and the Nature of East Coast Settlement," 316.

56 Carl Velten, "Erklärung einiger ostafrikanischer Ortsnamen," *Mittheilungen des Seminars für orientalische Sprachen an der Königlichen Friedrich Wilhelms-Universität zu Berlin* 1 (1898): 199.

57 "Lindi und die Handelsverhältnisse im Süden von Deutsch-Ostafrika," *Deutsches Kolonialblatt*, November 15, 1892, 578–83.

58 See, among others, "Vita vya kwanza," and Makanda bin Mwenyi Mkuu, "Utenzi wa habari za mrima na za bara," in Velten, *Suaheli-Gedichte*, 10–12, 27.

59 Sleman bin Mwenyi Tshande bin Mwenyi Hamisi esh-Shirazi in Carl Velten, *Schilderungen der Suaheli* (Göttingen: Vandenhoeck & Ruprecht, 1901), 1–55.

60 Mtoro bin Mwinyi Bakari, "Meine Reise nach Udoe bis Uzigua sowie Geschichtliches über die Wadoe und Sitten und Gebräuche derselben," in Carl Velten, *Schilderungen der Suaheli* (Göttingen: Vandenhoeck & Ruprecht, 1901), 138–44.

61 "Der weite und der nahe Weg," in *Märchen und Erzählungen der Suaheli*, ed. Carl Velten (Stuttgart: W. Spemann, 1898), 96–7.

62 Fahad Ahmad Bishara, *A Sea of Debt: Law and Economic Life in the Western Indian Ocean, 1780–1950* (Cambridge: Cambridge University Press, 2017), 25.

63 Jeremy Prestholdt, "On the Global Repercussions of East Africa Consumerism," *American Historical Review* 109, no. 3 (June 2004): 760.

64 Abdul Sheriff, *Slaves, Spices & Ivory in Zanzibar: Integration of an East African Commercial Empire into the World Economy, 1770–1873* (London: James Currey, 1987), 1.

65 Prestholdt, "On the Global Repercussions of East Africa Consumerism," 755–81.

66 Jonathon Glassman, *Feasts and Riot: Revelry, Rebellion, and Popular Consciousness on the Swahili Coast, 1856–1888* (Portsmouth, NH: Heinemann, 1995), 35–54.

67 Pekka Hämäläinen, *The Comanche Empire* (New Haven: Yale University Press, 2008), 4.

68 Alice Moore-Harell, *Egypt's African Empire: Samuel Baker, Charles Gordon & the Creation of Equatoria* (Brighton: Sussex Academic Press, 2012).

69 Neville Chittick, *Kilwa: An Islamic Trading City on the East African Coast*, vol. 2 (Nairobi: The British Institute in Eastern Africa, 1974).

70 Georg Lieder, Zur Kenntniß der Karawanenwege im südlichen Teile des ostafrikanischen Schutzgebietes, April 28, 1894, TNA G 1/35.

71 Georg Lieder, Zur Kenntniß der Karawanenwege im südlichen Teile des ostafrikanischen Schutzgebietes, April 28, 1894, TNA G 1/35.

72 "Lindi und die Handelsverhältnisse im Süden von Deutsch-Ostafrika," *Deutsches Kolonialblatt*, November 15, 1892, 578–83.

73 Felix Driver, *Geography Militant: Cultures of Exploration and Empire* (London: Blackwell, 2001).

74 Imre Josef Demhardt, *Deutsche Kolonialgrenzen in Afrika. Historisch-geographische Untersuchungen ausgewählter Grenzräume von Deutsch-Südwestafrika und Deutsch-Ostafrika* (Hildesheim: Georg Olms, 1997), 16.

75 Nipperdey, *Deutsche Geschichte, 1866–1918*, vol. 1, Arbeitswelt und Bürgergeist (Munich: Beck, 1990), 654.

76 Kathrin Fritsch, " 'You Have Everything Confused and Mixed Up…!' Georg Schweinfurth, Knowledge and Cartography of Africa in the 19th Century," *History in Africa* 36 (2009): 87.

77 Carsten Graebel, *Die Erforschung der Kolonien. Expeditionen und koloniale Wissenskultur deutscher Geographen, 1884–1919* (Bielefeld: Transfer, 2015), 31–5.

78 Johannes Fabian, *Out of Our Minds: Reason and Madness in the Exploration of Central Africa* (Berkeley: University of California Press, 2000), 221.

79 "Skizze einer Karte eines Theils von Ost u. Central Afrika mit Angabe der wahrscheinlichen Lage u. Ausdehnung des See's von Uniamesi nebs Bezeichnung der Grenzen u. Wohnsitze der Verschiedenen Völker sowie der Caravanen Strassen nach dem Innern," *Mittheilungen aus Justus Perthes' Geographischer Anstalt über wichtige neue Erforschungen auf dem Gesammtgebiete der Geographie von Dr. A. Petermann* (1856): plate 1.

80 John Phillip Short, *Magic Lantern Empire: Colonialism and Society in Germany* (Ithaca, NY: Cornell University Press, 2012), 127.

81 "Geschichte der Entstehung der Gesellschaft," *Mittheilungen der afrikanischen Gesellschaft in Deutschland* 1 (1878): 1–2. Among the organization's members were not only famous explorers like Gerhard Rohlfs and Gustav Nachtigal, but leading financial figures like Gerson Bleichröder, Bismarck's banker, and members of the von der Heydt family, the chemist Robert Bunsen, Werner von Siemens, and the geographer Friedrich Ratzel. "Verzeichnis der Stifter und Mitglieder," *Mittheilungen der afrikanischen Gesellschaft in Deutschland* 1 (1879): 51–5; "Verzeichnis der Stifter und Mitglieder," *Mittheilungen der afrikanischen Gesellschaft in Deutschland* 2 (1880): 66.

82 "Thätigkeit d. Gesellschaft vom Juni bis Oktober 1879," *Mittheilungen der afrikanischen Gesellschaft in Deutschland* 1 (1879): 170; "Thätigkeit der Gesellschaft seit dem Beginn des Jahres 1880," *Mittheilungen der afrikanischen Gesellschaft in Deutschland* 2 (1880): 123.

83 Gerhard Rohlfs, "Die deutsche Expedition zur Erforschung Centralafrikas," *Illustrirte Zeitung*, June 21, 1873, 477.

84 Anthony Kirk-Greene, "Heinrich Barth: An Exercise in Empathy," in *Africa and Its Explorers: Motives, Methods, and Impact*, ed. Robert Rotberg (Cambridge: Harvard University Press, 1970), 13–38; Heinrich Barth, *Reisen und Entdeckungen in Nord- und Central-Afrika in den Jahren 1849 bis 1855*, 5 vols. (Gotha: Justus Perthes, 1857–8). The degree to which German explorers formed a community with their British peers is attested to by the *Edinburgh Review*'s recommendation of Barth's books over David Livingstone's because of Livingstone's working-class background: Clare Pettit, *Dr. Livingstone, I Presume? Missionaries, Journalists, Explorers, and Empire* (Cambridge: Harvard University Press, 2007), 34.

85 Matthew Unangst, "Men of Science and of Action: The Celebrity of Explorers and German National Identity," *Central European History* 50, no. 3 (September 2017): 305–27.

86 Ratzel, "Sahara und Sudan," *Nord und Süd* 12, no. 34 (January 1880): 121.

87 Ratzel, *Anthropogeographie*, vol. 1, *Grundzüge der Anwendung der Erdkunde auf die Geschichte* (Stuttgart: Engelhorn, 1882).

88 "Verzeichnis der Stifter und Mitglieder," *Mittheilungen der afrikanischen Gesellschaft in Deutschland* 2 (1880): 66.

89 From the first issue of the *Deutsche Kolonialzeitung*: "Land und Leute in Argentinien," *Deutsche Kolonialzeitung*, January 1, 1884, 9; for an example of the relationship between *Land* and *Leute*, see Charles New, *Life, Wanderings, and Labours in Eastern Africa* (London: Hodder and Stoughton, 1874), in Charles Richards and James Place, eds., *East African Explorers* (London: Oxford University Press, 1960), 57–8; Adelheid von Pless, *Baron Carl Claus von der Decken's Reisen in Ost-Afrika in den Jahren 1859 bis 1865* (Leipzig and Heidelberg, Germany: C.F. Winter, 1869), 235.

90 Georg Schweinfurth, *The Heart of Africa. Three Years' Travels and Adventures in the Unexplored Regions of Central Africa from 1868–1871*, vol. 1, trans. Ellen E. Frewer (Chicago: Afro-Am Press, 1969), 119.

91 Schweinfurth, *The Heart of Africa*, vol. 2, 34.

92 Jeff Bowersox, "Classroom Colonialism: Race, Pedagogy, and Patriotism in Imperial Germany," in *German Colonialism in a Global Age*, ed. Bradley Naranch and Geoff Eley (Durham, NC: Duke University Press, 2014), 171–4.

93 Adam Jones and Isabel Voigt, "'Just a First Sketchy Makeshift': German Travellers and Their Cartographic Encounters in Africa, 1850–1914," *History in Africa* 39 (2012): 24–5.

94 Wilhelm Junker, *Travels in Africa during the Years 1882–1886*, trans. A.H. Keane (London: Chapman and Hall, 1892), 275.

95 Fritsch, "You Have Everything Confused and Mixed Up," 99; Driver, *Geography Militant*; Adrian Wisnicki, "Charting the Frontier: Indigenous Geography, Arab-Nyamwezi Caravans, and the East African Expedition of 1856–59," *Victorian Studies* 51, no. 1 (2008): 106, 124; Isabel Voigt,

"Die 'Schneckenkarte' – Mission, Kartographie und transkulturelle Wissensaushandlung in Ostafrika um 1850," *Cartographica Helvetica* 45 (2012): 27–38.

96 Pless, *Baron Carl Claus von der Decken's Reisen*; Henry M. Stanley, *How I Found Livingstone* (London: Sampson Lowe, 1872), in *East African Explorers*, 92–3; Ludwig von Höhnel, *Discovery of Lakes Rudolf and Stefanie* (London: Longmans, 1894), in *East African Explorers*, 231–2; Thomson, *Through Masai-Land*, 2, 62.

97 Schweinfurth, *Heart of Africa*, vol. 1, 459.

98 Georg Schweinfurth, "Ueber die Art des Reisens in Afrika," *Deutsche Rundschau* 2 (January-March, 1875): 246–50.

99 John Hanning Speke, *Journal of the Discovery of the Source of the Nile* (London and Edinburgh: William Blackwood & Sons, 1863), in *East African Explorers*, 138; New, *Life, Wanderings, and Labours* 39; Étienne Baur and Alexandre Le Roy, *À Travers le Zanguebar. Voyage dans l'Oudoé, l'Ouzigua, l'Oukwèré, l'Oukami et l'Ousaraga* (Tours: Alfred Mame et Fils, 1886), 201, 215; Duff MacDonald, *Africana; or, the Heart of Heathen Africa, vol. II – Mission Life* (London: Simpkin Marshall, 1882), 145–6.

100 Gustav Nachtigal, "Die Afrikaforschung und Henry M. Stanley's Zug durch den schwarzen Continent," *Deutsche Rundschau* 21 (1879): 208.

101 Schweinfurth, *Heart of Africa*, vol. 2, 123–5.

102 Michael Schubert, *Der schwarze Fremde: Das Bild des Schwarzafrikaners in der parlamentarischen und publizistischen Kolonialdiskussion in Deutschland von der 1870er bis in die 1930er Jahre* (Stuttgart: Steiner, 2003), 153.

103 Ethan R. Sanders, "Missionaries and Muslims in East Africa before the Great War" (paper presented at the Henry Martyn Seminar, Westminster College, Cambridge, March 9, 2011), 1–11.

104 Patrick Harries, "Roots of Ethnicity," *African Affairs* 87 (1988): 39–40.

105 C.G. Büttner, "Mission und Kolonien: Vortrag auf der sächsischen Missionskonferenz in Halle, 1885," *Allgemeine Missions-Zeitschrift* 12 (1885): 98–110.

106 Cameron, *Across Africa*, 118; MacDonald, *Africana*, 290; W. Claus, *Dr. Ludwig Krapf, weil. Missionar in Ostafrika* (Basel: C.F. Spittler, 1882), 119.

107 The British government bullied Zanzibar into signing a series of treaties creating increasing restrictions on the slave trade, while also justifying greater British policing of the Indian Ocean and the Sultanate's internal affairs. Sheriff, *Slaves, Spices & Ivory*, 201, 235. On the movement more generally, see Suzanne Miers, *Britain and the Ending of the Slave Trade* (New York: Africana, 1975); Amalia Ribi Forclaz, *Humanitarian Imperialism: The Politics of Anti-Slavery Activism, 1880–1940* (Oxford: Oxford University Press, 2015); Kevin Grant, *A Civilised Savagery: Britain and the New Slaveries in Africa* (New York: Routledge, 2004); Frederick Cooper,

"Conditions Analogous to Slavery: Imperialism and Free Labor Ideology in Africa," in *Beyond Slavery: Explorations of Race, Labor and Citizenship in Postemancipation Societies*, ed. Cooper, Thomas C. Holt, and Rebecca J. Scott (Chapel Hill: University of North Carolina Press, 2000), 127.

108 Richard Allen has shown that European ships carried around 200,000 African slaves on the Indian Ocean between 1800 and 1850 alone. Richard B. Allen, *European Slave Trading in the Indian Ocean, 1500–1850* (Athens: Ohio University Press, 2014), 16–19.

109 Jan-Georg Deutsch, *Emancipation without Abolition in German East Africa c. 1884–1914* (Oxford: James Currey, 2006); Martin Klein, *Slavery and Colonial Rule in French West Africa* (Cambridge: Cambridge University Press, 1998).

110 Abdul Sheriff, *Slaves, Spices & Ivory in Zanzibar: Integration of an East African Commercial Empire into the World Economy, 1770–1873* (London: James Currey, 1987), 228–30; Cooper, *Plantation Slavery on the East Coast of Africa* (New Haven: Yale University Press, 1977).

111 Paul E. Lovejoy, *Transformations in Slavery: A History of Slavery in Africa*, 3rd ed. (New York: Cambridge University Press, 2012), 140; Joseph C. Miller, *Way of Death: Merchant Capitalism and the Angolan Slave Trade 1730–1830* (Madison: University of Wisconsin Press, 1988), 141.

112 Lovejoy, *Transformations in Slavery*, 150.

113 H. Hd., "Quer durch Afrika. (Skizze des 'Berliner Tageblatts')," *Berliner Tageblatt*, November 24, 1882, 5.

3 The Introduction of a Land-Based Approach to Colonization, 1884–1885

1 Ostafrika was the fourth and final of Germany's long-term African colonies to be founded. The DOAG *Schutzbrief* followed Bismarck's recognition of Southwest Africa, Togo, and Cameroon in 1884.

2 The most prominent theories about Bismarck's change of heart are theories of social imperialism (that Bismarck wanted to divert socialist sentiment overseas) and that he wanted to draw Germany and France closer together against Britain. A good summary of the prevailing theories can be found in Sebastian Conrad, *German Colonialism: A Short History*, trans. Sorcha O'Hagan (Cambridge: Cambridge University Press, 2012), 21–7. Ideas of social imperialism come primarily from Hans-Ulrich Wehler, *Bismarck und der Imperialismus* (Cologne: Kiepenheuer & Witsch, 1969). Matthew Fitzpatrick argues that German liberals pressured Bismarck to pursue overseas empire as a means of globalizing German trade. Matthew Fitzpatrick, *Liberal Imperialism in Germany: Expansionism and Nationalism, 1848–1884* (New York: Berghahn, 2008), 116–31.

3 Mary Louise Pratt, *Imperial Eyes: Travel Writing and Transculturation* (London: Routledge, 1992), 7, describes earlier travelogues as "anti-conquest," the establishment of control through totalizing frameworks of knowledge rather than violence.

4 Michael Pesek and Jan-Georg Deutsch have discussed the treaties' function in creating an imagined precolonial political reality out of an East African political environment for European audiences that bore little resemblance to the reality of East African politics. Michael Pesek, *Koloniale Herrschaft in Deutsch-Ostafrika: Expeditionen, Militär und Verwaltung seit 1880* (Frankfurt am Main: Campus, 2005); Jan-Georg Deutsch, "Inventing an East African Empire: The Zanzibar Delimitation Commission of 1885/1886," in *Studien zur Geschichte des deutschen Kolonialismus in Afrika: Festschrift zum 60. Geburtstag von Peter Sebald*, ed. Peter Heine and Ulrich van der Heyden (Pfaffenweiler: Centaurus, 1995), 210; John Iliffe, *A Modern History of Tanganyika* (Cambridge: Cambridge University Press, 1979); Fritz Ferdinand Müller, *Deutschland-Zanzibar-Ostafrika: Geschichte einer deutschen Kolonialeroberung 1884–1890* (Berlin: Rütten & Loening, 1959).

5 Steven Press, *Rogue Empires: Contracts and Conmen in Europe's Scramble for Africa* (Cambridge: Harvard University Press, 2017).

6 Allen Roberts, *Dance of Assassins: Performing Early Colonial Hegemony in the Congo* (Bloomington: Indiana University Press, 2013), 33.

7 J. Wagner, *Deutsch-Ostafrika: Geschichte der Gesellschaft für deutsche Kolonisation und der Deutsch-Ostafrikanischen Gesellschaft nach den amtlichen Quellen* (Berlin: Verlag der Engelhardt'schen Landkartenhandlung, 1886), 22.

8 Fitzpatrick, *Liberal Imperialism in Germany*, 109–11, 153; Michael Schubert, *Der schwarze Fremde: Das Bild des Schwarzafrikaners in der parlamentarischen und publizistischen Kolonialdiskussion in Deutschland von der 1870er bis in die 1930er Jahre* (Stuttgart: Steiner, 2003); 89, 102; Woodruff Smith, *The German Colonial Empire* (Chapel Hill: University of North Carolina Press, 1978), 8–13, 19; Horst Gründer, *Die Geschichte der deutschen Kolonien* (Paderborn: Ferdinand Schöningh, 1985), 42; Mary Evelyn Townsend, *The Rise and Fall of Germany's Colonial Empire, 1884–1918* (New York: Howard Fertig, 1966), 82–4.

9 Carl Peters, "Alltagspolemik und Kolonialpolitik," in *Gesammelte Schriften*, vol. 1 (Munich and Berlin: C.H. Beck, 1943), 340.

10 Peters, *Gesammelte Schriften*, vol. 1 (Munich and Berlin: C.H. Beck, 1943), 19.

11 Pfeil, *Zur Erwerbung*, 47.

12 Pfeil, *Zur Erwerbung*, 31.

13 Peters, *Die Gründung von Deutsch-Ostafrika* (Berlin: C.A. Schwetschke und Sohn, 1906), 55.

14 Pfeil, *Zur Erwerbung*, 68.

15 Peters, *Die Gründung*; Pesek, *Koloniale Herrschaft*; Peters, "Die Usagara-Expedition," in *Gesammelte Schriften*, vol. 1, 290–1.

16 Peters, "Die Usagara-Expedition," 292–3, 295, 299, 306–8, 311; *Die Gründung*, 69.

17 Michael Pesek, "Eine Gründungsszene des deutschen Kolonialismus – Peters' Expedition nach Usagara, 1884," in *Die (koloniale) Begegnung: AfrikanerInnen in Deutschland 1880–1945, Deutsche in Afrika 1880–1918*, ed. Marianne Bechhaus-Gerst and Reinhard Klein-Arendt (Frankfurt am Main: Peter Lang, 2003), 261; Peters, "Die Usagara-Expedition," 37–8.

18 Peters, "Die Usagara-Expedition," 301, 317–18.

19 Press, *Rogue Empires*; John C. Weaver, *The Great Land Rush and the Making of the Modern World, 1650–1900* (Montreal: McGill-Queen's University Press, 2003).

20 Amy S. Greenberg, *Manifest Manhood and the Antebellum American Empire* (Cambridge: Cambridge University Press, 2005), 3.

21 Peters, *Gesammelte Schriften*, vol. 1, 64; Greenberg, *Manifest Manhood*, 3, 61. Alexander von Humboldt constructed the Americas for Americans as well as Europeans.

22 Greenberg, *Manifest Manhood*, 45–6, 61–2, 70, 87.

23 Before beginning his colonial work, Peters had written a thesis (*Promotion*) on the 1177 Treaty of Venice between the Holy Roman Emperor Friedrich I, and Pope Alexander III and his allies.

24 I will use "sultans" without the quotes to denote the figures Peters labelled as such for the sake of brevity, with an understanding that they would not have used the title themselves.

25 Wagner, *Deutsch-Ostafrika*, 51–6; Peters, "Die Usagara-Expedition," 303.

26 This investigation, a joint German-British-French commission to delimit Zanzibar's borders, will be discussed in chapter 4.

27 Wagner, *Deutsch-Ostafrika*, 53.

28 Peters, *Untersuchung zur Geschichte des Friedens von Venedig* (Hannover: Hahn'sche Buchhandlung, 1879).

29 Narratives of European progress from medieval sovereignties to modern were important to European claims to authority in colonial conquests from the sixteenth century. See Kathleen Davis, *Periodization & Sovereignty: How Ideas of Feudalism & Secularization Govern the Politics of Time* (Philadelphia: University of Pennsylvania Press, 2008); Kathleen Biddick, *The Shock of Medievalism* (Durham: Duke University Press, 1998).

30 Wagner, *Deutsch-Ostafrika*, 51–6; Peters, "Die Usagara-Expedition," 303.

31 Peters, "Die Usagara-Expedition," 302, 304; Wagner, *Deutsch-Ostafrika*, 43; Pfeil, *Zur Erwerbung*, 70–1; Harald Sippel, "Recht und Herrschaft in kolonialer Frühzeit: Die Rechtsverhältnisse in den Schutzgebieten der Deutsh-Ostafrikanishen Gesellschaft (1885–1890)," in Heine and an der Heyden, eds., *Studien zur Geschichte des deutschen Kolonialismus*, 467; Michael Pesek has described Peters's methods as "mimicry" of an imagined African potentate. Pesek, "Eine Gründungsszene des deutschen

Kolonialismus," 257–8; Allen F. Roberts, *A Dance of Assassins: Performing Early Colonial Hegemony in the Congo* (Bloomington: Indiana University Press, 2013), 7, details the inability to "hear" between Europeans and Africans in the colonial encounter. Richard White's "middle ground" is a process by which two sides in an intercultural encounter attempted to "justify their own actions in terms of what they perceived to be their partner's cultural premises." Richard White, *The Middle Ground: Indians, Empires, and Republics in the Great Lakes Region, 1650–1815*, 20th anniversary ed. (New York: Cambridge University Press, 2011), xii.

32 Peters, "Die Usagara-Expedition," 306.

33 A. Altenberg, "Dr. G.A. Fischer, der rechte Arzt zur rechten Zeit," *Deutsche Kolonialzeitung*, May 15, 1885, 310–12; "Koloniale Aufgaben," *Deutsche Kolonialzeitung*, September 1, 1885, 538–50.

34 Pfeil, *Zur Erwerbung*, 74–5; Peters, "Deutsche Gegner der Deutsch-Ostafrikanischen Gesellschaft," in *Gesammelte Schriften*, vol. 1, 342–5.

35 Wagner, *Deutsch-Ostafrika*, 59.

36 Wagner, *Deutsch-Ostafrika*, 59; Peters, *Die Gründung*, 76–7.

37 No title, *National Zeitung*, May 16, 1885, BArch R 8023/265, pag. 15.

38 "Redaktionelle Korrespondenz," *Deutsche Kolonialzeitung*, September 15, 1885, 593–4.

39 Bruno Kurtze, *Die Deutsch-Ostafrikanische Gesellschaft: Ein Beitrag zum Problem der Schutzbriefgesellschaften und zur Geschichte Deutsch-Ostafrikas* (Jena: Gustav Fischer, 1913), 45–6; Antheilsschein der Deutsch-ostafrikanischen Gesellschaft, BArch R 8124/1, pag. 10.

40 Kurtze, *Die Deutsch-Ostafrikanische Gesellschaft*, 10–12, 40–4; "Erster Schutzbrief für Deutsch-Ostafrika vom 27. February 1885," accessed 16 March 2014, http://www.deutsche-schutzgebiete.de/ostafrika-schutzvertrag.htm.

41 Kurtze, *Die Deutsch-Ostafrikanische Gesellschaft*, 50–1.

42 Kurtze, *Die Deutsch-Ostafrikanische Gesellschaft*, 52–4; Ernst Vohsen to Gerhard Rohlfs, no date, BArch R 8023/265, pag. 104; Herbert Bismarck to Freiherr von Plessen, August 26, 1885, BArch R 8124/1, pag. 35.

43 Wagner, *Deutsch-Ostafrika*, 41, 50; Peters, "Die Usagara-Expedition," 305; *Frankfurter Zeitung*, August 1885, BArch R 8023/265, pag. 26.

44 Friedrich Back, "Die Unterbringung befreiter Sklaven," *Deutsche Kolonialzeitung*, February 15, 1890, 46; GfdK treaty with Quafungo, PA AA 372/537–9.

45 Wagner, *Deutsch-Ostafrika*, 53, 56; GfdK treaties, PA AA London 372/537–77.

46 No title, *National Zeitung*, August 6, 1885, BArch R 8023/265, pag. 29; GfdK treaties, PA AA London 372/537–77.

47 Steven Feierman, *The Shambaa Kingdom: A History* (Madison: University of Wisconsin Press, 1974).

48 Wagner, *Deutsch-Ostafrika*, 93; Said Barghash to Gerhard Rohlfs, June 24, 1885, PA AA London 372/536.

49 No title, *Times*, October 14, 1885, BArch R 8023/265, pag. 55, No title, *Weser-Zeitung*, November 26, 1885, BArch R 8023/265, pag. 62.

50 Ernst Vohsen to Gerhard Rohlfs, no date, BArch R 8023/265, pag. 103–7; Pfeil, *Zur Erwerbung*, 121–2.

51 No title, *National Zeitung*, May 16, 1885, BArch R 8023/265, pag. 14, No title, *Deutsche Tageblatt*, August 6, 1885, BArch R 8023/265, pag. 31.

52 Peters, *Die Gründung*, 76; Peters, "Die neuesten Erwerbungen der Deutsch-Ostafrikanischen Gesellschaft und die Presse," *Gesammelte Schriften*, vol. 1, 348.

53 Ernst Vohsen to Gerhard Rohlfs, no date, BArch R 8023/265, pag. 107–12.

54 Aylward Shorter, "Nyungu-Ya-Mawe and the 'Empire of the Ruga-Ruga," *Journal of African History* 9, no. 2 (1968): 243.

55 Iliffe, *Modern History of Tanganyika*, 62.

56 Norman R. Bennett, "Isike, *Ntemi* of Unyanyembe," in *African Dimensions: Essays in Honor of William O. Brown*, ed. Mark Karp (Brookline: African Studies Center, Boston University, 1975), 67.

57 Pfeil, "Schwierigkeiten bei der Zusammenstellung einer Expedition," BArch N 2225/5, pag. 1.

58 Pfeil, "Wanderungen in Afrika," August 17, 1886, BArch N 2225/21, pag. 3.

59 Detail from Paul Engelhardt and I. von Wenzierski, "Karte von Central-Ost-Afrika" (Berlin: Engelhardtischen Landkartenhandlung, 1886).

60 Wagner, *Deutsch-Ostafrika*, 92–5; "Kolonialpolitsche Vorgänge," *Deutsche Kolonialzeitung*, September 1, 1885, 533; No title, *Berliner Börse Zeitung*, August 27, 1885, BArch R 8023/265, pag. 37.

61 No title, *National Zeitung*, April 11, 1885, BArch R 8023/265, pag. 11. The Sultans of Zanzibar appointed a *liwali*, an administrative official with responsibility for customs and civil affairs, in each coastal town. It is unclear whether the marines met with a real *liwali* or someone else they gave the title to or someone who gave the title to himself.

62 "Freundschafts-, Handels- und Schiffahrtsvertrag zwischen dem Deutschen Reich und dem Sultan von Zanzibar. 20 December 1885," *Reichs-Gesetzblatt* 28 (1886): 261–84.

4 Inventing *Hinterland*: Historiography and Cultural Geography in the DOAG's Takeover of the Indian Ocean Coast, 1886–1888

1 "Kiloas Besitzergreifung durch die Portugiesen. (1500 bis 1502)," *Deutsche Kolonialzeitung*, January 28, 1888, 28.

2 "Der portugiesische Vizekönig Don Franzisko d'Almeida, sowie die Augsburger Kaufleute Balthasar Sprenger und Hans Mayer im Jahr 1505 in Kiloa," *Deutsche Kolonialzeitung*, February 18, 1888, 52.

3 "Land und Leute in Kiloa im Jahre 1505. Nach der Schilderung des Balthasar Sprenger und Hans Mayr," *Deutsche Kolonialzeitung*, March 10, 1888, 74–6.

4 In the 1920s, both Frederick Lugard and M.F. Lindley credited the Hinterland Theory as the central framework for drawing African borders in the 1880s. Lindley, *The Acquisition and Government of Backward Territory in International Law* (London: Longmans, Green and Co., 1926), 32–4, 143–6; Frederick Lugard, *The Dual Mandate in British Tropical Africa* (Edinburgh: William Blackwood and Sons, 1922), 14–15. Among the works dating the Hinterland Theory to the Berlin Conference and describing it as a landmark in the European partition of Africa are G.N. Uzoigwe, "Spheres of Influence and the Doctrine of the Hinterland in the Partition of Africa," *Journal of African Studies* 3, no. 2 (1976): 194; S. Akweenda, *International Law and the Protection of Namibia's Territorial Integrity: Boundaries and Territorial Claims* (The Hague: Kluwer Law International, 1997), 8; Simon Katzenellenbogen, "It Didn't Happen at Berlin: Politics, Economics and Ignorance in the Setting of Africa's Colonial Boundaries," in *African Boundaries*, ed. Paul Nugent and A.I. Asiwaju, (London: Pinter, 1996), 21–2; H.L. Wesseling, *Divide and Rule: The Partition of Africa, 1880–1914*, trans. Arnold Pomerans (Westport: Praeger, 1996), 118.

5 The scholarship on the Indian Ocean World is too copious to discuss fully here. Notable syntheses include Michael Pearson, *Port Cities and Intruders: The Swahili Coast, India, and Portugal in the Early Modern Era* (Baltimore: Johns Hopkins University Press, 1998); Pearson, *The Indian Ocean* (London: Routledge, 2003); Ashin Das Gupta, *India and the Indian Ocean World: Trade and Politics* (Oxford: Oxford University Press, 2004); Edward Alpers, *The Indian Ocean in World History* (Oxford: Oxford University Press, 2014); K.N. Chaudhuri, *Trade and Civilisation in the Indian Ocean: An Economic History from the Rise of Islam to 1750* (New York: Cambridge University Press, 2005); Abdul Sheriff, *Dhow Cultures of the Indian Ocean: Cosmopolitanism, Commerce and Islam* (New York: Columbia University Press, 2010); Milo Kearney, *The Indian Ocean in World History* (New York: Routledge, 2004); Satish Chandra, ed., *The Indian Ocean: Explorations in History, Commerce and Politics* (New Delhi: Sage, 1987); Denys Lombard and Jean Aubin, eds., *Marchands et hommes d'affaires asiatiques dans l'Océan Indien et la Mer de Chine 13e-20e siècles* (Paris: EHESS, 1988); Roderich Ptak and Dietmar Rothermund, eds., *Emporia, Commodities and Entrepreneurs in Asian Maritime Trade, c. 1400–1750* (Stuttgart: Franz Steiner, 1992); Anthony Reid, *Southeast Asia in the Age of Commerce, 1450–1680*, vol. 1, The Lands below the Winds (New Haven: Yale University Press, 1988).

6 The conference has been the subject of much scholarly discussion. For a useful overview, see M.E. Chamberlain, *The Scramble for Africa*, 3rd ed. (London: Routledge, 2010); for in-depth discussions of elements of the

conference, see Stig Förster, Wolfgang Mommsen, and Ronald Robinson, eds., *Bismarck, Europe and Africa: The Berlin Africa Conference, 1884–1885, and the Onset of Partition* (London: Oxford University Press, for the German Historical Institute, 1988).

7 Wesseling, *Divide and Rule*, 118; G.N. Uzoigwe, "Spheres of Influence and the Doctrine of the Hinterland in the Partition of Africa," *Journal of African Studies* 3, no. 2 (1976): 186; Simon Katzenellenbogen, "It Didn't Happen at Berlin: Politics, Economics and Ignorance in the Setting of Africa's Colonial Boundaries," in *African Boundaries*, ed. Nugent and Asiwaju, 21–2.

8 Daniel Boorstin, *The Americans: The National Experience* (New York: Knopf, 2010), 242–4.

9 For example, see Michael Geistbeck, *Grundzüge der Geographie für Mittelschulen sowie zum Selbstunterricht* (Munich: Oldenbourg, 1885), 174, http://gei-digital.gei.de/viewer/resolver?urn=urn%3Anbn%3Ade %3A0220-gd-8174959.

10 Said Barghash to Otto von Bismarck, May 12, 1885 (26 Rajab 1302), John Kirk to Earl Granville, July 6, 1885, in *Correspondence Relating to Zanzibar*, January 1886, in *Irish University Press Series of British Parliamentary Papers. The Zanzibar Papers 1841–98*. Colonies Africa 68 (Shannon: Irish University Press, 1971), 58.

11 Zanzibar Delimitation Commission Annex to procès-verbal No. 11, May 18, 1886, BArch R 1001/600, pag. 101–5.

12 Mohammed ben Salem to Zanzibar Delimitation Commission, May 17, 1886, BArch R 1001/600, pag. 98–100.

13 Carl Grimm, *Der wirthschaftliche Werth von Deutsch-Ostafrika* (Berlin: Walther & Apolant, 1886), 41, 57, 159.

14 Enclosure with Salisbury to Herbert Kitchener, October 17, 1885, in *Correspondence Relating to Zanzibar*, January 1886, in *The Zanzibar Papers 1841–98*, 98–9.

15 Kirk to Granville, July 6, 1885, in *Correspondence Relating to Zanzibar*, January 1886, in *The Zanzibar Papers 1841–98*, 67–9.

16 Salisbury to C.S. Scott, July 20, 1885, *Correspondence Relating to Zanzibar*, January 1886, in *The Zanzibar Papers 1841–98*, 63–4.

17 Declaration of the Sultans of Chagga and Kilimanjaro, May 30, 1885, *Correspondence Relating to Zanzibar*, January 1886, in *The Zanzibar Papers 1841–98*, 66.

18 Notiz betr. Die den Delegirten von Großbrit. Zu ertheilende Instruktion für den Zusammentritt der mit der Abgrenzung des Sultanats von Zanzibar auf der ostafrik. Küste zu betrauenden Kommission von Deutschland, Großbrit. u. Frankreich, September 21, 1885, BArch R 1001/596, pag. 41–3.

19 Gutachten über den S5 des englischen Instruktionsentwurfs aufgestellten Grundsatz des Völkerrecht, daß die Besitzergreifung des Küstengebiets

diej. des dahinter liegenden Flußgebiets in sich schließt, no date, BArch R 1001/596, pag. 39–40.

20 Georg Herbert zu Münster to Granville, May 5, 1885, in *Correspondence Relating to Zanzibar*, January 1886, in *The Zanzibar Papers 1841–98*, 38–9.

21 Otto von Bismarck to Münster, June 2, 1885, in *Correspondence Relating to Zanzibar*, January 1886, in *The Zanzibar Papers 1841–98*, 49.

22 Klaus von Anderten, "Die Galla-Länder," *Kolonial-Politische Korrespondenz*, August 14, 1886, 228–31.

23 Otto Schmidt-Leda to Gustav Krauel, February 15, 1886, BArch R 1001/598, pag. 65–7; Schmidt-Leda to Otto von Bismarck, February 13, 1886, BArch R 1001/598, pag. 5–7.

24 Schmidt-Leda to Otto von Bismarck, January 17, 1886, BArch R 1001/596, 126–31.

25 *Aide mémoire* to Paul von Hatzfeldt, June 2, 1886, BArch R 1001/600, pag. 3–4.

26 Schmidt-Leda to Otto von Bismarck, February 13, 1886, BArch R 1001/598, pag. 8–15.

27 Informations recueillies par la Commission pendant son voyage sur la côte de Zanzibar du 19 Janvier an 6 Février 1886, BArch R 1001/598, pag. 18–34.

28 Herbert Bismarck instructions to Schmidt-Leda, no date (September 1885), BArch R 1001/596, pag. 59–61.

29 Instructions to Schmidt-Leda, no date (August 1885), BArch R 1001/596, pag. 7–8.

30 Salisbury to Kitchener, October 17, 1885, in *Correspondence Relating to Zanzibar*, January 1886, in *The Zanzibar Papers 1841–98*, 97–8. This parallels contemporaneous discussions in the British Empire in Natal over whether chiefly authority was based in territory or people. David Welsh, *The Roots of Segregation; Native Policy in Colonial Natal, 1845–1910* (New York: Oxford University Press, 1971); Thomas V. McClendon, *White Chief, Black Lords: Shepstone and the Colonial State in Natal, South Africa, 1845–1878* (Rochester: University of Rochester Press, 2010).

31 Schmidt-Leda to Otto von Bismarck, March 12, 1886, BArch R 1001/598, pag. 127–40.

32 Zanzibar Delimitation Commission, Procès-Verbal No. 6, Séance du 15 Février 1886, BArch R 1001/598, pag. 141–60.

33 Zanzibar Delimitation Commission, Procès-Verbal No. 6, Séance du 15 Février 1886, BArch R 1001/598, pag. 141–60.

34 Informations recueillies par la Commission pendant son voyage sur la côte de Zanzibar du 19 janvier à 6 février 1886, BArch R 1001/598, pag. 18–34.

35 Zanzibar Delimitation Commission, Procès-Verbal No. 6, Séance du 15 Février 1886, BArch R 1001/598, pag. 141–60.

36 "Kolonialpolitische Vorgänge und geographische Erforschungen. Die ostafrikanische Grenzregulierung," *Deutsche Kolonialzeitung* 4, no. 1 (1887): 4.

37 As Jan-Georg Deutsch has argued, the delimitation commission invented "a particular concept of Zanzibari sovereignty … in order to afford the British and German occupation of the coastal areas of the East African mainland an appearance of legality and legitimacy." Jan-Georg Deutsch, "Inventing an East African Empire," 210–19.

38 Gregory Bracken, "Treaty Ports in China: Their Genesis, Development, and Influence," *Journal of Urban History* 45, no. 1 (2019): 168–76.

39 Foreign Office to Otto Arendt, November 18, 1886, BArch R 1001/605, pag. 17–19.

40 Ibid. The United Kingdom also benefited from the agreement. It received the rights to trade at two ports, Mombasa and Tana, the eventual basis for its Kenyan colony.

41 Gerhard Rohlfs, "Die Vorgänge in Ostafrika," *Kölnische Zeitung*, September 30, 1888, BArch R 1001/688, pag. 64.

42 "Geographischer Monatsbericht. Afrika. Äquatoriale Gebiete," *Dr. A. Petermanns Mitteilungen aus Justus Perthes' geographischer Anstalt* (1887): 58.

43 "Sansibar und der benachbarte Teil von Deutsch-Ostafrika," *Meyers Konversations-Lexicon*, 4th ed., vol. 14, 300a, https://commons.wikimedia .org/wiki/Category:Old_maps_of_Tanzania#/media/File:Meyers_b14 _s0300a.jpg.

44 "Kolonialpolitische Vorgänge und geographische Erforschungen. Erweiterung des Stationennetzes der Deutsch-Ostafrikanischen Gesellschaft," *Deutsche Kolonialzeitung* 4, no. 3 (1887): 71.

45 Foreign Office report, no date (1887), PA AA London 376.142; "Kolonialpolitische Vorgänge und geographische Erforschungen. Neue Erwerbungen der Deutsch-Ostafrikanischen Gesellschaft und deren Stationen," *Deutsche Kolonialzeitung* 3, no. 23 (1886): 786.

46 Foreign Office report, no date (1887), PA AA London 376.147. "Dampfersubvention für Ostafrika," *DKZ*, October 1, 1887. 575–76.

47 Peters, *Die Gründung von Deutsch-Ostafrika*, 179–80, 188. Peters lamented later that the plan had fallen apart because German capital had focused on Mexico and Brazil instead of East Africa.

48 Marcia Wright, *German Missions in Tanganyika, 1891–1941: Lutherans and Moravians in the Southern Highlands* (Oxford: Clarendon, 1971), 38–9.

49 Michael Pesek, *Koloniale Herrschaft in Deutsch-Ostafrika: Expeditionen, Militär und Verwaltung seit 1880* (Frankfurt am Main: Campus, 2005), 147–8.

50 Peters to DOAG, June 3,1887, printed in Peters, *Die Gründung von Deutsch-Ostafrika*, 202–3.

51 The German gold mark was pegged at 20.43 to the British pound. Conversion to today's money; via https://www.measuringworth.com /ppoweruk/.

52 Joachim von Pfeil speech at Allgemeiner Deutscher Kongress zur
   Förderung überseeischer Interessen, August 13, 1886, BArch N 2225/7,
   pag. 9–10.

53 F.M. Zahn, "Die Mission auf dem allgemeinen deutschen Kongreß zur
   Förderung überseeischer Interessen," *Allgemeine Missions-Zeitschrift,*
   1887, 31.

54 M. Ittameier, *Die evangelisch-lutherische Mission in Ostafrika in den ersten zwei
   Jahren ihres Bestehens* (Rothenburg: J.P. Peters, 1888), 13, 17.

55 E. Reichel, "Was haben wir zu thun, damit die deutsche Kolonialpolitik
   nicht zur Schädigung, sondern zur Förderung der Mission ausschlage?,"
   Allgemeine Missions-Zeitschrift (January 1886): 46; Dr. Schreiber, "Besetzung
   deutscher Kolonien mit deutschen Missionaren," *Allgemeine Missions-
   Zeitschrift* (January 1886): 56–8.

56 Wilhelm Hübbe-Schleiden, "Deutsche Kultivation Afrikas," *Deutsche
   Kolonialzeitung,* August 1, 1887, 459–61.

57 As reprinted in "Modernste Missionsgeschichtschreibung," *Allgemeine
   Missions-Zeitschrift* (July 1886): 297–311.

58 "Aus unserer Gesellschaft," *Nachrichten aus der ostafrikanischen Mission,*
   April 1888, 51–2.

59 Friedrich Ratzel, "Eine neue Spezialkarte von Afrika," *Dr. A. Petermanns
   Mitteilungen aus Justus Perthes' geographischer Anstalt,* 1886, 161–2.

60 "Ein Ausflug nach Dar-es-Salaam," *Deutsche Kolonialzeitung* 3, no. 1
   (1886): 8–11.

61 Foreign Office report, no date (1887), PA AA London 376.149.

62 DOAG report to Foreign Office, November 1886, BArch R 8023/265,
   pag. 114.

63 Foreign Office memorandum, no date (1887), PA AA London 376.149.

64 Bericht über die Deutsch-Ostafrikanische Gesellschaft, January 27, 1887,
   BArch R 8023/265, pag. 126–7.

65 Peters to DOAG, printed in Peters, *Die Gründung von Deutsch-Ostafrika,* 215.

66 Peters, Berichterstattung über meine Wirksamkeit in Ostafrika im Jahre
   1887, March 26, 1888, BArch R 8124/2, 106–7.

67 Carl Grimm, *Die Pharaonen in Ostafrika. Eine kolonialpolitische Studie*
   (Karlsruhe: Grimm, 1887).

68 Grimm, *Die Pharaonen in Ostafrika,* 175–9.

69 Georg G. Iggers, *The German Conception of History: The National Tradition
   of Historical Thought from Herder to the Present* (Middletown: Wesleyan
   University Press, 1968); Wolfgang J. Mommsen, ed., *Leopold von Ranke und
   die moderne Geschichtswissenschaft* (Stuttgart: Klett-Cotta, 1988); Theodore
   Ziolkowski, *Clio the Romantic Muse: Historicizing the Faculties in Germany*
   (Ithaca: Cornell University Press, 2004); Benedikt Stuchtey and Peter
   Wende, eds., *British and German Historiography 1750–1950: Traditions,*

Perceptions and Transfers (Oxford: Oxford University Press, 2000); Bonnie G. Smith, *The Gender of History: Men, Women, and Historical Practice* (Cambridge: Harvard University Press, 1998).

70 By the 1880s, many German historians had adopted a consciously political tone, to the point that several were elected to the Reichstag or regional legislatures. Stefan Berger, *The Search for Normality: National Identity and Historical Consciousness in Germany since 1800* (New York: Berghahn, 2007), 29–34; Iggers, *German Conception of History*.

71 Suzanne L. Marchand, *Down from Olympus: Archaeology and Philhellenism in Germany, 1750–1970* (Princeton: Princeton University Press, 2003).

72 Contrast the German idea of reliable history with the African "homespun historians" discussed in Derek R. Peterson and Giacomo Macola, eds., *Recasting the Past: History Writing and Political Work in Modern Africa* (Athens: Ohio University Press, 2009), who sought out alternative sources of evidence.

73 "Wandlungen," *Deutsche Kolonialzeitung*, February 18, 1888, 49–50.

74 John Hanning Speke, *Journal of the Discovery of the Source of the Nile* (London: William Blackwood & Sons, 1863). See Mahmood Mamdani, *When Victims Become Killers* (Princeton: Princeton University Press, 2001) on the Hamitic hypothesis' use in colonial rule.

75 "Europas Aufgaben und Aussichten im tropischen Afrika. Vortrag des Prof. Dr. Schweinfurth in der 59. Naturforscher- und Ärzte-Versammlung," *Deutsche Kolonialzeitung* 3, no. 20 (1886): 695–702.

76 "Die vierte ordentliche Generalversammlung des Deutschen Kolonialvereins zu Dresden am 6. Und 7. Mai 1887," *Deutsche Kolonialzeitung*, June 1, 1887, 345.

77 DOAG report, November 1886, BArch R 8023/265, pag. 114.

78 "Die Veränderung der politischen Lage in Ostafrika," *Nachrichten aus der ostafrikanischen Mission*, August 1888, 115–16.

79 Hermann Wissmann, *My Second Journey through Equatorial Africa. From the Congo to the Zambesi, in the Years 1886 and 1887*, trans. Minna J.A. Bergmann (London: Chatto & Windus, 1891), frontispiece.

80 Wissmann, *My Second Journey*, 186ff.

81 Wissmann, *My Second Journey*, 202ff.

82 Wissmann, *My Second Journey*, 244ff.

83 Tim Jeal, *Livingstone*, revised ed. (New Haven: Yale University Press, 2013).

84 Edward Berenson, *Heroes of Empire: Five Charismatic Men and the Conquest of Africa* (Berkeley: University of California Press, 2011), 83, 112–15.

85 Daniel Laqua, "The Tensions of Internationalism: Transnational Anti-Slavery in the 1880s and 1890s," *The International History Review* 33, no. 4 (December 2011): 705–26; Suzanne Miers, *Britain and the Ending of the Slave Trade* (New York: Africana, 1975); Amalia Ribi Forclaz, *Humanitarian*

Imperialism: The Politics of Anti-Slavery Activism, 1880–1940 (Oxford: Oxford University Press, 2015).

86 "Kiloas Besitzergreifung durch die Portugiesen. (1500 bis 1502)," *Deutsche Kolonialzeitung*, January 28, 1888, 28; "Der portugiesische Vizekönig Don Franzisko d'Almeida, sowie die Augsburger Kaufleute Balthasar Sprenger und Hans Mayer im Jahr 1505 in Kiloa," *Deutsche Kolonialzeitung*, February 18, 1888, 52.

87 "Land und Leute in Kiloa im Jahre 1505. Nach der Schilderung des Balthasar Sprenger und Hans Mayr," *Deutsche Kolonialzeitung*, March 10, 1888, 74–6.

88 Harald Sippel, "Recht und Herrschaft in kolonialer Frühzeit," in *Studien zur Geschichte des deutschen Kolonialismus in Afrika*, ed. Heine and van der Heyden, 474.

89 Charles Euan-Smith to Salisbury, June 1, 1888, in *The Zanzibar Papers 1841–98*, 207–9.

90 Preliminary treaty between DOAG and Muhamed bin Salim, July 30, 1887, BArch R 8124/2, pag. 46.

91 Preliminary treaty between DOAG and Muhamed bin Salim, July 30, 1887, BArch R 8124/2, pag. 43–4.

92 Preliminary treaty between DOAG and Muhamed bin Salim, July 30, 1887, BArch R 8124/2, pag. 44.

93 Peters, *Die Gründung von Deutsch-Ostafrika*, 163, 170.

94 Otto Arendt, "Die Übernahme der Küstenverwaltung in Ostafrika durch die Deutsch-Ostafrikanische Gesellschaft," *Deutsche Kolonialzeitung*, August 25, 1888, 265–6.

5 The Emin Pasha Expedition, New Forms of Political Mobilization, and Reimagining East African Space, 1888–1890

1 While the British expedition, led by Henry Morton Stanley, has been the subject of several books, the German expedition's importance has been overlooked because of its failure. Books on the Stanley expedition include Alan Caillou, *South from Khartoum: The Story of Emin Pasha* (New York: Hawthorn, 1974); Iain Smith, *The Emin Pasha Relief Expedition, 1886–1890* (Oxford: Clarendon, 1972); Roger Jones, *The Rescue of Emin Pasha: The Story of Henry M. Stanley and the Emin Pasha Relief Expedition, 1887–1889* (London: Allison and Busby, 1972); Daniel Liebowitz and Charles Pearson, *The Last Expedition: Stanley's Mad Journey through the Congo* (New York: Norton, 2005); Olivia Manning, *The Reluctant Rescue: The Story of Stanley's Rescue of Emin Pasha from Equatorial Africa* (Garden City: Doubleday, 1947); as well as parts of Tim Jeal, *Stanley: The Impossible Life of Africa's Greatest Explorer* (London: Faber and Faber, 2007).

2 James Retallack, *The German Right, 1860–1920* (Toronto: University of Toronto Press, 2006); Heinz Hagenlücke, *Deutsche Vaterlandspartei: Die nationale Rechte am Ende des Kaiserreiches* (Düsseldorf: Droste, 1997); Geoff Eley, *Reshaping the German Right: Radical Nationalism and Political Change after Bismarck* (New Haven: Yale University Press, 1980); Roger Chickering, *We Men Who Feel Most German: A Cultural Study of the Pan-German League* (Boston: Allen & Unwin, 1984); Thomas Rohkrämer, *A Single Communal Faith? The German Right from Conservatism to National Socialism* (New York: Berghahn, 2007); Rainer Hering, *Konstruierte Nation: Der Alldeutsche Verband 1890 bis 1939* (Hamburg: Hans Christians, 2003); Uwe Puschner, *Die völkische Bewegung im wilhelminischen Kaiserreich. Sprache – Rasse – Religion* (Darmstadt: Wissenschaftliche Buchgesellschaft, 2001); Peter Walkenhorst, *Nation – Volk – Rasse. Radikaler Nationalismus im Deutschen Kaiserreich 1890–1914* (Göttingen: Vandenhoeck & Ruprecht, 2007).

3 I follow Sebastian Conrad in *Globalisation and the Nation in Imperial Germany*, trans. Sorcha O'Hagan (Cambridge: Cambridge University Press, 2010), as well as Peter Walkenhorst in *Nation – Volk – Rasse*, and Dennis Sweeney in "Pan-German Conceptions of Colonial Empire," in *German Colonialism in a Global Age*, ed. Bradley Naranch and Eley (Durham: Duke University Press, 2014), 265–82, that the encounter with the global was central to the formation of German nationalism in this period. Hagenlücke, *Deutsche Vaterlandspartei* argues that the Right developed in a top-down process directed by elites; Chickering, *We Men Who Feel Most German* argues that it was primarily Germany's respectable middle classes who participated; Eley, *Reshaping the German Right*, maintains that elites did not manage or manipulate events, but that the Right only really developed after the turn of the century. I argue that the new methods of right-wing populist mobilization developed through the interaction of different levels of German society.

4 Georg Schweitzer, *Emin Pascha. Eine Darstellung seines Lebens und Wirkens mit Benutzung seiner Tagebücher, Briefe und wissentschaftlichen Aufzeichnungen* (Berlin: Walther, 1898), 1. Oppeln is now Opole, Poland.

5 Schweitzer, *Emin Pascha*, 33.

6 Schnitzer was just one of many Germans, both academic and practical, who devoted themselves to Orientalist study. See, for example, Todd Kontje, *German Orientalisms* (Ann Arbor: University of Michigan Press, 2004); Ursula Wokoeck, *German Orientalism: The Study of the Middle East and Islam from 1800 to 1945* (London: Routledge, 2009); Suzanne Marchand, *German Orientalism in the Age of Empire: Religion, Race, and Scholarship* (Washington: German Historical Institute, 2009); Nicholas Germana, *The Orient of Europe: The Mythical Image of India and Competing Images of German National Identity* (Newcastle upon Tyne: Cambridge Scholars Publishing, 2009).

7 Schweitzer, *Emin Pascha*, 84; Alice Moore-Harrell, *Egypt's African Empire: Samuel Baker, Charles Gordon & the Creation of Equatoria* (Brighton: Sussex Academic Press, 2012), 99.

8 Schweitzer, *Emin Pascha*, 99–100, 110.

9 Moore-Harrell, *Egypt's African Empire*, 12, 76, 133–4, 181–2.

10 Schweitzer, *Emin Pascha*, 110, 145, 146.

11 Schweitzer, *Emin Pascha*, 235.

12 Lytton Strachey, *Eminent Victorians: Cardinal Manning, Florence Nightingale, Dr. Arnold, General Gordon* (London: Continuum, 2002). "Die Aufsuchung der Reisenden Junker, Schnitzler und Casati," *Deutsche Kolonialzeitung* 3, no. 14 (1886): 427.

13 "Dr. Junker," *Neueste Mittheilungen*, March 17, 1887, 3.

14 Wilhelm Junker, *Reisen in Afrika, 1875–1886*, vol. 3 (1882–1886) (Vienna: Eduard Hölzel, 1891), 347, 351, 354, 448, 365.

15 Schweitzer, *Emin Pascha*, 388–9.

16 Liebowitz and Pearson, *The Last Expedition*, 16–17.

17 August Leue to Emin Pasha, March 5, 1891, BArch N 2063/8, pag. 5.

18 Karl von den Sthiren to Emin Pasha, April 25, 1890, BArch N 2063/8, pag. 4.

19 Otto Schmidt-Leda to Otto von Bismarck, December 6, 1886, PA AA London 375.439.

20 No title, *Standard*, December 12, 1888, BArch R 1001/713, pag. 23.

21 "Mr. Stanley's Expedition," *Times of London*, February 9, 1887, in BArch R 1001/247, pag. 28.

22 Liebowitz and Pearson, *The Last Expedition*.

23 "Eine deutsche Emin Pascha-Expedition," *Kölnische Zeitung*, September 20, 1888, BArch R 1001/249c, pag. 81. Stanley tended to shed members of the expedition whenever they or their loads began to slow it down, so he was left with a skeleton crew when he did finally find Emin in April 1888, over a year after he had landed on the west coast. His "rescue force" was so small by that point, however, that Stanley decided to turn back and try to bring more men and supplies forward so as not to embarrass himself when he met Emin. When the two did finally meet, on 29 April 1888, Stanley's forces were too weak to bring Emin out to Indian Ocean coast.

24 "Kleine Mitteilungen," *Deutsche Kolonialzeitung*, August 11, 1888, 255.

25 DOAG to Otto von Bismarck, June 10, 1887, BArch R 1001/247, pag. 57.

26 Herbert Bismarck to Paul von Hatzfeldt, June 18, 1887, BArch R 1001/607, pag. 21–4; Edward Malet to Herbert Bismarck, July 8, 1887, BArch R 1001/247, pag. 67–71.

27 Lecture by Postdirektor Sachse on the Emin Pasha Expedition, Wiesbaden, September 11, 1888, BArch R 1001/249c, pag. 85.

28 "Antrag der Abteilung Nürnberg," *Deutsche Kolonialzeitung*, September 25, 1888, 312–16.

29 Carl Peters, *New Light on Dark Africa: Being the Narrative of the German Emin Pasha Expedition*, trans. H.W. Dulcken (London: Ward, Lock, and Co., 1891), 3. I quote from the English translation of the travelogue, as it captures the flowery prose of the late nineteenth century. The translation is mostly true to the original; Peters was fluent in English and had lived in the UK, so could review the text.

30 Emin Pasha Committee to Vorstand der Abtheilung der Deutschen Kolonialgesellschaft, September 23, 1888, BArch R 8023/852, 24–5.

31 Peters, *New Light in Dark Africa*, 4–5.

32 Schreiben des Dr. Peters v. 7 Juli 1888 betr. Bildung eines Comité's für die deutsche Expedition zum Entsatze von Emin Pascha, BArch R 1001/249c, pag. 5–6.

33 Bokemeyer to Emin Pasha Committee, September 28, 1888, BArch R 8023/852, pag. 33.

34 Emin Pasha Committee to Otto von Bismarck, July 20, 1888. BArch R 1001/249c, pag. 8–9.

35 No title, *Kölnische Zeitung*, August 28, 1889, in BArch R 1001/250, pag. 90.

36 Matthias Ittameier to Otto von Bismarck, August 23, 1888, BArch R 1001/607, pag. 79–86.

37 Carl Peters, *Gesammelte Schriften*, vol. 2, 514.

38 Schweitzer, *Emin Pascha*, 501.

39 No title, *Militär-Wochenblatt*, November 14, 1888, in BArch R 1001/249d, pag. 57.

40 "Kreuz und Halbmond in Central-Afrika," *Berliner Tageblatt*, January 26, 1889, 1–2.

41 Frederick Cooper, "Conditions Analogous to Slavery: Imperialism and Free Labor Ideology in Africa," in *Beyond Slavery: Explorations of Race, Labor and Citizenship in Postemancipation Societies*, ed. Cooper, Thomas C. Holt, and Rebecca J. Scott (Chapel Hill: University of North Carolina Press, 2000), 127.

42 William Gervase Clarence-Smith, "The British 'Official Mind' and Nineteenth-Century Islamic Debates over the Abolition of Slavery," in *Slavery, Diplomacy and Empire: Britain and the Suppression of the Slave Trade, 1807–1975*, ed. Keith Hamilton and Patrick Salmon (Brighton: Sussex Academic Press, 2013), 125.

43 Norman Robert Bennett, *Arab versus European: Diplomacy and War in Nineteenth-Century East Central Africa* (New York: Africana Publishing Company, 1986), 12–15.

44 Daniel Laqua, *The Age of Internationalism and Belgium, 1880–1930: Peace, Progress and Prestige* (Manchester: University of Manchester Press, 2013), 54.

45 Jean Bodin, *Six Books of the Commonwealth*, trans. M.J. Tooley (Oxford: Basil Blackwell, 1955), accessed June 23, 2016, http://www.constitution .org/bodin/bodin.txt; Henry Heller, "Bodin on Slavery and Primitive Accumulation," *The Sixteenth Century Journal* 25, no. 1 (1994): 53–65;

Kathleen Davis, *Periodization and Sovereignty: How Ideas of Feudalism and Secularization Govern the Politics of Time* (Philadelphia: University of Pennsylvania Press, 2017).

46 "Die Colonial- und die Sklavenjagd-Gebiete von Centralafrika," *Illustrirte Zeitung*, November 24, 1888, 526, 540.

47 Georg Schweinfurth, "Deutschlands Verpflichtungen gegen Emin Pascha," lecture in Abteilung Berlin der Deutschen Kolonialgesellschaft, August 17, 1889, BArch R 1001/250, pag. 116.

48 Georg Schweinfurth, et al., *Emin-Pascha. Eine Sammlung von Reisebriefen und Berichten Dr. Emin-Pascha's aus den ehemals ägyptischen Aequatorialprovinzen und deren Grenzländern* (Leipzig: Brockhaus, 1888), viii, xi–xii.

49 Schweinfurth, et al., *Emin-Pascha*, 507–8.

50 Friedrich Ratzel, *Völkerkunde*, vol. 1, Die Naturvölker Afrikas (Leipzig: Bibliographisches Institut, 1885), 478, 546.

51 Friedrich Ratzel, *Anthropogeographie*, vol. 2 Die geographische Verbreitung des Menschen (Stuttgart: J. Engelhorns Nachf., 1912), 384–5.

52 Woodruff Smith, "Friedrich Ratzel and the Origins of Lebensraum," *German Studies Review* 3, no. 1 (February 1980): 59 identifies this transition in Ratzel's politics.

53 Friedrich Ratzel to Ernst Hasse, October 10, 1888, Nachlass Ernst Hasse, Institut für Länderkunde Archiv für Geographie (IfL), K 437.

54 Ratzel to Hasse, August 6, 1889, IfL, K 437.

55 "Aus Emin Paschas Leben. Nach persönlichen Erlebnissen," *Berliner Tageblatt*, January 9, 1889, 1–3.

56 No title, *Kölnische Zeitung*, July 1, 1888, in BArch R 1001/249c, pag. 7.

57 "Lokal-Nachrichten und Vermischtes. In der Audienz des Afrikaforschers Wißmann," *Berliner Tageblatt*, August 8, 1888, 5.

58 Karl Rottenberger to the German government, Oct. 8, 1888; Paul Paatz to the Auswärtiges Amt, Nov. 13, 1888; Wilhelm Heuser to Otto von Bismarck, Dec. 6, 1888; Adolf Mauthner to German government, Sept. 5, 1888; L. Lahn to Otto von Bismarck, Nov. 15, 1888; BArch R 1001/254.

59 BArch R 1001/254, letter from Karl Rottenberger to the German government, Oct. 8, 1888; letter from Paul Paatz to the Auswärtiges Amt, Nov. 13, 1888; letter from Wilhelm Heuser to Bismarck, Dec. 6, 1888.

60 BArch R 1001/254, letter from Wilhelm Heuser to Bismarck, Dec. 6, 1888.

61 Eva Giloi, *Monarchy, Myth, and Material Culture in Germany 1750–1950* (Cambridge: Cambridge University Press, 2011), 295–6.

62 No title, *Berliner Tageblatt*, September 1, 1888, BArch R 1001/249c, pag. 38.

63 "Die Emin-Pascha-Expedition," *Schlesische Zeitung*, October 9, 1888, in BArch R 1001/249d, pag. 15.

64 Eugen Wolf, *Vom Fürsten Bismarck und seinem Haus. Tagebuchblätter*, 2nd ed. (Berlin: Egon Fleischel, 1904), 16.

65 Otto von Bismarck promemoria, September 14, 1888, BArch R 1001/249c, pag. 74.

66 Otto von Bismarck promemoria, September 14, 1888, BArch R 1001/249c, pag. 74.

67 Otto von Bismarck directive, August 11, 1889. BArch R 1001/250, pag. 26–7.

68 Foreign Office to the German Emin Pasha Committee, August 15, 1888, BArch R 1001/249c, pag. 31.

69 Records of German Pasha Comitteee, BArch R 8023/853, pag. 3–105, 124–9, 132–6, 139–41, 147–8, 157–89. On interest groups as a site of political power in the *Kaiserreich*, see Chickering, *We Men Who Feel Most German*; David Blackbourn and Geoff Eley, *The Peculiarities of German History: Bourgeois Society and Politics in Nineteenth-Century Germany* (Oxford: Oxford University Press, 1984).

70 Steifensand to Otto von Bismarck, July 1, 1889, BArch R 1001/249e, pag. 115.

71 "Koloniales," *Deutsches Tageblatt*, July 19, 1889, in BArch R 1001/249e, pag. 111.

72 No title, *Allgemeine Reichs-Correspondenz*, November 4, 1891, in BArch R 1001/253, pag. 36; No title, *Vossische Zeitung*, March 11, 1892, BArch R 1001/253, pag. 50.

73 On text in travelogues, see Adam Jones and Isabel Voigt, "'Just a First Sketchy Makeshift': German Travellers and Their Cartographic Encounters in Africa, 1850–1914," *History in Africa* 39 (2012): 24; the idea of "fugitive landscapes" comes from Raymond Craib, *Cartographic Mexico: A History of State Fixations and Fugitive Landscapes* (Durham: Duke University Press, 2004), 12; Simon Ryan, *The Cartographic Eye: How Explorers Saw Australia* (Cambridge: Cambridge University Press, 1996), 18.

74 Michel de Certeau, *The Practice of Everyday Life*, trans. Steven F. Rendall (Berkeley: University of California Press, 1984).

75 Tiedemann, *Tana-Baringo-Nil*, 166. As Simon Ryan has noted, marking that boundary between civilization and wilderness was a strategy to assert power over territory. Ryan, *Cartographic Eye*, 149.

76 Peters, *New Light on Dark Africa*, 203.

77 Peters, *New Light on Dark Africa*, 213.

78 Peters, *New Light on Dark Africa*, 203.

79 Tiedemann, *Tana-Baringo-Nil*, 240.

80 Peters, *New Light on Dark Africa*, 168

81 Peters, *New Light on Dark Africa*, 192–3.

82 Peters, *New Light on Dark Africa*, 242.

83 Carl Velten, *Schilderungen der Suaheli* (Göttingen: Vandenhoeck & Ruprecht, 1901), 7.

84 Ibid., 40, 52.

85 Ibid., 7.

86 Peters, *New Light on Dark Africa*, 64.

87 Peters, *New Light on Dark Africa*, 191, 217.

88 Peters, *New Light on Dark Africa*, 389.

89 Liebowitz and Pearson, *The Last Expedition*, 177–8.

90 Hermann Wissmann operational report, October 13, 1889, BArch R 1001/742, pag. 53.

91 Rochus Schmidt to Hermann Wissmann, November 10. 1889, BArch R 1001/249a, pag. 59–60, 73.

92 Schmidt to Wissmann, November 18, 1889, BArch R 1001/249a, pag. 66–7.

93 Admiral Fohs to Admiralty, December 16, 1889, BArch R 1001/249a, pag. 88.

94 Emin to Otto von Bismarck, April 8, 1890, BArch R 1001/249b, pag. 51–3.

95 Schmidt to Wissmann, November 18, 1889, BArch R 1001/249a, pag. 67.

96 These fears were a common element in discourses about colonialism and the instability of racial categories. See Ann Stoler, "Making Empire Respectable: The Politics of Race and Sexual Morality in 20th-Century Colonial Cultures," *American Ethnologist* 16, no. 4 (November 1989): 634–60.

97 Arthur Levysohn, "Politische Wochenschau," *Berliner Tageblatt*, November 25, 1889, 1.

98 "Stanley und Emin," *Berliner Tageblatt*, November 28, 1889, 1–2.

99 "Emin Paschas Heimkehr. (Von unserem Korrespondenten)," *Berliner Tageblatt*, November 6, 1889, 1.

100 "Rundschau. Deutsches Reich," *Teltower Kreisblatt*, December 10, 1889, 1.

101 No title, *Schlesische Zeitung*, November 26, 1889, BArch N 2063/12, pag. 5, Extrablatt, *Neisser Zeitung*, December 6, 1889, BArch N 2063/12, pag. 13.

102 On new ethnographic mapping, Jason D. Hansen, *Mapping the Germans: Statistical Science, Cartography, and the Visualization of the German Nation, 1848–1914* (Oxford: Oxford University Press, 2015), 10; on the Pan-German League's membership, Chickering, *We Men Who Feel Most German*; on the League's brand of nationalism, Peter Walkenhorst, *Nation – Volk – Rasse: radikaler Nationalismus im Deutschen Kaiserreich 1890–1914* (Göttingen: Vandenhoeck & Ruprecht, 2007).

6 The Bushiri War and Anti-Arab Internationalism

1 DOAG to Otto von Bismarck, March 22, 1889, BArch R 8124/8, pag. 88. Even a toll collector working for the DOAG, Baniame Jiram Sewjee, expressed grievances with the DOAG's administration, as it meant they could no longer collect customary tips or side payments. BArch R 8124/8, pag. 97.

2 DOAG to Otto von Bismarck, January 12, 1889, BArch R 8124/8, pag. 17.

3 The DOAG somehow hoped that the Sultan's forces, of which he still had many in the coastal cities, would assist in their operations to undermine

his economic power. Barghash of course refused, claiming that the Askaris would not listen to him because they had not been paid, at the DOAG's request. Charles Euan Smith to Said Khalifa BArch R 8124/8, pag. 84–5.

4 "Die kolonialpolitischen Reichstagverhandlungen," *Deutsche Kolonialzeitung*, May 24, 1890: 130.

5 Klaus J. Bade, "Antisklavereibewegung in Deutschland und Kolonialkrieg in Deutsch-Ostafrika, 1888–1890: Bismarck und Friedrich Fabri," *Geschichte und Gesellschaft* 3, no. 1 (1977): 32. Bade also argued that the Reichstag agreement was a "kind of crusade of big capital for the systematic 'opening' of East Africa," a sentiment belied by the reluctance of the largest German banks to invest in the DOAG.

6 Terence Ranger, "Connexions between 'Primary Resistance' Movements and Modern Mass Nationalism in East and Central Africa," *Journal of African History* 9 (1968): 437–53, 631–41; G.C.K. Gwassa, "The German Intervention and African Resistance in Tanzania," in *A History of Tanzania*, ed. I.N. Kimambo and A.J. Temu (Nairobi: East African Publishing House, 1969), 85–122; Robert D. Jackson, "Resistance to the German Invasion of the Tanganyikan Coast," in *Protest and Power in Black Africa*, ed. A.A. Mazuri and R.I. Rotberg (New York: Oxford University Press, 1970), 37–79.

7 Jonathon Glassman, *Feasts and Riot: Revelry, Rebellion, and Popular Consciousness on the Swahili Coast, 1856–1888* (Portsmouth: Heinemann, 1995); Norman Robert Bennett, *Arab versus European: Diplomacy and War in Nineteenth-Century East Central Africa* (New York: Africana Publishing Company, 1986), 175.

8 For recentring imperial narratives, Daniel K. Richter, *Facing East from Indian Country: A Native History of Early America* (Cambridge: Harvard University Press, 2001), 8–9.

9 "Oriental Despotism" also served as a foil for British concepts of colonial law. Nasser Hussain, *The Jurisprudence of Emergency: Colonialism and the Rule of Law* (Ann Arbor: University of Michigan Press, 2003), 7.

10 Said Khalifa to Otto von Bismarck, August 22, 1888, BArch R 1001/770, pag. 5–6; Gustav Michahelles to Otto von Bismarck, August 24, 1888, BArch R 1001/770, pag. 27–33.

11 Makanda bin Mwenyi Mkuu, "Vita vya Bagamoyo," in *Suaheli-Gedichte*, ed. Carl Velten (Berlin: Reichsdruckerei, 1918), 16–19.

12 Glassman, *Feasts and Riot*, 199.

13 Glassman, *Feasts and Riot*, 226, 242–3.

14 Michahelles to Otto von Bismarck, August 27, 1888, BArch R 1001/770, pag. 59–60.

15 Maximilian von Berchem report, September 17, 1888, BArch R 1001/687, pag. 54.

16 Michahelles to Foreign Office, September 24, 1888, BArch R 1001/687, pag. 75.

17 Said Khalifa to Otto von Bismarck, October 3, 1888, BArch R 1001/770, pag. 83–4.

18 "Die Unruhen an der ostafrikanischen Küste," *Nachrichten aus der ostafrikanischen Mission*, October 1888, 147–8.

19 Michahelles to Foreign Office, October 3, 1888, in *Stenographische Berichte über die Verhandlungen des Reichstages*, 1888/89, 4, http://www.reichstagsprotokolle.de/Band3_k7_bsb00018657.html, 403–4.

20 Michahelles to Otto von Bismarck, December 2, 1888, BArch R 1001/694, pag. 82.

21 Michahelles to Foreign Office, September 24, 1888, in *Stenographische Berichte über die Verhandlungen des Reichstages*, 1888/89, 4, http://www.reichstagsprotokolle.de/Band3_k7_bsb00018657.html, 401–2.

22 Oscar Baumann, *In Deutsch-Ostafrika während des Aufstandes* (Vienna: Eduard Hölzel, 1890), 138–40.

23 Tippu Tip was a slave trader and plantation owner who became well known in Europe for his assistance to David Livingstone and Henry Morton Stanley, then as governor of the Stanley Falls District in the Congo Free State. Mbaruk was the leader of the powerful Mazrui clan in Mombasa who was sometimes an official working for Khalifa and sometimes opposed his rule.

24 On Tippu Tip, see Bennett, *Arab versus European*, 222–30.

25 DOAG meeting minutes, October 12, 1888, BArch R 8124/7, pag. 111.

26 Otto von Bismarck to Michahelles, October 6, 1888, BArch R 1001/770, pag. 98–9.

27 Kusserow to Otto von Bismarck, December 22, 1888, BArch R 1001/694, pag. 67–9.

28 Foreign Office to Paul von Hatzfeldt, October 22, 1888, BArch R 1001/707, pag. 5–6.

29 "Das Weißbuch über Ostafrika," *Deutsche Kolonialzeitung*, January 19, 1889, 20–2.

30 Paul Reichard, "Koloniale Rundschau," *Deutsche Kolonialzeitung*, April 27, 1889, 133.

31 "Die Colonial- und die Sklavenjagd-Gebiete von Centralafrika," *Illustrirte Zeitung*, November 24, 1888, 526.

32 DOAG Denkschrift, November 24, 1888, BArch R 8124/2, pag. 171–5.

33 Friedrich Ratzel, *Völkerkunde*, vol. 3, Die Kulturvölker der alten und neuen Welt (Leipzig: Bibliographisches Institut, 1888), 105–7.

34 Ratzel, *Völkerkunde*, vol. 3, 12–14.

35 Otto von Bismarck to Paul von Hatzfeldt, October 21, 1888, BArch R 1001/706, pag. 89; 99–100.

36 Auswärtiges Amt to Paul von Hatzfeldt, October 23, 1888, BArch R 1001/707, pag. 23–8.

37 "Kardinal Lavigerie und die Sklaverei in Afrika," *Berliner Tageblatt*, August 28, 1888, 1.

38 "Kleine Mitteilungen," *Deutsche Kolonialzeitung*, May 18, 1889, 160.

39 Instructions to Paul von Hatzfeldt, October 21, 1888, BArch R 1001/706, pag. 100.

40 German Foreign Office to British Foreign Office, October 17, 1888, BArch R 1001/706, pag. 83–7.

41 British Foreign Office memorandum to German Foreign Office, September 29, 1888, BArch R 1001/706, pag. 14–18.

42 Robert Gascoyne-Cecil to Edward Malet, November 5, 1888, *Parliamentary Papers. Africa* No. 6. Correspondence Respecting Suppression of Slave Trade in East African Waters (London: Harrison and Sons, 1888), 3–4.

43 Paul von Hatzfeldt to Foreign Office, October 19, 1888, BArch R 1001/707, pag. 5–7.

44 Foreign Office to Georg Herbert zu Münster, November 10, 1888, BArch R 1001/709, pag. 46.

45 Draft of French agreement with the United Kingdom, BArch R 1001/707, pag. 61.

46 Otto von Bismarck to Ludwig von Wäcker-Gotter, October [25?], 1888, BArch R 1001/707, pag. 46–8.

47 Henrique de Barros Gomes to Wäcker-Gotter, November 16, 1888, BArch R 1001/714, pag. 8–9.

48 Foreign Office to Georg Herbert zu Münster, November 8, 1888, BArch R 1001/708, pag. 83–6.

49 No title, *Deutscher Reichsanzeiger*, November 17, 1888, BArch R 1001/710, pag. 24.

50 Carl Falkenhorst, "Hermann Wißmann," *Die Gartenlaube*, 1889, no. 6, 84.

51 Hermann Wissmann, *Unter deutscher Flagge quer durch Afrika von West nach Ost*, 7th ed. (Berlin: Walther & Apolant, 1890), 145; Wissmann, Ludwig Wolf, Curt von François, and Hans Mueller, *Im Innern Afrikas. Die Erforschung des Kassai während der Jahre 1883, 1884 und 1885*, 3rd ed. (Leipzig: Brockhaus, 1891).

52 Reichstag – 27. Sitzung. Sonnabend den 26. Januar 1889. *Stenographische Berichte über die Verhandlungen des Reichstages*, 1888/89, vol. 1, 604–6. http://www.reichstagsprotokolle.de/Band3_k7_bsb00018653.html.

53 Michael Schubert, *Der schwarze Fremde: Das Bild des Schwarzafrikaners in der parlamentarischen und publizistischen Kolonialdiskussion in Deutschland von der 1870er bis in die 1930er Jahre* (Stuttgart: Steiner, 2003), 152–3, 200.

54 Herbert von Bismarck report, November 5, 1888, BArch R 1001/730, pag. 11.

55 Foreign Office to Hatzfeldt, October 22, 1888, BArch R 1001/707, pag. 27.

56 Reichstag – 27. Sitzung. Sonnabend den 26. Januar 1889. *Stenographische Berichte über die Verhandlungen des Reichstages*, 1888/89, vol. 1, 603–4. http://www.reichstagsprotokolle.de/Band3_k7_bsb00018653.html.

57 Reichstag – 27. Sitzung. Sonnabend den 26. Januar 1889. *Stenographische Berichte über die Verhandlungen des Reichstages*, 1888/89, vol. 1, 615–17. http://www.reichstagsprotokolle.de/Band3_k7_bsb00018653.html.

58 Reichstag – 15. Sitzung. Freitag den 14. Dezember 1888. BArch R 1001/730, pag. 100–7.

59 Reichstag – 30. Sitzung. Mittwoch den 30. Januar 1889. *Reichstagsprotokolle*, 1888/89, vol. 2, 681, http://www.reichstagsprotokolle.de/Band3_k7 _bsb00018654.html.

60 Reichstag – 29. Sitzung. Sonnabend den 29. Januar 1889. *Reichstagsprotokolle*, 1888/89, vol. 2, 670–671. http://www. reichstagsprotokolle.de/Band3_k7_bsb00018653.html.

61 Diana K. Davis, *Resurrecting the Granary of Rome: Environmental History and French Colonial Expansion in North Africa* (Athens: Ohio University Press, 2009); Yves Lacoste, *Ibn Khaldun: The Birth of History and the Past of the Third World*, trans. David Macey (New York: Verso, 1984); Abdelmajid Hannoum, "Translation and the Colonial Imaginary: Ibn Khaldûn Orientalist," *History and Theory* 42 (February 2003): 61–81.

62 Reichstag – 27. Sitzung. Sonnabend den 26. Januar 1889. *Stenographische Berichte über die Verhandlungen des Reichstages*, 1888/89, vol. 1, 618–22. http://www.reichstagsprotokolle.de/Band3_k7_bsb00018653.html.

63 Appointment of Wissmann as Reichskommissar, February 8, 1889, BArch R 1001/735, pag. 81.

64 Otto von Bismarck to Wissmann, February 12, 1889, BArch R 1001/735, pag. 42–3.

65 Wissmann to Foreign Office, December 16, 1889, BArch R 1001/743, pag. 5.

66 Karl von Gravenreuth to Otto von Bismarck, October 31, 1889, BArch R 1001/712, pag. 86–90.

67 Wissmann to Foreign Office, December 27, 1889, BArch R 1001/744, pag. 6 8.

68 Bennett, *Arab versus European*, 171.

69 Instructions to Wissmann, November 25, 1889, BArch R 1001/742, pag. 94.

70 Paul Reichard, "Die Bedeutung von Tabora für Deutsch-Ostafrika," *Deutsche Kolonialzeitung*, March 15, 1890, 67–8.

71 Berchem report, December 27, 1889, BArch R 1001/743, pag. 82.

72 Berchem report, December 27, 1889, BArch R 1001/743, pag. 82.

73 Franz Johannes von Rottenburg to Herbert von Bismarck, December 1, 1889, BArch R 1001/742, pag. 109.

74 Berchem to Wissmann, November 25, 1889, BArch R 1001/742, pag. 94.

75 "Mitteilungen aus der Gesellschaft," *Deutsche Kolonialzeitung*, January 4, 1890, 14.

76 See R. v. Spalding, "Araberaufstand," *Deutsches Kolonial-Lexikon*, vol. 1 (Leipzig: Quelle & Meyer, 1920), 68.

77 "An dem Commers zu Ehren des Reichskommissars v. Wißmann," *Neueste Mittheilungen*, July 1, 1890, 3.

7 Rethinking the Spread of *Kultur* West

1 British Foreign Office, September 29, 1890, BArch R 1001/760, pag. 36. The treaty also conveyed the island of Heligoland in the Baltic Sea to Germany and recognized a British protectorate over Zanzibar.

2 Ralph Austen describes the setting of international borders as a turning point for German colonialism in East Africa. German attention turned away from acquiring land to other means of ensuring obedience. I argue that this perspective misses how concerned German colonialists remained with the establishment of control over space and remaking it. Ralph A. Austen, *Northwest Tanzania under German and British Rule: Colonial Policy and Tribal Politics, 1889–1939* (New Haven: Yale University Press, 1968), 32.

3 As Michael Shapiro has noted in the cased of the European conquest of North America, violence against Indigenous peoples turns from war into "police action" after territory is bound within the international state system, as Ostafrika was with the Heligoland-Zanzibar Treaty. Michael J. Shapiro, *Violent Cartographies: Mapping Cultures of War* (Minneapolis: University of Minnesota Press, 1997), 30.

4 Koponen divides German colonists involved in the development of *Ostafrika* into two camps. One, the "emigrationist," promoted settler colonialism and the creation of new economic structures in *Ostafrika*; the other, the "economic" camp, promoted development primarily through trade with existing economic structures. Juhani Koponen, *Development for Exploitation: German Colonial Policies in Mainland Tanzania, 1884–1914* (Helsinki: Tiedekirja, 1994), 89; Michael Pesek, *Koloniale Herrschaft in Deutsch-Ostafrika: Expeditionen, Militär und Verwaltung seit 1880* (Frankfurt am Main: Campus, 2005), 190.

5 Caprivi replaced Bismarck as chancellor in March 1890 as Wilhelm II tried to take greater control of German politics. See John C.G. Röhl, *Wilhelm II: The Kaiser's Personal Monarchy, 1888–1900*, trans. Sheila de Bellaigue (Cambridge: Cambridge University Press, 2004).

6 "Soden, Julius Freiherr v.," *Deutsches Kolonial-Lexikon*, vol. 3 (Leipzig: Quelle & Meyer, 1920), 369.

7 Julius von Soden, September 24, 1890, BArch R 1001/762, pag. 55–9.

8 Circular Erlaß No. 15, April 28, 1891, TNA G 1/1.

9 Kolonialabteilung to DOAG, August 22, 1891, BArch R 8124.13, pag. 61.

10 On the importance of emigration worries to the colonial movement, see Woodruff D. Smith, *The German Colonial Empire* (Chapel Hill: University of North Carolina Press, 1978), 12–19.

11 Denkschrift betreffend Deutsch-Ostafrika, no date [1892], BArch R 1001/765, pag. 28.

12 "Die Denkschrift über Ostafrika und die Araber," *Deutsche Kolonialzeitung*, January 7, 1893, 4–5.

13 Soden to Leo von Caprivi, April 4, 1892, BArch R 1001/764, pag. 67–70.

14 Denkschrift betreffend Deutsch-Ostafrika, no date [1892], BArch R 1001/765, pag. 28.

15 Paul Kayser, "Eindrücke aus Sansibar. Bericht eines unbekannten deutschen Reisenden," BArch N 2139/60, pag. 7–8.

16 Eugen Wolf, "Unsere Kulturerfolge in Deutschostafrika. Bagamoyo, Saadani und Mkiodja," *Berliner Tageblatt*, January 27, 1891, BArch R 1001/750, pag. 5.

17 Sewa Haji testament, April 15, 1892, TNA G 32/8.

18 H. Hermann Graf von Schweinitz, *Deutsch-Ost-Afrika im Krieg und Frieden* (Berlin: Walther, 1894), 16.

19 "Wissmanns Expedition nach dem Kilimandjaro. (Von unserem Spezialberichterstatter). XI. Die Rückkehr nach der Küste," *Berliner Tageblatt*, April 21, 1891, BArch R 1001/750, pag. 29.

20 "Aus unserer Arbeit. Dar-es-Salaam. Hohenfriedeberg. Hoffnungshöhe bei Kisserawe," *Nachrichten aus der ostafrikanischen Mission 7*, no. 2 (February 1893): 18

21 Amur bin Nasur ilOmeiri, "Geschichte des erwähnten Knechtes des Propheten Gotttes," in *Lieder und Geschichten der Suaheli*, ed. Carl Büttner (Berlin: Emil Felber, 1894), 161–2.

22 Soden to Caprivi, September 1, 1891, BArch R 1001/756, pag. 15–18.

23 Soden to Caprivi, September 1, 1891, BArch R 1001/756, pag. 15–18.

24 Soden to Caprivi, July 25, 1891, BArch R 1001/755, pag. 26.

25 "Der Dampfer auf dem Viktoria-Nyanza," *Deutsche Kolonialzeitung*, September 20, 1890, 233.

26 Wolf, "Wie sind Handel und Verkehr in nördlichen Deutschostafrika zu heben?" *Berliner Tageblatt*, June 3, 1891, BArch R 1001/750, pag. 45–6.

27 German Anti-Slavery Committee to Wilhelm II, April 15, 1891, BArch R 1001/1007a, pag. 40.

28 Both the Wissmann and Peters steamer campaigns received few donations from Catholic regions of Germany, to the point that the *Kolonialgesellschaft* celebrated the first donation from Bavaria late in 1890. Deutsche Kolonialgesellschaft, Abtheilung Nürnberg, November 3, 1890, BArch R 8023/855, pag. 167–9.

29 Soden to Caprivi, July 25, 1891, BArch R 1001/755, pag. 26; Soden to Caprivi, September 1, 1891, BArch R 1001/756, pag. 15–18.

30 Jan-Georg Deutsch, *Emancipation without Abolition in German East Africa c. 1884–1914* (Oxford: James Currey, 2006), 105.

31 Contributions for Wissmann steamer, 1891, BArch R 8023/856, pag. 29–47, 55–64.

32 Franz Stuhlmann, *Mit Emin Pascha ins Herz von Afrika. Ein Reisebericht* (Berlin: Dietrich Reimer, 1894), 40.

33 Korvettenkapitän Rüdiger, Bericht über die Ereignisse auf der ostafrikanischen Station während des Monats January 1891, January 31, 1891, BArch R 1001/747, pag. 199–206.

34 Stuhlmann, *Mit Emin Pascha*, 235–6.

35 Abschrift zwischen Schweinfurth als Vorsitzender des Gesammtcomité's der Petersstiftung und Bergrat Dr. Busse als I. stellvertretenden Vorsitzenden der Ausführungs Kommission der Deutschen Antisklaverei-Lotterie, BArch R 1001/755, pag. 23–4; Geschäftsführende Ausschuß der Karl Peters-Stiftung, May 13, 1891, BArch R 8023/855, pag. 71.

36 Verein für Handelsgeographie und Colonial-Politik zu Leipzig to geschäftsführenden Ausschluß für die Carl Peters-Stiftung, November 3, 1890; Carl Peters-Stiftung. Stand d. Sammlung am 21. Oct. 1890, IfL K 437.

37 Carl Peters, "Gefechtsweise und Expeditionsführung in Afrika," 1892, in *Gesammelte Schriften*, vol. 2 (Munich and Berlin: C.H. Beck, 1943), 519–28.

38 Peters to Kayser, September 5, 1892, BArch R 1001/755, pag. 95–6.

39 Franco Moretti, *Atlas of the European Novel, 1800–1900* (London: Verso, 1998), 58–60.

40 Hermann Wissmann to Caprivi, February 11, 1891, BArch R 1001/748, pag. 8–13.

41 Peters to Soden, September 8, 1891, BArch R 1001/755, pag. 37–43.

42 Peters, *Das Deutsch-Ostafrikanische Schutzgebiet* (Munich: Oldenbourg, 1895), 120.

43 Stuhlmann, *Mit Emin Pascha*, v.

44 Peters to Soden, September 8, 1891, BArch R 1001/755, pag. 48–58.

45 Stuhlmann, *Mit Emin Pascha*, 62.

46 Stuhlmann, *Mit Emin Pascha*, 74.

47 Stuhlmann, *Mit Emin Pascha*, 410.

48 Pechmann, sketch of Kilimanjaro station, September 1891, BArch R 1001/755, pag. 45.

49 Kurt Johannes to Soden, February 7, 1893, TNA G 1/18.

50 Reichstag – 188. Sitzung. Sonnabend den 5. März 1892, *Reichstagsprotokolle*, 1890–1892, vol. 7, 4581. http://www.reichstagsprotokolle.de/Band3_k8 _bsb00018670.html.

51 Tom von Prince, *Gegen Araber und Wahehe. Erinnerungen aus meiner ostafrikanischen Leutnantszeit 1890–1895* (Berlin: Mittler und Sohn, 1914), 79.

52 Alison Redmayne, "Mkwawa and the Hehe Wars," *Journal of African History* 9, no. 3 (1968): 419.

53 "Neues aus Deutsch-Ostafrika. (Von unserem Spezialberichterstatter)," *Berliner Tageblatt*, October 26, 1891, BArch R 1001/750, pag. 75.

54 "Die Zersprengung der Expedition v. Zelewski durch die Wahehe," *Deutsches Kolonialblatt*, October 1, 1891, 409–11.

55 Franz Leopold von Sonnenschein to Kayser, October 24, 1892, BArch N 2139/68, pag. 1.

56 Rochus Schmidt to Kolonialabteilung, November 29, 1892, BArch R 1001/765, pag. 20–7.

57 "Von der Expedition des Lieutenants Herrmann," *Deutsches Kolonialblatt*, July 1, 1892, 357–9.

58 Prince, *Gegen Araber und Wahehe*, 115–16.

59 Prince, *Gegen Araber und Wahehe*, 199.

60 Schweinitz, *Deutsch-Ost-Afrika im Krieg und Frieden*, 65–6.

61 Prince, *Gegen Araber und Wahehe*, 209–10.

62 Prince, *Gegen Araber und Wahehe*, 220, 223.

63 "Koloniale Kämpfe," *Deutsche Kolonialzeitung*, September 16, 1893, 119–21.

64 Denkschrift betreffend Deutsch-Ostafrika, no date [1892], BArch R 1001/765, pag. 28.

65 Eugen Krenzler to Soden, May 14, 1891, BArch R 8124/13, pag. 17–18.

66 On the growth of socialism as a political force in the late nineteenth century, see James Retallack, *Red Saxony: Election Battles and the Spectre of Democracy in Germany, 1860–1918* (Oxford: Oxford University Press, 2017); Margaret Lavinia Anderson, *Practicing Democracy: Elections and Political Culture in Imperial Germany* (Princeton: Princeton University Press, 2000).

67 Peters, *Deutsch-Ostafrikanische Schutzgebiet*, 29–31.

68 Peters, *Deutsch-Ostafrikanische Schutzgebiet*, 133–4.

69 Oskar Baumann, *Usambara und seine Nachbargebiete* (Berlin: Dietrich Reimer, 1891), 96.

70 Schweinitz, *Deutsch-Ost-Afrika im Krieg und Frieden*, 163.

71 Ann Laura Stoler, "Rethinking Colonial Categories: European Communities and the Boundaries of Rule," *Comparative Studies in Society and History* 31, no. 1 (January 1989): 134–61; Julia Clancy-Smith and Frances Gouda, eds., *Domesticating the Empire: Race, Gender, and Family Life in French and Dutch Colonialism* (Charlottesville: University Press of Virginia, 1998).

72 Verein für Handelsgeographie und Kolonialpolitik zu Leipzig minutes, May 28, 1891, IfL, Nachlass Ernst Hasse, K 437/2.

73 Kurt Weiß, Hauptmann à la suite der II. Ingenieur-Inspektion, "Die Eisenbahnfrage in Ostafrika," *Deutsche Kolonialzeitung*, January 10, 1891, 20.

74 Minutes, DOAG Verwaltungsrat meeting, May 5, 1891, BArch R 8124/4, pag. 52.

75 Baumann, *Usambara und seine Nachbargebiete*, 309.

76 Karl von der Heydt, January 1, 1892, BArch R 8023/279, pag. 45–51.

77 Protokoll über die am 5. März 1891 auf Veranlassung der Deutsch-Ostafrikanischen Gesellschaft zu Berlin in ihren Geschäftsräumen veranstaltete Beratung in Sachen des Eisenbahnbaues in Deutsch-Ostafrika, BArch R 8124/12, pag. 196–204.

78 Protokoll über die am 5. März 1891 auf Veranlassung der Deutsch-Ostafrikanischen Gesellschaft zu Berlin in ihren Geschäftsräumen veranstaltete Beratung in Sachen des Eisenbahnbaues in Deutsch-Ostafrika, BArch R 8124/12, pag. 196–204.

79 Ausführungen des Mitglieds des Verwaltungsrats, Herrn Geheimen Kommerzienrat [Friedrich] Lenz, vom 25. Juni 1891 über die von Tanga nach Usambara zu bauende Eisenbahn, BArch R 8124/4, pag. 67.

80 Geschäfts-Bericht der Deutsch-Ostafrikanischen Gesellschaft für das Jahr 1890, BArch R 8124/4, pag. 76–7.

81 Geschäfts-Bericht der Deutsch-Ostafrikanischen Gesellschaft für das Jahre 1891, BArch R 8124/5, pag. 18.

82 Peters, *Deutsch-Ostafrikanische Schutzgebiet*, 64–5.

83 Richard White, *Railroaded: The Transcontinentals and the Making of Modern America* (New York: W.W. Norton, 2011); Michael Adas, *Dominance by Design: Technological Imperatives and America's Civilizing Mission* (Cambridge: Harvard University Press, 2009).

84 Francis W. Wcislo, *Tales of Imperial Russia: The Life and Times of Sergei Witte, 1849–1915* (New York: Oxford University Press, 2011); Sean McMeekin, *The Berlin-Baghdad Express: The Ottoman Empire and Germany's Bid for World Power* (Cambridge: Belknap, 2010).

85 Sonnenschein to Kayser, July 4, 1892, BArch N 2139/68, pag. 2.

86 Denkschrift betreffend Deutsch-Ostafrika, no date [1892], BArch R 1001/765, pag. 28.

87 Peters, *Deutsch-Ostafrikanische Schutzgebiet*, 16.

88 Peters, *Deutsch-Ostafrikanische Schutzgebiet*, 403, 40.

89 Alexander Merensky, *Was lehren uns die Erfahrungen der andere Völker bei Kolonisationsversuchen in Afrika gemacht haben?* (Berlin: Matthies, 1890), 13.

90 Johannes, November 16, 1892, TNA G 1/18.

91 Geschäfts-Bericht der Deutsch-Ostafrikanischen Gesellschaft für das Jahr 1891, BArch R 8124/5, pag. 18–19.

92 Reintrock to Kolonialabteilung, April 28, 1892, TNA G 1/10.

93 "Bodenkultur Ostafrika," BArch N 2139/69, pag. 1–3.

94 Chinese workers were also part of German economic plans in Samoa, but not until Governor Wilhelm Solf lifted a ban on the importation of Chinese workers in 1903. Stewart Firth, "Governors versus Settlers: The Dispute over Chinese Labour in German Samoa," *New Zealand Journal of History* 11, no. 2 (1977): 155–79.

95 Report from Singapore consulate, July 5, 1892, BArch R 8124/14, pag. 10,
    Report from Batavia general consulate, August 9, 1892, BArch R 8124/14,
    pag. 22, BArch R 8124/14, Report from Singapore consulate, August 17,
    1892, BArch R 8124/14, pag. 37, DOAG to Foreign Office, December 7,
    1892, BArch R 8124/14, pag. 194; Kolonaialabtheilung to DOAG, March 8,
    1892, BArch R 8023/273, pag. 57.

96 DOAG to Kolonialabteilung, February 27, 1892, BArch R 8023/274, pag. 53.

97 Denkschrift betreffend Deutsch-Ostafrika, no date [1892], BArch R
    1001/765, pag. 28.

98 DOAG to Kolonialabtheilung, February 27, 1892, BArch R 8023/273,
    pag. 53.

99 DOAG to Kayser, January 30, 1894, BArch N 2139/65, pag. 4.

100 "Bodenkultur Ostafrika," BArch N 2139/69, pag. 4.

101 "Lindi und die Handelsverhältnisse im Süden von Deutsch-Ostafrika,"
    Deutsches Kolonialblatt, November 15, 1892, 578–83.

102 Frederick Cooper, *Plantation Slavery on the East Coast of Africa* (New
    Haven, CT: Yale University Press, 1977).

103 Joachim von Pfeil, "East Africa again," BArch N 2225/147, pag. 328.

104 Joachim von Pfeil, "East Africa again," BArch N 2225/147, pag. 329.

105 Geschäfts-Bericht der Deutsch-Ostafrikanischen Gesellschaft für das Jahre
    1892, BArch R 8124/5, pag. 55–6.

106 DOAG Verwaltungsrat to Herr von Burchard, February 22, 1893, BArch R
    8023/279, pag. 123–4.

107 Denkschrift betreffend Deutsch-Ostafrika, no date [1892], BArch R
    1001/765, pag. 28.

108 Maximilian von Rode to Soden, December 16, 1892, BArch R 8023/274,
    pag. 69.

109 Kayser to Deutsch Ostafrikanische Plantagen Gesellschaft, February 20,
    1893, BArch R 8023/274, pag. 70.

110 "Lindi und die Handelsverhältnisse im Süden von Deutsch-Ostafrika,"
    Deutsches Kolonialblatt, November 15, 1892, 578–83.

111 "Kulturpolitik in Ostafrika. II. Der Kaffee von Mrogoro," *Deutsche
    Kolonialzeitung*, January 7, 1893, 3–4.

112 Schweinitz, *Deutsch-Ost-Afrika im Krieg und Frieden*, 164–5.

113 John Hanning Speke, *Journal of the Discovery of the Source of the Nile
    (London: Blackwoods, 1863)*. See Mahmood Mamdani, *When Victims Become
    Killers* (Princeton: Princeton University Press, 2001) on the Hamitic
    hypothesis' use in colonial rule.

114 Joseph Thomson, *Through Masai-Land: A Journey of Exploration among the
    Snowclad Volcanic Mountains and Strange Tribes of Eastern Equatorial Africa*
    (Boston: Houghton, Mifflin, and Company: 1885), 337, 443.

115 Thomson, *Through Masai-Land*, 411.
116 Thomson, *Through Masai-Land*, 415.
117 Peters, *Deutsch-Ostafrikanische Schutzgebiet*, 39.
118 Peters, *Deutsch-Ostafrikanische Schutzgebiet*, 142.
119 "Die Expedition des Majors v. Wissmann nach dem Kilimandscharo," *Deutsches Kolonialblatt*, April 1, 1891, 151.
120 Baumann, *Usambara und seine Nachbargebiete*, 244–5.
121 Friedrich Ratzel, *Völkerkunde*, vol. 1, book 1: Grundzüge der Völkerkunde, (Leipzig: Bibliographisches Institut, 1885), 86.
122 Ratzel, *Völkerkunde*, vol. 1, book 1, 95.
123 Ratzel, *Völkerkunde*, vol. 1, book 2, 409.
124 Carl Holst, "Die Kulturen der Waschambaa," *Deutsche Kolonialzeitung*, February 4, 1893, 23–4.
125 Holst, "Der Landbau der Eingeborenen von Usambara," *Deutsche Kolonialzeitung*, August 19, 1893, 113–14.
126 Peters, *Deutsch-Ostafrikanische Schutzgebiet*, 241.
127 Peters, *Deutsch-Ostafrikanische Schutzgebiet*, 266, 348.
128 Peters, *Deutsch-Ostafrikanische Schutzgebiet*, 338.
129 Stuhlmann, *Mit Emin Pascha*, 855.
130 Stuhlmann, *Mit Emin Pascha*, 410, 423.
131 Stuhlmann, *Mit Emin Pascha*, 472–3.
132 Stuhlmann, *Mit Emin Pascha*, 855.
133 Stuhlmann, *Mit Emin Pascha*, 866.
134 Peters, *Deutsch-Ostafrikanische Schutzgebiet*, 161.
135 Peters, *Deutsch-Ostafrikanische Schutzgebiet*, 226–8.

8 Creating Familiar Landscapes: *Heimatkunde* and African Spatialities, 1893–1900

1 "Wangemannshöh," in Martha Eitner, *Berliner Mission im Njaßa-Lande (Deutsch Ostafrika)* (Berlin: Evangelische Missiongesellschaft, 1897), 51; "Kirche in Hohenfriedeberg," *Nachrichten aus der ostafrikanischen Mission*, August 1894, 121.
2 Celia Applegate, *A Nation of Provincials: The German Idea of Heimat* (Berkeley: University of California Press, 1990). According to Alon Confino, Germans of the *Kaiserreich* built a cohesive national identity out of disparate parts by imagining "nationhood as a form of localness." Alon Confino, *The Nation as a Local Metaphor: Württemberg, Imperial Germany, and National Memory, 1871–1918* (Chapel Hill: University of North Carolina Press, 1997), 188.
3 Timothy Mitchell, *Colonising Egypt* (Berkeley: University of California Press, 1988) describes a similar process at work with images of Egypt.

4 On "domesticating" empire for metropolitan audiences, see Julia Clancy-Smith and Frances Gouda, eds., *Domesticating the Empire: Race, Gender, and Family Life in French and Dutch Colonialism* (Charlottesville, VA: University Press of Virginia, 1998); Antoinette Burton, *Burdens of History: British Feminists, Indian Women, and Imperial Culture, 1865–1915* (Chapel Hill: The University of North Carolina Press, 1994).

5 From September 1893 until 1901, all three permanent governors in Ostafrika – Friedrich von Schele, Hermann Wissmann, and Eduard von Liebert – possessed military backgrounds and made their names fighting in the colony. In addition to the continued war with Mkwawa and the Hehe, smaller military actions were ubiquitous. Albert Wirz, Andreas Eckert, and Katrin Bromber, eds., *Alles unter Kontrolle: Disziplinierungsprozesse im kolonialen Tansania (1850–1960)* (Cologne: Rüdiger Köppe, 2003); Michael Pesek, *Koloniale Herrschaft in Deutsch-Ostafrika: Expeditionen, Militär und Verwaltung seit 1880* (Frankfurt am Main: Campus, 2005).

6 Frederick Cooper, *Africa since 1940: The Past of the Present* (Cambridge: Cambridge University Press, 2002).

7 Schele came from a different background than had Soden. He was a Prussian army officer rather than a diplomat, indicating the importance of the militarization of the interior to plans for Ostafrika.

8 Schele, Runderlaß an die Bezirks- und Bezirksnebenämter sowie die Stationen im Innern, May 22, 1894, BArch R 1001/786/1, pag. 17–18.

9 Adolf Görz to Kayser, October 24, 1893, BArch R 1001/774, pag. 2–7.

10 Arthur Abraham, "Bai Bureh, the British, and the Hut Tax War," *The International Journal of African Historical Studies* 7, no. 1 (1974): 99–106.

11 Keletso Atkins, *The Moon Is Dead! Give Us Our Money! The Cultural Origins of an African Work Ethic, Natal, South Africa, 1843–1900* (Portsmouth: Heinemann, 1993).

12 "Zur Träger- und Arbeiterfrage," March 3, 1900, TNA G 1/35.

13 Konstantin Kramer to DOA administration, September 16, 1900, TNA G 1/35.

14 Karl Rannenberg, "Bericht über die in Bezug auf die Häuser und Hüttensteuer gemachten Erfahrungen," October 2, 1898, TNA G 3/48.

15 "Karte der Ortschaften im Bezirk Mpapua welche im Sommerhalbjahr 1898 Steuern gebracht, bezw. Steuerarbeit geleistet haben, nach den Volkstämmen durch Farben unterschieden," TNA G 3/48.

16 Schele, "Denkschrift über den Werth und die Entwicklungsfähigkeit der Kolonie Deutsch Ostafrika, sowie über die Verwaltung derselben," no date [1894], BArch R 1001/774, pag. 11–32.

17 Allgemeiner Bericht unsere Plantagen betreffend, November 1, 1893, BArch R 8124/5, pag. 71.

18 See Atkins, *The Moon Is Dead!*

19 Schele, "Denkschrift über den Werth und die Entwicklungsfähigkeit der Kolonie Deutsch Ostafrika, sowie über die Verwaltung derselben," no date [1894], BArch R 1001/774, pag. 11–32.

20 DOAG to AA, November 5, 1897, BArch R 1001/7850, pag. 22–3.

21 Reiseprogramm, BArch R 1001/7850, pag. 55.

22 Ferdinand Wohltmann, Bericht über die Ergebnisse seiner Reise nach Deutsch-Ost-Afrika im Auftrage der Kolonial-Abtheilung des Auswärtigen Amtes, September 2, 1898, BArch R 1001/7850, pag. 89–160.

23 Ibid.

24 Ibid.

25 Ibid.

26 Ibid.

27 Ibid.

28 Runderlass an die Bezirks- und Bezirksnebenämter sowie die Stationen im Innern, September 19, 1898, TNA G 1/35.

29 August Leue to DOA administration, November 15, 1900, October 11, 1900, TNA G 1/35.

30 Soliman bin Nasr to Decken, December 20, 1898, BArch R 1001/796, pag. 171–9.

31 Soliman bin Nasr to Bezirksamtmann von Strantz, December 22, 1898, BArch R 1001/796, pag. 180–7.

32 Karl Ewerbeck to DSM government, June 24, 1899, BArch R 1001/796, pag. 201; Eduard von Liebert to Kolonialabteilung, July 21, 1899, BArch R 1001/796, pag. 209–10.

33 Georg von Prittwitz und Gaffron to DOA administration, April 17, 1902, TNA G 1/35.

34 Robert von Beringe to DOA administration, February 3, 1900; Albrecht von Rechenberg to DOA administration, March 29, 1900, TNA G 1/35.

35 August Leue to DOA administration, February 11, 1899, TNA G 1/35.

36 Bezirksamt Pangani to DOA administration, August 9, 1895, TNA G 1/35.

37 Georg Lieder, Zur Kenntniß der Karawanenwege im südlichen Teile des ostafrikanischen Schutzgebietes, April 28, 1894, TNA G 1/35; Bezirksamt Lindi to DOA administration, June 14, 1895, TNA G 1/35.

38 Carl Velten, ed., *Sitten und Gebräuche der Suaheli* (Göttingen: Vandenhoeck & Ruprecht, 1903), 294.

39 Karl von Ewerbeck to DOA administration, November 21, 1900, TNA G 1/35.

40 Military stations did at least reflect increased attention to important locations for spatial control, unlike the DOAG stations of the previous decade. Michelle R. Moyd, *Violent Intermediaries: African Soldiers, Conquest, and Everyday Colonialism in German East Africa* (Athens: Ohio University Press, 2014), 150.

41 Cultural diffusion and borrowing in the missionary encounter in Africa has been analysed in detail with a framework laid out in Jean Comaroff and John L. Comaroff, *Of Revelation and Revolution*, vol. 1, Christianity, Colonialism, and Consciousness in South Africa (Chicago: University of Chicago Press, 1991).

42 "Nachrichten der Wakamba-Mission. Mitgeteilt von Senior Ittameier in Erlangen," *Evangelisch-Lutherisches Missionsblatt*, May 1, 1895, 165–8.

43 Mike Davis, *Late Victorian Holocausts: El Niño Famines and the Making of the Third World* (London: Verso, 2000).

44 "Hungersnot in Ostafrika," *Nachrichten aus der ostafrikanischen Mission*, August 1894, 124–6.

45 "Heuschrecken und kein Ende," *Nachrichten aus der ostafrikanischen Mission*, May 1895, 70–3.

46 "Licht und Dunkel in der Wakambamission. Bericht aus der Station Jimba von Miss. Karl Kämpf in Jimba," *Evangelisch-Lutherisches Missionsblatt*, April 1, 1894, 130–3.

47 Carl Paul, *Die Mission in unsern Kolonien*, vol. 2, Deutsch-Ostafrika (Leipzig: Fr. Richter, 1900), 6.

48 Paul, *Die Mission in unsern Kolonien*, vol. 2, 58.

49 Paul, *Die Mission in unsern Kolonien*, vol. 2, 291–2.

50 Paul, *Die Mission in unsern Kolonien*, vol. 2, 298.

51 Paul, *Die Mission in unsern Kolonien*, vol. 2, 301.

52 Tom von Prince, *Gegen Araber und Wahehe. Erinnerungen aus meiner ostafrikanischen Leutnantszeit 1890–1895* (Berlin: Mittler und Sohn, 1914), 246–7.

53 Prince, *Gegen Araber und Wahehe*, 290.

54 Moyd, *Violent Intermediaries*, 154.

55 Kompagnie Führer Fromm to Lothar von Trotha, April 28, 1895, BArch R 1001/698, pag. 188–9.

56 "Ein Besuch auf unserer neuen Station Kwarango in Madschame. Nach dem Tagebuch von Miss. Päsler," *Evangelisch-Lutherisches Missionsblatt*, March 1, 1894, 89–93.

57 Miss. Faßmann, "Eine vornehme Frau in Madschame," *Evangelisch-Lutherisches Missionsblatt*, May 1, 1895, 168–71.

58 Georg von Prittwitz und Gaffron diary, July 11, 1897, IfL, Box 245, folder 1.

59 Georg von Prittwitz und Gaffron diary, May 9–29, 1898, IfL, Box 245, folder 2.

60 Kompagnie Führer Fromm to Lothar von Trotha, April 28, 1895, BArch R 1001/698, pag. 188–9.

61 "Kisserawe – ein Heim für befreite Sklaven," *Nachrichten aus der ostafrikanischen Mission*, May 1894, 66–7.

62 Georg von Prittwitz und Gaffron diary, January 11, 1898, IfL, Box 245, folder 2.

63 Georg von Prittwitz und Gaffron diary, November 9, 1898, IfL, Box 248, folder 6.

64 Hans Meyer journal, July 19, 1898, IfL, Box 180/1, book 9.

65 Hans Meyer journal, July 22, 1898, Institut für Länderkunde, Leipzig, Box 180/1, folder 9, 10.

66 Hans Meyer journal, July 24, 1898, Institut für Länderkunde, Leipzig, Box 180/1, folder 9, 10.

67 Hans Meyer journal, August 4–8, 1898, Institut für Länderkunde, Leipzig, Box 180/1, folder 9, 10.

68 Hans Meyer journal, August 9, 1898, Institut für Länderkunde, Leipzig, Box 180/1, folder 9, 10.

69 Hans Meyer journal, July 18, 1898, IfL, Box 180/1, book 9.

70 Ernst Albrecht von Eberstein to DOA administration, March 16, 1899, TNA G 1/35.

71 Tabora station to DOA administration, April 1, 1899, TNA G 1/35.

72 Georg von Elpons to DOA administration, April 20, 1899, TNA G 1/35.

73 "Deutsch-Ostafrika," *Missionsblatt der Brüdergemeine*, January 1896, 10.

74 "Deutsch-Ostafrika," *Missionsblatt der Brüdergemeine*, May 1896, 145–6.

75 "Ein Besuch bei den Massai. Bericht von Miss. Althaus in Karangwo (5. Febr. 1894)," *Evangelisch-Lutherisches Missionsblatt*, July 1, 1894, 237–45.

76 G. Althaus, "Sprachstudien im Dschagga-Lande," *Evangelisch-Lutherisches Missionsblatt*, September 1, 1894, 325–30.

77 "Das Kilimandjaro-Gebirge in Ostafrika," *Evangelisch-Lutherisches Missionsblatt*, January 1, 1894, 8–13.

78 Eitner, *Berliner Mission im Njaßa-Lande*, 93, 95.

79 Eitner, *Berliner Mission im Njaßa-Lande*, 23, 38.

80 Lothar von Trotha report, May 8, 1898, BArch R 1001/774, pag. 57–197.

81 Lothar von Trotha report, May 8, 1898, BArch R 1001/774, pag. 57–197.

82 Lothar von Trotha, "Meine Bereisung von Deutsch-Ostafrika," lecture to Gesellschaft für Erdkunde (Berlin: Brigl, 1897), 33.

83 Trotha, "Meine Bereisung von Deutsch-Ostafrika," 37.

84 Trotha, "Meine Bereisung von Deutsch-Ostafrika," 77–9.

85 Trotha, "Meine Bereisung von Deutsch-Ostafrika," 56.

86 Lothar von Trotha report, May 8, 1898, BArch R 1001/774, pag. 57–197.

87 G. Althaus, "Eine Besichtigung der neuen Station Mamba und ihrer Umgebung," *Evangelisch-Lutherisches Missionsblatt*, January 15, 1895, 28–35.

88 Friedrich Ratzel, *Deutschland. Einführung in die Heimatkunde*, 7th ed. (Berlin: Walter de Gruyter & Co., 1943), v.

89 Ratzel, *Deutschland*, 158–9.

90 Ratzel, *Deutschland*, 215–16.

91 Ratzel, *Der Staat und Sein Boden*, 99.

92 Ratzel, *Der Staat und Sein Boden*, 102–5.

93 Ratzel, *Der Staat und Sein Boden*, 107.

94 Friedrich Ratzel, *Der Staat und Sein Boden geographisch betrachtet* (Leipzig: Hirzel, 1896), 58.

95 Ratzel, *Der Staat und Sein Boden*, 18.

96 Ratzel, *Der Staat und Sein Boden*, 19.

97 Ratzel, *Der Staat und Sein Boden*, 65.

98 Ratzel, *Der Staat und Sein Boden*, 70.

99 Hans Meyer journal, September 18, 1898, Institut für Länderkunde, Leipzig, Box 180/1, folder 9, 10.

100 Ratzel, *Der Staat und Sein Boden*, 64.

101 Ratzel, *Der Staat und Sein Boden*, 99.

102 Ratzel, *Der Staat und Sein Boden*, 21.

103 Ule, *Lehrbuch der Erdkunde*, II. Teil, 378.

104 *Volksschulen* were the mandatory primary and lower secondary institutions all Germans had to attend. On the development of *Heimatkunde* as a school subject, see Ulrich Schubert, *Das Schulfach Heimatkunde im Spiegel von Lehrerhandbüchern der 20er Jahre* (Hildesheim: Olms, 1987).

105 J.M. Blaut, *The Colonizer's Model of the World* (New York: Guilford Press, 1993), 6.

106 Jeff Bowersox, *Raising Germans in the Age of Empire: Youth and Colonial Culture, 1871–1914* (Oxford: Oxford University Press, 2013), 54–5.

107 Willi Ule, *Lehrbuch der Erdkunde für höhere Schulen*, I. Teil, Für die unteren Klassen (Leipzig: Freytag, 1897), v–vi.

108 Ule, *Lehrbuch der Erdkunde*, vol. I, 19.

109 Ule, *Lehrbuch der Erdkunde*, vol. I, 16.

110 Ule, *Lehrbuch der Erdkunde*, vol. I, 145.

111 Willi Ule, *Lehrbuch der Erdkunde für höhere Schulen*, II. Teil, Für die mittleren und oberen Klassen (Leipzig: Freytag, 1896), 243–4.

112 Ule, *Lehrbuch der Erdkunde*, vol. II, 258.

113 Ule, *Lehrbuch der Erdkunde*, vol. II, 382–3.

114 Johannes Bumüller and Ignaz Schuster, *Erdkunde* (Freiburg im Breisgau: Herder, 1900), 26.

115 Paul Gockisch, *E. v. Seydlitzsche Geographie*. Ausgabe E. Für höhere Mädchenschulen. Heft 3: Die außereuropäischen Erdteile (Breslau: Ferdinand Hirt, 1896), 10.

116 Franz Tschauder, *Die deutschen Kolonien* (Breslau: Heinrich Handel, 1900), 8.

117 Tschauder, *Die deutschen Kolonien*, 12.

118 Rudolf Langenbeck, *Leitfaden der Geographie für höhere Lehranstalten*, vol. 2, Lehrstoff der mittleren und oberen Klassen (Leipzig: Engelmann, 1897), 102.

119 Otto Sommer, *Leitfaden der Erdkunde* (Braunschweig: Appelhans, 1899), 79–80.

120 Reichstag – 63. Sitzung. Montag den 18. März 1895. *Reichstagsprotokolle*, 1894/95, vol. 2, 1566. https://www.reichstagsprotokolle.de/Band3_k9 _bsb00018724.html.
121 Ibid., 1562.

9 Biogeography and Resettlement for *Kultur*

1 Eduard von Liebert to Kolonialabteilung, October 4, 1899, BArch R 1001/27, pag. 71–5.
2 Aga Khan, "The Aga Khan on German East Africa," *The Times of India*, November 25, 1899, BArch R 1001/27, pag. 84.
3 Jagdamba Prosad Verma to Gustav von Götzen, February 25, 1901, TNA G 8/59.
4 James C. Scott, *Seeing Like a State: How Certain Schemes to Improve the Human Condition Have Failed* (New Haven: Yale University Press, 1998).
5 Woodruff Smith, "Friedrich Ratzel and the Origins of Lebensraum," *German Studies Review* 3, no.1 (February 1980): 51–68; *The Ideological Origins of Nazi Imperialism* (Oxford: Oxford University Press, 1989), 83–105; Karl Lange, "Der Terminus 'Lebensraum' in Hitlers 'Mein Kampf,'" *Vierteljahrsheft für Zeitgeschichte* 13, no. 4 (1965): 426–37; Harriet Wanklyn, *Friedrich Ratzel: A Biographical Memoir and Bibliography* (London: Cambridge University Press, 1961); Shelly Baranowski, *Nazi Empire: German Colonialism and Imperialism from Bismarck to Hitler* (Cambridge: Cambridge University Press, 2010), 64; Eberhard Jäckel, *Hitler's World View: A Blueprint for Power* (Cambridge: Harvard University Press, 1981); Richard Evans, *The Coming of the Third Reich* (London: Penguin, 2004), 34–5; Holger Herwig, "Geopolitik: Haushofer, Hitler and Lebensraum," *Journal of Strategic Studies* 22, no. 2–3 (1999): 218–41; Horst Dreier, "Wirtschaftsraum – Großraum – Lebensraum. Facetten eines belasteten Begriffs," in *Raum und Recht. Festschrift 600 Jahre Würzburger Juristenfakultät*, eds. Dreier, Hans Forkel, and Klaus Laubenthal (Berlin: Duncker & Humblot, 2002), 47–84; Gearoid O'Tuathail, *Critical Geopolitics: The Politics of Writing Global Space* (Minneapolis: University of Minnesota Press, 1996), 37–8.
6 J.A. Hobson, *Imperialism: A Study* (New York: J. Pott & Co., 1902). On Hobson's anti-Semitism, see Norman Etherington, *Theories of Imperialism: War, Conquest and Capital*, Routledge Revivals ed. (London: Routledge, 2014), 70.
7 Ernst Haeckel, *Die Welträthsel. Gemeinverständliche Studien über Monistische Philosophie* (Bonn: Emil Strauß, 1899).
8 Ratzel, *Der Lebensraum*, 51.

9 Ratzel, *Der Lebensraum*, 60.

10 Jens-Uwe Guettel, "From the Frontier to German South-West Africa: German Colonialism, Indians, and American Westward Expansion," *Modern Intellectual History* 7, no. 3 (2010): 523–52.

11 Patrick Wolfe, "Settler Colonialism and the Elimination of the Native," *Journal of Genocide Research* 8, no. 4 (December 2006): 387–409.

12 Friedrich Ratzel, *Der Lebensraum. Eine biogeographische Studie* (Tübingen: H. Laupp'schen Buchhandlung, 1901), 44.

13 Friedrich Ratzel, *Die Erde und das Leben. Eine vergleichende Erdkunde*, vol. 2 (Leipzig: Bibliographisches Institut, 1902), 550.

14 Ratzel, *Die Erde und das Leben*, vol. 2, 626.

15 Ratzel, *Die Erde und das Leben*, vol. 2, 627.

16 Ratzel, *Die Erde und das Leben*, vol. 2, 628.

17 Ratzel, *Die Erde und das Leben*, vol. 2, 665.

18 Ratzel, *Die Erde und das Leben*, vol. 2, 670.

19 Ratzel, *Die Erde und das Leben*, vol. 2, 652.

20 Ratzel, *Die Erde und das Leben*, vol. 2, 627.

21 Ratzel, *Die Erde und das Leben*, vol. 2, 633.

22 Ratzel, *Die Erde und das Leben*, vol. 2, 643.

23 Ratzel, *Die Erde und das Leben*, vol. 2, 669.

24 Ratzel, *Die Erde und das Leben*, vol. 2, 579.

25 Ratzel, *Die Erde und das Leben*, vol. 2, 670.

26 John Phillip Short, *Magic Lantern Empire: Colonialism and Society in Germany* (Ithaca: Cornell University Press, 2012), 33–4.

27 Woodruff D. Smith, "Friedrich Ratzel and the Origins of Lebensraum," *German Studies Review* 3, no. 1 (February 1980): 51–2. See also David Thomas Murphy, "'Retroactive Effects': Ratzel's Spatial Dynamics and the Expansionist Imperative In Interwar Germany," *Journal of Historical Geography* 61 (2018): 86–90.

28 Guettel, "From the Frontier to German South-West Africa."

29 Ratzel, *Die Erde und das Leben*, vol. 2, 285.

30 Ratzel, *Die Erde und das Leben*, vol. 2, 286.

31 Ratzel, *Die Erde und das Leben*, vol. 1, 452.

32 Ratzel, *Die Erde und das Leben*, vol. 1, 459.

33 Th. Bechler, *Eine Karawanenreise ins Innere Afrikas* (Herrnhut: Verlag der Missionsbuchhandlung, 1903), 4.

34 Bechler, *Eine Karawanenreise ins Innere Afrikas*, 12.

35 Georg von Prittwitz und Gaffron journal, February 4, 1902, IfL, Nachlass Prittwitz, Folder 2/247.

36 Contract between Commune Bagamoyo and Golek bin Laschko, June 3, 1902, among others, TNA G 32/1.

37 Ratzel, *Die Erde und das Leben*, vol. 2, 638.

38 Gustav von Götzen, *Deutsch-Ostafrika im Aufstand 1905–1906* (Berlin: Dietrich Reimer, 1909), 23.

39 Gustav von Götzen, *Deutsch-Ostafrika im Aufstand 1905–1906* (Berlin: Dietrich Reimer, 1909), 24.

40 Gustav von Götzen, *Deutsch-Ostafrika im Aufstand 1905–1906* (Berlin: Dietrich Reimer, 1909), 25.

41 Gustav von Götzen, *Deutsch-Ostafrika im Aufstand 1905–1906* (Berlin: Dietrich Reimer, 1909), 29.

42 Lothar von Trotha to Chlodwig Hohenlohe-Schillingsfürst, June 22, 1895, BArch R 1001/27, pag. 5–8.

43 Karl Perrot to Oswald von Richthofen, November 15, 1897, BArch R 1001/27, pag. 48–51.

44 E.W., "Gelbe Gefahr," *Deutsch-ostafrikanische Zeitung*, September 5, 1903, 9.

45 Heinz Gollwitzer, *Die Gelbe Gefahr. Geschichte eines Schlagworts. Studien zum imperialistischen Denken* (Göttingen: Vandenhoeck & Ruprecht, 1962); Ute Mehnert, *Deutschland, Amerika und die "Gelbe Gefahr"* (Stuttgart: Steiner, 1998).

46 "Die gelbe Gefahr für Deutsch-Ostafrika," *Deutsch-ostafrikanische Zeitung*, August 9, 1902, 1.

47 "Zur Inderfrage," *Deutsch-ostafrikanische Zeitung*, July 18, 1903, 2. *Pesa* means "money" in Swahili.

48 Gustav von Götzen, *Deutsch-Ostafrika im Aufstand 1905–1906* (Berlin: Dietrich Reimer, 1909), 35.

49 Gustav von Götzen, *Deutsch-Ostafrika im Aufstand 1905–1906* (Berlin: Dietrich Reimer, 1909), 36.

50 Gustav von Götzen, *Deutsch-Ostafrika im Aufstand 1905–1906* (Berlin: Dietrich Reimer, 1909), 32.

51 On cotton agriculture in Togo, see Andrew Zimmerman, *Alabama in Africa: Booker T. Washington, the German Empire and the Globalization of the New South* (Princeton: Princeton University Press, 2010); on Southwest Africa settlement, see Daniel Joseph Walther, *Creating Germans Abroad: Cultural Policies and National Identity in Namibia* (Athens: Ohio University Press, 2002).

52 Gustav von Götzen, *Deutsch-Ostafrika im Aufstand 1905–1906* (Berlin: Dietrich Reimer, 1909), 13.

53 Gustav von Götzen, *Deutsch-Ostafrika im Aufstand 1905–1906* (Berlin: Dietrich Reimer, 1909), 16.

54 Franz Stuhlmann, "Übersicht über Land- und Forstwirtschaft in Deutsch-Ostafrika im Berichtsjahre vom 1. July 1900 bis 30. Juni 1901," *Berichte über Land- und Forstwirtschaft in Deutsch-Ostafrika* 1 (1903): 20–1.

55 Götzen, *Deutsch-Ostafrika im Aufstand*, 17.

56 Zimmerman, *Alabama in Africa*.

57 "Auszüge aus den Jahresberichten der Bezirksämter und Militärstationen für die Zeit vom 1. Juli 1900 bis 30. Juni 1901," *Berichte über Land- und Forstwirtschaft in Deutsch-Ostafrika* 1 (1903): 23.

58 "Auszüge aus den Berichten der Bezirksämter, Militärstationen und anderer Berichtsstellen über die wirtschaftliche Entwicklung im Berichtsjahre vom 1. April 1901 bis 31. März 1902," *Berichte über Land- und Forstwirtschaft in Deutsch-Ostafrika* 1 (1903): 218.

59 Meyer to Johann Albrecht zu Mecklenburg, November 6, 1902, BArch R 1001/11, pag. 86–92.

60 Liebert to Kolonialabteilung, October 4, 1899, BArch R 1001/27, pag. 71–5.

61 Victor Eschke to Hohenlohe-Schillingsfürst, June 9, 1900, BArch R 1001/27, pag. 106–9.

62 Victor Eschke to Hohenlohe-Schillingsfürst, July 10, 1900, BArch R 1001/27, pag. 112–14.

63 Piya Chatterjee discusses how this process unfolded in Northern India's tea industry, where British planters recruited labour from Southern India based on idealized archetypes of labourers after both local and Chinese workers refused to do work they considered unskilled. Piya Chatterjee, "'Secure This Excellent Class of Labour': Gender and Race in Labour Recruitment for British Indian Tea Plantations," *Bulletin of Concerned Asian Scholars* 27, no. 3 (1995): 43–56.

64 Götzen to Kolonialabteilung, September 10, 1901, BArch R 1001/28, pag. 12–16.

65 Detlev von Winterfeld to Gustav von Götzen, October 11, 1901, TNA G 8/59.

66 Detlev von Winterfeld to Gustav von Götzen, October 17, 1901, TNA G 8/59.

67 A. Zimmermann, "Erster Jahresbericht des Kaiserl. Biologisch-Landwirtschaftlichen Instituts Amani," *Berichte über Land- und Forstwirtschaft in Deutsch-Ostafrika* 1 (1903): 435.

68 Ludwig Meyer to Gustav von Götzen, September 30, 1901, TNA G 8/59.

69 The *Sachsengänger* were Polish labourers who provided seasonal labour on Saxon estates before returning home and animated much social concern in late nineteenth-century Germany as a threat to a conservative ideal of society based in settled agriculture. Most prominently, Max Weber wrote about the *Sachsengänger* as part of a study on labour. Max Weber, "Die ländliche Arbeitsverfassung" and "Entwickelungstendenzen in der Lage der ostelbischen Landarbeiter," in *Gesammelte Aufsätze zur Sozial- und Wirtschaftgeschichte*, ed. Marianne Weber (Tübingen: J.C.B. Mohr, 1924), 444–69, 470–507.

70 Ludwig Meyer to Gustav von Götzen, October 28, 1901, TNA G 8/59.

71 Gustav von Götzen, October 29, 1901, TNA G 8/59.

72 Götzen to Meyer, February 27, 1902, TNA G 8/59; Götzen to Freudenberg, March 25, 1902, BArch R 1001/11, pag. 49–50.

73 Eric Ames, *Carl Hagenbeck's Empire of Entertainments* (Seattle: University of Washington Press, 2008). Like Ostafrika's administrators imagined the Sinhalese migrants as "Indian," Hagenbeck advertised their appearance as an "Indian village."

74 Meyer to Götzen, April 8, 1902, TNA G 8/59.

75 Götzen to Kolonialabteilung, February 27, 1902, TNA G 8/59.

76 Meyer to Götzen, April 8, 1902, TNA G 8/59.

77 Wilhelm Sperling to Götzen, January 31, 1903, TNA G 8/60.

78 "Auszüge aus den Berichten der Bezirksämter, Militärstationen und anderer Berichtsstellen über die wirtschaftliche Entwicklung im Berichtsjahre vom 1. April 1901 bis 31. März 1902," *Berichte über Land- und Forstwirtschaft in Deutsch-Ostafrika* 1 (1903): 219.

79 Freudenberg to DOA government, April 26, 1902, BArch R 1001/11, pag. 81.

80 Ludwig Meyer to Gustav von Götzen, March 9, 1902, TNA G 8/59.

81 "Auszüge aus den Berichten der Bezirksämter, Militärstationen und anderer Dienststellen über die wirtschaftliche Entwicklung im Berichtsjahre vom 1. April 1902 bis 31. März 1903," *Berichte über Land- und Forstwirtschaft* 2 (1904–6): 40.

82 Wilhelm Sperling to Götzen, January 31, 1903, TNA G 8/60.

83 Meyer to Johann Albrecht zu Mecklenburg, November 6, 1902, BArch R 1001/11, pag. 86–92.

84 "Auszüge aus den Berichten der Bezirksämter, Militärstationen und anderer Berichtsstellen über die wirtschaftliche Entwicklung im Berichtsjahre vom 1. April 1901 bis 31. März 1902," *Berichte über Land- und Forstwirtschaft in Deutsch-Ostafrika* 1 (1903): 219–20.

85 Wilhelm Sperling to Götzen, January 31, 1903, TNA G 8/60.

86 Stuhlmann to Kolonialabteilung, February 24, 1903, BArch R 1001/12, pag. 13.

87 "Auszüge aus den Berichten der Bezirksämter, Militärstationen und anderer Berichtsstellen über die wirtschaftliche Entwicklung im Berichtsjahre vom 1. April 1901 bis 31. März 1902," *Berichte über Land- und Forstwirtschaft in Deutsch-Ostafrika* 1 (1903): 219.

88 Meyer to Kolonialabteilung, January 4, 1903, BArch R 1001/11, pag. 93–8.

89 "Auszüge aus den Berichten der Bezirksämter, Militärstationen und anderer Dienststellen über die wirtschaftliche Entwicklung im Berichtsjahre vom 1. April 1902 bis 31. März 1903," *Berichte über Land- und Forstwirtschaft* 2 (1904–6): 40.

90 Detlev von Winterfeld to Götzen, March 10, 1904, TNA G 8/60; Götzen to Winterfeld, March 26, 1904, TNA G 8/60.

91 Meyer to Götzen, September 29, 1904, TNA G 8/60.

92 Meyer to Götzen, January 3, 1905, TNA G 8/60.

93 Meyer to Götzen, June 2, 1905, TNA G 8/60.

94 *West-Usambara und seine Besiedelungsfähigkeit* (Berlin: Deutsche Kolonialgesellschaft, 1902), BArch R 1001/11, pag. 59.

95 "Auskunft für Ansiedler im Bezirk Langenburg (Nordufer des Nyassa-Sees)," June 1904, BArch R 1001/12, pag. 124–5; "Auskunft für Ansiedler im Bezirk Moschi, Deutsch-Ostafrika (Kilimandscharo und Meruland)," August 1905, BArch R 1001/13, pag. 49–50.

96 "Europäische Handwerker in Deutsch-Ostafrika," *Deutsch-ostafrikanische Zeitung*, April 13, 1901, 1.

97 Kurt Toeppen, "Die Besiedlung von Deutsch-Ostafrika," *Deutsche Kolonialzeitung*, July 10, 1902, 269–71.

98 Theobald Schütze, "Deutsch-Ost-Afrika und die Südafrikanischen Boeren," September 1901, BArch R 1001/25, pag. 10–19.

99 Theobald Schütze, "Deutsch-Ost-Afrika und die Südafrikanischen Boeren," September 1901, BArch R 1001/25, pag. 10–19.

100 W. Janke to Vorstand des Allgemeinen deutschen Verbandes, October 20, 1902, BArch R 1001/25, pag. 21–2.

101 August Leue, "Die Besiedlung von Deutsch-Ostafrika," *Deutsche Kolonialzeitung*, June 26, 1902, 253–6.

102 Roloff Abraham Zeederberg-Chiappini to Langenburg station, February 27, 1903, BArch R 1001/25, pag. 53.

103 Chiappini to Langenburg station, February 28, 1903, BArch R 1001/25, pag. 54.

104 "Buren-Ansiedlung in Deutsch-Ostafrika," July 12, 1905, TNA G 54/5.

105 Louis Alberts to C.J. Zorn, July 14, 1904, BArch R 1001/25, pag. 86–8.

106 Theobald Schütze, "Praktische Massregeln zur Einleitung einer Boerenauswanderung nach D.O. Afrika," BArch R 1001/25, pag. 29–34.

107 Stuhlmann to Kolonialabteilung, December 9, 1902, BArch R 1001/25, pag. 24–6.

108 Stuhlmann to Kolonialabteilung, February 21, 1905, BArch R 1001/25, pag. 129–30.

109 Haber to Kolonialabteilung, June 8, 1905, BArch R 1001/25, pag. 145–6.

110 Wolfgang Wippermann, *Wie die Zigeuner: Antisemitismus und Antiziganismus im Vergleich* (Berlin: Elefanten, 1997); Marion Bonillo, *"Zigeunerpolitik" im Deutschen Kaiserreich 1871–1918* (Frankfurt: Peter Lang, 2001).

111 "Deutsche Schutzgebiete. Die Burenansiedlung," *Kölnische Zeitung*, July 4, 1905, BArch R 1001/25, pag. 188.

112 Ann Laura Stoler, "Rethinking Colonial Categories: European Communities and the Boundaries of Rule," *Comparative Studies in Society and History* 31, no. 1 (January 1989): 134–61. George Steinmetz has detailed these concerns in the case of German Samoa in *The Devil's*

Handwriting: The German Colonial State in Qingdao, Samoa, and Southwest Africa (Chicago: University of Chicago Press, 2007), 317–18.

113 Franz Stuhlmann to Owen Joseph and J. Seth Johnson, August 7, 1901, TNA G 8/56.

114 Ernst Steinweder to Gustav von Götzen, January 12, 1902, TNA G 8/56.

115 Hans-Ulrich Wehler, *The German Empire 1871–1918*, trans. Kim Traynor (Dover: Berg, 1985), 173–7.

116 "Ein Beitrag zur deutsch-ostafrikanischen Kommunal-Frage," *Deutsch-ostafrikanische Zeitung*, June 22, 1901, 1.

117 Eugen Styx to Dar es Salaam, July 6, 1905, TNA G 3/48.

118 Eugen Styx to Dar es Salaam, May 18, 1905, TNA G 3/48; "Plan der Militärstation Mpapua über den friedlichen Machbereich derselben für das Steuerjahr 1905," TNA G 3/48.

119 "Auszüge aus den Berichten der Bezirksämter, Militärstationen und anderer Dienststellen über die wirtschaftliche Entwicklung im Berichtsjahre vom 1. April 1902 bis 31. März 1903," *Berichte über Land- und Forstwirtschaft* 2 (1904–6): 94–5.

120 James Giblin and Jamie Monson, "Introduction," in *Maji Maji: Lifting the Fog of War*, eds. James Giblin and Jamie Monson (Leiden: Brill, 2010), 5–8.

121 Walter Nuhn, *Flammen über Deutschost: Der Maji-Maji-Aufstand in Deutsch-Ostafrika, 1905–1906: die erste gemeinsame Erhebung schwarzafrikanischer Völker gegen weisse Kolonialherrschaft: ein Beitrag zur deutschen Kolonialgeschichte* (Bonn: Bernard & Graefe, 1998), 9; Terence O. Ranger, "Connexions between 'primary resistance' movements and modern mass nationalism in East and Central Africa," *Journal of African History* 9 (1968): 437–53, 631–41.

122 "Aufstands-Ursachen und Schmuggel," *Deutsch-Ostafrikanische Zeitung*, November 25, 1905, n.p. "Die Ruinierung der ostafrikanischen angesessenen Bevölkerung," *Deutsch-Ostafrikanische Zeitung*, January 13, 1906, n.p.

123 "Indische Völkerwanderung nach Deutsch-Ostafrika," *Deutsch-Ostafrikanische Zeitung*, October 28, 1905, n.p.

124 "Eine indische Preßstimme über den Aufstand in unserer Kolonie," *Deutsch-Ostafrikanische Zeitung*, September 9, 1905, n.p.

125 "Aufstands-Ursachen und Schmuggel," *Deutsch-Ostafrikanische Zeitung*, November 25, 1905, n.p.

126 No title, *Deutsch-Ostafrikanische Zeitung*, March 31, 1906, n.p.

127 "Verlorene Liebesmüh," *Deutsch-Ostafrikanische Zeitung*, February 24, 1906, n.p.

128 No title, *Deutsch-Ostafrikanische Zeitung*, March 31, 1906, n.p.

129 "Die Inderfrage in der letzten Gouvernementsratssitzung," *Deutsch-Ostafrikanische Zeitung*, June 16, 1906, n.p.

130 "Soll Deutsch-Ostafrika eine deutsche Kolonie werden oder eine Hamburg-indische Domäne bleiben? 5. Die Inderfrage und die Behandlung der Farbigen," *Deutsch-Ostafrikanische Zeitung*, October 7, 1905, n.p.

131 "Aus unseren Kolonien. Ostafrika. Die Inderfrage," *Deutsche Kolonialzeitung*, March 24, 1906, 114.

132 "Aus unseren Kolonien. Ostafrika. Die Inderfrage," *Deutsche Kolonialzeitung*, May 5, 1906, 178–9.

133 "Aus Deutsch-Ostafrika," *Deutsche Zeitung*, November 5, 1905, in *Neueste Nachrichten aus unseren Kolonien in Afrika. Pressemeldungen von den Aufständen in Deutsch-Ostafrika und Deutsch-Südwestafrika, 1905–1906*, ed. Reinhard Schneider (Berlin: Carola Hartmann Miles, 2010), 204–5.

134 Hans Paasche, *Im Morgenlicht: Kriegs-, Jagd- und Reise-Erlebnisse in Ostafrika* (Berlin: C.A. Schwetschke und Sohn, 1907), 106; The *DOAZ* also described Indians as "yellow parasites" attacking both Germans and Africans, "Aus der Kolonie. Indische Schachzüge," *Deutsch-Ostafrikanische Zeitung*, August 11, 1906, n.p.

135 Paasche, *Im Morgenlicht*, 321.

136 Paasche, *Im Morgenlicht*, 348.

137 "Nachrichten aus den deutschen Schutzgebieten. Deutsch-Ostafrika. Über die in Deutsch-Ostafrika ausgebrochenen Unruhen," *Deutsches Kolonialblatt*, September 1, 1905, 525–6.

138 "Denkschrift über die Ursachen des Aufstandes in Deutsch-Ostafrika 1905," *Deutsch-Ostafrikanische Zeitung*, March 3, 1906, n.p.

139 "Ostafrika. Eine Unterredung mit Graf Götzen," *Deutsche Zeitung*, September 3, 1905, in Schneider, *Neueste Nachrichten aus unseren Kolonien in Afrika*, 165–7.

140 "Ein Inderfreund," *Deutsch-Ostafrikanische Zeitung*, February 17, 1906, n.p.

141 Fritz Langheld, "Ist die Hüttensteuer als Grund des ostafrikanischen Aufstandes anzusehen?," *Deutsche Kolonialzeitung*, November 25, 1905, 493.

142 "Aus unseren Kolonien. Ostafrika. Aufstand im englischen Nyassagebiet," *Deutsche Kolonialzeitung*, January 6, 1906, 6.

143 The National Archives of the UK: FO 800/19. Page 31. March 9, 1906, letter from Frank Lascelles to Charles Hardinge.

144 The National Archives of the UK: FO 800/13. Pages 181–203. August 16, 1906, Sir C. Hardinge to Sir E. Grey, Meeting of King and Emperor and Cronberg. Account of Anglo-German Relations. Conversation with Tschirschky. Conversation with Kaiser. Reports.

145 Gustav von Götzen, *Deutsch-Ostafrika im Aufstand 1905–1906* (Berlin: Dietrich Reimer, 1909), 24–5.

146 Mary Evelyn Townsend, *The Rise and Fall of Germany's Colonial Empire, 1884–1918* (New York: Howard Fertig, 1966), 246; Helmuth Stoecker, ed.,

German Imperialism in Africa: From the Beginnings until the Second World War (London: Hurst, 1986), 197–203.

147 The origins of that narrative are explored in Inka Chall and Sonja Mezger, "Die Perspektive der Sieger: Der Maji-Maji-Krieg in der Kolonialen Presse," in *Der Maji-Maji-Krieg in Deutsch-Ostafrika, 1905–1907*, ed. Felicitas Becker and Jigal Beez (Berlin: Ch. Links, 2005), 143–53.

148 James Giblin and Jamie Monson, "Introduction," in *Maji Maji: Lifting the Fog of War*, eds. James Giblin and Jamie Monson (Leiden: Brill, 2010), 19–21; Marcia Wright, "Maji Maji Prophecy and Historiography," in *Revealing Prophets: Prophecy in Eastern African History*, ed. David Anderson and Douglas H. Johnson (London: James Currey, 1995), 125; Felicitas Becker and Jigal Beez, eds., *Der Maji-Maji-Krieg in Deutsch-Ostafrika, 1905–1907* (Berlin: Links, 2005).

10 Conclusion

1 On the Lettow-Vorbeck mythos, see Uwe Schulte-Varendorff, *Kolonialheld für Kaiser und Führer. General Lettow-Vorbeck – Mythos und Wirklichkeit* (Berlin: Links, 2011).

2 Jeff Bowersox, *Raising Germans in the Age of Empire: Youth and Colonial Culture, 1871–1914* (Oxford: Oxford University Press, 2013); John Phillip Short, *Magic Lantern Empire: Colonialism and Society in Germany* (Ithaca: Cornell University Press, 2012); David Ciarlo, *Advertising Empire: Race and Visual Culture in Imperial Germany* (Cambridge: Harvard University Press, 2011); Matthew Unangst, "Men of Science and of Action: The Celebrity of Explorers and German National Identity, 1870–1895" *Central European History* 50, no. 3 (September 2017): 305–27.

3 Woodruff Smith, *The Ideological Origins of Nazi Imperialism* (Oxford: Oxford University Press, 1989).

4 Shelly O. Baranowski, *The Sanctity of Rural Life: Nobility, Protestantism, and Nazism in Weimar Prussia* (Oxford: Oxford University Press, 1995); Robert G. Moeller, *German Peasants and Agrarian Politics, 1914–1924: The Rhineland and Westphalia* (Chapel Hill: University of North Carolina Press, 2010).

5 Woodruff Smith, *The Ideological Origins of Nazi Imperialism* (Oxford: Oxford University Press, 1989).

6 Vejas Gabriel Liulevicius, *War Land on the Eastern Front: Culture, National Identity, and German Occupation in World War I* (Cambridge: Cambridge University Press, 2000); Kristin Kopp, *Germany's Wild East: Constructing Poland as Colonial Space* (Ann Arbor: University of Michigan Press, 2012).

7 Leo Frobenius, *Der Ursprung der afrikanischen Kulturen*, vol. 1 (Berlin: Gebrüder Borntraeger, 1898).

8 Fritz Graebner, *Methode der Ethnographie* (Heidelberg: Carl Winter, 1911).
9 Joseph C. Miller, "History and Africa/Africa and History," *American Historical Review* 104, no. 1 (February 1999): 1–32; Suzanne Marchand, "Leo Frobenius and the Revolt against the West," *Journal of Contemporary History* 32, no. 2 (1997): 153–70.

Bibliography

Archival Sources

Auswärtiges Amt, Berlin, Germany (AA)
  - London consular records
Bundesarchiv, Berlin-Lichterfelde, Germany (BArch)
    - Deutsche Kolonialgesellschaft (R 8023)
    - Deutsch-Ostafrikanische Gesellschaft (R 8124)
    - Nachlass Paul Kayser (N 2139)
    - Nachlass Emin Pascha (N 2063)
    - Nachlass Carl Peters (N 2223)
    - Nachlass Joachim von Pfeil (N 2225)
    - Reichskolonialamt (R 1001)
Institut für Länderkunde, Leipzig, Germany (IfL)
    - Nachlass Ernst Hasse
    - Nachlass Hans Meyer
    - Nachlass Georg von Prittwitz und Gaffron
National Archives of the United Kingdom
    - FO 800, Private Offices: Various Ministers' and Officials' Papers
Tanzania National Archives, Dar es Salaam, Tanzania (TNA)
    - German Records

Published Primary Sources

Periodicals
Allgemeine Missions-Zeitschrift
Berichte über Land- und Forstwirtschaft in Deutsch-Ostafrika
Berliner Tageblatt
Deutsches Kolonialblatt

Deutsche Kolonialzeitung

Deutsche Rundschau

Deutsch-ostafrikanische Zeitung

Evangelisch-Lutherisches Missionsblatt

Die Gartenlaube

Illustrirte Zeitung

Die katholischen Missionen

Kolonial-Politische Korrespondenz

Missionsblatt der Brüdergemeine

Mittheilungen aus Justus Perthes' Geographischer Anstalt über wichtige neue Erforschungen auf dem Gesammtgebiete der Geographie von Dr. A. Petermann/Dr. A. Petermanns Mitteilungen aus Justus Perthes' geographischer Anstalt

Mittheilungen der afrikanischen Gesellschaft in Deutschland

Le Mouvement Géographique

Nachrichten aus der ostafrikanischen Mission

Neue freie Presse

Neueste Mittheilungen

Nord und Süd

Reichs-Gesetzblatt

Stenographische Berichte über die Verhandlungen des Reichstages

BOOKS

Barth, Heinrich. *Reisen und Entdeckungen in Nord- und Central-Afrika in den Jahren 1849 bis 1855.* 5 vols. Gotha: Justus Perthes, 1857–8.

Baumann, Oskar. *Usambara und seine Nachbargebiete.* Berlin: Dietrich Reimer, 1891.

Baumgarten, Johannes. *Ostafrika, der Sudan und das Seeengebiet: Land und Leute, Naturschilderungen, charakterische Reisebilder: die Antisklavereibewegung, ihre Ziele und ihr Ausgang, kolonialpolitische Fragen der Gegenwart: nach den neuesten und besten Quellen.* Gotha: F.A. Perthes, 1890.

Baur, Étienne, and Alexandre Le Roy. *À Travers le Zanguebar. Voyage dans l'Oudoé, l'Ouzigua, l'Oukwèré, l'Oukami et l'Ousaraga.* Tours: Alfred Mame et Fils, 1886.

Bechler, Th. *Eine Karawanenreise ins Innere Afrikas.* Herrnhut: Verlag der Missionsbuchhandlung, 1903.

Bodin, Jean. *Six Books of the Commonwealth.* Translated by M.J. Tooley. Oxford: Basil Blackwell, 1955.

Bumüller, Johannes, and Ignaz Schuster. *Erdkunde.* Freiburg im Breisgau: Herder, 1900.

Büttner, Carl, ed. *Lieder und Geschichten der Suaheli.* Berlin: Emil Felber, 1894.

Cameron, Verney Lovett. *Across Africa.* New York: Harper & Brothers, 1877.

Claus, W. *Dr. Ludwig Krapf, weil. Missionar in Ostafrika*. Basel: C.F. Spittler, 1882.

Deutsches Kolonial-Lexikon. 3 vols. Leipzig: Quelle & Meyer, 1920.

Eitner, Martha. *Berliner Mission im Njaßa-Lande (Deutsch Ostafrika)*. Berlin: Evangelische Missiongesellschaft, 1897.

Engelhardt, Paul, and I. von Wenzierski. *Karte von Central-Ost-Afrika*. Berlin: Engelhardtischen Landkartenhandlung, 1886.

Frobenius, Leo. *Der Ursprung der afrikanischen Kulturen*. Vol. 1. Berlin: Gebrüder Borntraeger, 1898.

Geistbeck, Michael. *Grundzüge der Geographie für Mittelschulen sowie zum Selbstunterricht*. Munich: Oldenbourg, 1885.

Gockisch, Paul. *E. v. Seydlitzsche Geographie*. Ausgabe E. Für höhere Mädchenschulen. Heft 3: Die außereuropäischen Erdteile. Breslau: Ferdinand Hirt, 1896.

Götzen, Gustav von. *Deutsch-Ostafrika im Aufstand 1905–1906*. Berlin: Dietrich Reimer, 1909.

Graebner, Fritz. *Methode der Ethnographie*. Heidelberg: Carl Winter, 1911.

Grimm, Carl. *Der wirthschaftliche Werth von Deutsch-Ostafrika*. Berlin: Walther & Apolant, 1886.

– *Die Pharaonen in Ostafrika. Eine kolonialpolitische Studie*. Karlsruhe: Grimm, 1887.

Haeckel, Ernst. *Die Welträthsel. Gemeinverständliche Studien über Monistische Philosophie*. Bonn: Emil Strauß, 1899.

Hobson, J.A. *Imperialism: A Study*. New York: J. Pott & Co., 1902.

Höhnel, Ludwig von. *Discovery of Lakes Rudolf and Stefanie*. London: Longmans, 1894.

Humboldt, Alexander von, and Aimé Bonpland. *Essay on the Geography of Plants*. Translated by Sylvie Romanowski. Chicago: University of Chicago Press, 2009.

Irish University Press Series of British Parliamentary Papers. The Zanzibar Papers 1841–98. Colonies, Africa. Vol. 68. Shannon: Irish University Press, 1971.

Ittameier, M. *Die evangelisch-lutherische Mission in Ostafrika in den ersten zwei Jahren ihres Bestehens*. Rothenburg: J.P. Peters, 1888.

Junker, Wilhelm. *Reisen in Afrika, 1875–1886*. Volume 3 (1882–6). Vienna: Eduard Hölzel, 1891.

– *Travels in Africa during the Years 1882–1886*. Translated by A.H. Keane. London: Chapman and Hall, 1892.

Krapf, Johann Ludwig. *Travels, Researches, and Missionary Labours, during an Eighteen Years' Residence in Eastern Africa*. London: Trübner & Co., 1860.

Kurtze, Bruno. *Die Deutsch-Ostafrikanische Gesellschaft: Ein Beitrag zum Problem der Schutzbriefgesellschaften und zur Geschichte Deutsch-Ostafrikas*. Jena: Gustav Fischer, 1913.

Langenbeck, Rudolf. *Leitfaden der Geographie für höhere Lehranstalten.* Vol. 2, *Lehrstoff der mittleren und oberen Klassen.* Leipzig: Engelmann, 1897.

MacDonald, Duff. *Africana; or, the Heart of Heathen Africa.* 2 vols. London: Simpkin Marshall, 1882.

Merensky, Alexander. *Was lehren uns die Erfahrungen der andere Völker bei Kolonisationsversuchen in Afrika gemacht haben?* Berlin: Matthies, 1890.

New, Charles. *Life, Wanderings, and Labours in Eastern Africa.* London: Hodder and Stoughton, 1874.

Nigmann, Ernst. *Die Wahehe. Ihre Geschichte, Kult-, Rechts- Kriegs- und Jagd-Gebräuche.* Berlin: Mittler und Sohn, 1908.

Paasche, Hans. *Im Morgenlicht: Kriegs-, Jagd- und Reise-Erlebnisse in Ostafrika.* Berlin: C.A. Schwetschke und Sohn, 1907.

Parliamentary Papers. Africa No. 6 (1888). *Correspondence Respecting Suppression of Slave Trade in East African Waters.* London: Harrison and Sons, 1888.

Parliamentary Papers, Africa No. 10 (1888). *Further Correspondence Respecting Germany and Zanzibar.* London: Harrison and Sons, 1888.

Paul, Carl. *Die Mission in unsern Kolonien.* Vol. 2, *Deutsch-Ostafrika.* Leipzig: Fr. Richter, 1900.

Peters, Carl. *Untersuchung zur Geschichte des Friedens von Venedig.* Hannover: Hahn'sche Buchhandlung, 1879.

– *New Light on Dark Africa: Being the Narrative of the German Emin Pasha Expedition.* Translated by H.W. Dulcken. London: Ward, Lock, and Co., 1891.

– *Das Deutsch-Ostafrikanische Schutzgebiet.* Munich: Oldenbourg, 1895.

– *Die Gründung von Deutsch-Ostafrika.* Berlin: C.A. Schwetschke und Sohn, 1906.

– *Gesammelte Schriften.* 3 vols. Munich and Berlin: C.H. Beck, 1943.

Pfeil, Joachim Graf von. *Zur Erwerbung von Deutsch-Ostafrika.* Berlin: Karl Curtius, 1907.

Pless, Adelheid von. *Baron Carl Claus von der Decken's Reisen in Ost-Afrika in den Jahren 1859 bis 1865.* Leipzig and Heidelberg: C.F. Winter, 1869.

Prince, Tom von. *Gegen Araber und Wahehe. Erinnerungen aus meiner ostafrikanischen Leutnantszeit 1890–1895.* Berlin: Mittler und Sohn, 1914.

Pruen, S. Tristam. *The Arab and the African: Experiences in Eastern Equatorial Africa during a Residence of Three Years.* London: Seeley and Co., 1891.

Ratzel, Friedrich. *Anthropo-Geographie, oder Grundzüge der Anwendung der Erdkunde auf die Geschichte.* Vol. 1. Stuttgart: Engelhorn, 1882.

– *Völkerkunde.* Vol, 1, *Die Naturvölker Afrikas.* Leipzig: Bibliographisches Institut, 1885.

– *Völkerkunde.* Vol. 3, *Die Kulturvölker der alten und neuen Welt.* Leipzig: Bibliographisches Institut, 1888.

– *Der Staat und Sein Boden geographisch betrachtet.* Leipzig: Hirzel, 1896.

– *Der Lebensraum. Eine biogeographische Studie.* Tübingen: H. Laupp'schen Buchhandlung, 1901.

– *Die Erde und das Leben. Eine vergleichende Erdkunde.* Vol. 2. Leipzig: Bibliographisches Institut, 1902.

– "Geschichte, Völkerkunde und historische Perspektive." *Historische Zeitschrift* 93, no. 1 (1904): 1–46.

– *Anthropogeographie.* Vol. 2, *Die geographische Verbreitung des Menschen.* Stuttgart: J. Engelhorns Nachf., 1912.

– *Deutschland. Einführung in die Heimatkunde.* 7th ed. Berlin: Walter de Gruyter & Co., 1943.

– *Sketches of Urban and Cultural Life in North America.* Translated by Stewart A. Stehlin. New Brunswick: Rutgers University Press, 1988.

Richards, Charles, and James Place, eds. *East African Explorers.* London: Oxford University Press, 1960.

Riehl, Wilhelm Heinrich. *Die Naturgeschichte des Volkes als Grundlage einer deutschen Social-Politik.* Vol. 1, *Land und Leute.* Stuttgart and Tübingen: J.G. Cotta, 1854.

Schneider, Reinhard, ed. *Neueste Nachrichten aus unseren Kolonien in Afrika. Pressemeldungen von den Aufständen in Deutsch-Ostafrika und Deutsch-Südwestafrika, 1905–1906.* Berlin: Carola Hartmann Miles, 2010.

Schweinfurth, Georg. *The Heart of Africa. Three Years' Travels and Adventures in the Unexplored Regions of Central Africa from 1868–1871.* 2 vols. Translated by Ellen E. Frewer. Chicago: Afro-Am Press, 1969.

Schweinfurth, Georg, Friedrich Ratzel, Robert W. Felkin, and Gustav Hartlaub, eds. *Emin-Pascha. Eine Sammlung von Reisebriefen und Berichten Dr. Emin-Pascha's aus den ehemals ägyptischen Aequatorialprovinzen und deren Grenzländern.* Leipzig: Brockhaus, 1888.

Schweinitz, H. Hermann Graf von. *Deutsch-Ost-Afrika im Krieg und Frieden.* Berlin: Walther, 1894.

Schweitzer, Georg. *Emin Pascha. Eine Darstellung seines Lebens und Wirkens mit Benutzung seiner Tagebücher, Briefe und wissentschaftlichen Aufzeichnungen.* Berlin: Walther, 1898.

Sommer, Otto. *Leitfaden der Erdkunde.* Braunschweig: Appelhans, 1899.

Speke, John Hanning. *Journal of the Discovery of the Source of the Nile.* London and Edinburgh: William Blackwood & Sons, 1863.

Stanley, Henry M. *How I Found Livingstone.* London: Sampson Lowe, 1872.

Stuhlmann, Franz. *Mit Emin Pascha ins Herz von Afrika. Ein Reisebericht.* Berlin: Dietrich Reimer, 1894.

Sturz, J. *Land und Leute in Deutsch-Ost-Afrika. Erinnerungen aus der ersten Zeit des Aufstandes und der Blockade.* Berlin: Mittler und Sohn, 1894.

Thomson, Joseph. *Through Masai-Land: A Journey of Exploration among the Snowclad Volcanic Mountains and Strange Tribes of Eastern Equatorial Africa.* Boston: Houghton, Mifflin, and Company, 1885.

Tiedemann, Adolf von. *Aus Busch und Steppe.* Berlin: Winckelmann & Söhne, 1905.

290   Bibliography

– *Tana-Baringo-Nil.* Berlin: C.A. Schwetschke, 1907.

Trotha, Lothar von. "Meine Bereisung von Deutsch-Ostafrika." Lecture to Gesellschaft für Erdkunde. Berlin: Brigl, 1897.

Tschauder, Franz. *Die deutschen Kolonien.* Breslau: Heinrich Handel, 1900.

Ule, Willi. *Lehrbuch der Erdkunde für höhere Schulen.* II. Teil, *Für die mittleren und oberen Klassen.* Leipzig: Freytag, 1896.

– *Lehrbuch der Erdkunde für höhere Schulen.* I. Teil, *Für die unteren Klassen.* Leipzig: Freytag, 1897.

Velten, Carl. "Erklärung einiger ostafrikanischer Ortsnamen." *Mittheilungen des Seminars für orientalische Sprachen an der Königlichen Friedrich Wilhelms-Universität zu Berlin* 1 (1898): 199–204.

–, ed. *Märchen und Erzählungen der Suaheli.* Stuttgart: W. Spemann, 1898.

– *Schilderungen der Suaheli.* Göttingen: Vandenhoeck & Ruprecht, 1901.

–, ed. *Sitten und Gebräuche der Suaheli.* Göttingen: Vandenhoeck & Ruprecht, 1903.

–, ed. *Suaheli-Gedichte.* Berlin: Reichsdruckerei, 1918.

Wagner, J. *Deutsch-Ostafrika: Geschichte der Gesellschaft für deutsche Kolonisation und der Deutsch-Ostafrikanischen Gesellschaft nach den amtlichen Quellen.* Berlin: Verlag der Engelhardt'schen Landkartenhandlung, 1886.

West-Usambara und seine Besiedelungsfähigkeit. Berlin: Deutsche Kolonialgesellschaft, 1902.

Weule, Karl. "Zur Kartographie der Naturvölker." *Petermanns geographische Mitteilungen* 61 (1915): 18–21, 59–62.

Wissmann, Hermann. *Unter deutscher Flagge quer durch Afrika von West nach Ost.* 7th ed. Berlin: Walther & Apolant, 1890.

– *My Second Journey through Equatorial Africa. From the Congo to the Zambesi, in the Years 1886 and 1887.* Translated by Minna J.A. Bergmann. London: Chatto & Windus, 1891.

Wissmann, Hermann, Ludwig Wolf, Curt von François, and Hans Mueller. *Im Innern Afrikas. Die Erforschung des Kassai während der Jahre 1883, 1884 und 1885.* 3rd ed. Leipzig: Brockhaus, 1891.

Secondary Sources

Abraham, Arthur. "Bai Bureh, the British, and the Hut Tax War." *The International Journal of African Historical Studies* 7, no. 1 (1974): 99–106.

Adas, Michael. *Machines as the Measure of Men: Science, Technology, and Ideologies of Western Dominance.* Ithaca: Cornell University Press, 1989.

– *Dominance by Design.* Cambridge: Harvard University Press, 2009.

Ajayi, J.F.A. "The Continuity of African Institutions under Colonialism." In *Emerging Themes in African History,* edited by Terence Ranger, 189–200. Nairobi: East African, 1968.

Akerman, James R., ed. *The Imperial Map*. Chicago: University of Chicago Press, 2009.

Allen, James de Vere. "Swahili Culture and the Nature of East Coast Settlement." *The International Journal of African Historical Studies* 14, no. 2 (1981): 306–34.

Allen, Richard B. *European Slave Trading in the Indian Ocean, 1500–1850*. Athens: Ohio University Press, 2014.

Alpers, Edward. *The Indian Ocean in World History*. Oxford: Oxford University Press, 2014.

Ames, Eric. *Carl Hagenbeck's Empire of Entertainments*. Seattle: University of Washington Press, 2008.

Anderson, Margaret Lavinia. *Practicing Democracy: Elections and Political Culture in Imperial Germany*. Princeton: Princeton University Press, 2000.

Anghie, Antony. "Finding the Peripheries: Sovereignty and Colonialism in Nineteenth-Century International Law." *Harvard International Law Journal* 40, no. 1 (1999): 1–81.

Ansprenger, Franz. "Schulpolitik in Deutsch-Ostafrika." In *Studien zur Geschichte des deutschen Kolonialismus in Afrika: Festschrift zum 60. Geburtstag von Peter Sebald*, edited by Peter Heine and Ulrich van der Heyden, 59–93. Pfaffenweiler: Centaurus, 1995.

Appadurai, Arjun. *Modernity at Large: Cultural Dimensions of Globalization*. Minneapolis: University of Minnesota Press, 1996.

Applegate, Celia. *A Nation of Provincials: The German Idea of Heimat*. Berkeley: University of California Press, 1990.

Applegate, Celia, and Pamela Potter, eds. *Music & German National Identity*. Chicago: University of Chicago Press, 2002.

Asiwaju, A.I., ed. *Partitioned Africans: Ethnic Relations across Africa's International Boundaries 1884–1895*. London: C. Hurst & Company, 1985.

Atkins, Keletso. *The Moon Is Dead! Give Us Our Money! The Cultural Origins of an African Work Ethic, Natal, South Africa, 1843–1900*. Portsmouth: Heinemann, 1993.

Austen, Ralph. *Northwest Tanzania under German and British Rule: Colonial Policy and Tribal Politics, 1889–1939*. New Haven: Yale University Press, 1968.

Bade, Klaus J. "Antisklavereibewegung in Deutschland und Kolonialkrieg in Deutsch-Ostafrika, 1888–1890: Bismarck und Friedrich Fabri." *Geschichte und Gesellschaft* 3, no. 1 (1977): 31–58.

Balandier, Georges. "The Colonial Situation: A Theoretical Approach." In *Social Change: The Colonial Situation*, edited by Immanuel Wallerstein, 34–61. New York: Wiley, 1966.

Baltrusch, Ernst, Morten Hegewisch, Michael Meyer, Uwe Puschner, Christian Wendt, and Reinhard Wolters, eds. *2000 Jahre Varusschlacht: Geschichte, Archäologie, Legenden*. Berlin: De Gruyter, 2012.

Bandeira Jerónimo, Miguel. *The "Civilizing Mission" of Portuguese Colonialism, 1870–1930.* Translated by Stewart Lloyd-Jones. New York: Palgrave Macmillan, 2015.

Baranowski, Shelly O. *The Sanctity of Rural Life: Nobility, Protestantism, and Nazism in Weimar Prussia.* Oxford: Oxford University Press, 1995.

– *Nazi Empire: German Colonialism and Imperialism from Bismarck to Hitler.* Cambridge: Cambridge University Press, 2010.

Barkin, Kenneth. *The Controversy over German Industrialization, 1890–1902.* Chicago: University of Chicago Press, 1970.

Becker, Felicitas. *Becoming Muslim in Mainland Tanzania 1890–2000.* Oxford: Oxford University Press, 2008.

Becker, Felicitas, and Jigal Beez, eds. *Der Maji-Maji-Krieg in Deutsch-Ostafrika, 1905–1907.* Berlin: Links, 2005.

Beckert, Sven. *Empire of Cotton: A Global History.* New York: Penguin, 2014.

Bennett, Norman R. "Isike, *Ntemi* of Unyanyembe." In *African Dimensions: Essays in Honor of William O. Brown*, edited by Mark Karp, 53–68. Brookline: African Studies Center, Boston University, 1975.

– *Arab versus European: Diplomacy and War in Nineteenth-Century East Central Africa.* New York: Africana Publishing Company, 1986.

Bentley, Jerry H. Introduction to "Perspectives on Global History: Cultural Encounters between the Continents over the Centuries." In *Proceedings of the Nineteenth Congress of Historical Sciences*, edited by Anders Jolstad and Marianne Lunde, 29–45. Oslo: International Committee of Historical Sciences, 2000.

Benton, Lauren. *A Search for Sovereignty: Law and Geography in European Empires, 1400–1900.* Cambridge: Cambridge University Press, 2010.

Berdahl, Robert M. *Conservative Politics and Aristocratic Landholders in Bismarckian Germany.* Chicago: University of Chicago Press, 1972.

Berenson, Edward. *Heroes of Empire: Five Charismatic Men and the Conquest of Africa.* Berkeley: University of California Press, 2011.

Berger, Stefan. *Germany.* London: Arnold, 2004.

– *The Search for Normality: National Identity and Historical Consciousness in Germany since 1800.* New York: Berghahn, 2007.

Bergmann, Klaus. *Agrarromantik und Großstadtfeindschaft.* Meisenheim: Anton Hain, 1970.

Beusekom, Monica van. *Negotiating Development: African Farmers and Colonial Experts at the Office du Niger, 1920–1960.* Portsmouth: Heinemann, 2002.

Biddick, Kathleen. *The Shock of Medievalism.* Durham: Duke University Press, 1998.

Bishara, Fahad Ahmad. *A Sea of Debt: Law and Economic Life in the Western Indian Ocean, 1780–1950.* Cambridge: Cambridge University Press, 2017.

Blackbourn, David. *The Conquest of Nature: Water, Landscape, and the Making of Modern Germany*. London: Pimlico, 2007.

Blackbourn, David, and Geoff Eley. *The Peculiarities of German History: Bourgeois Society and Politics in Nineteenth-Century Germany*. Oxford: Oxford University Press, 1984.

Blaut, J.M. *The Colonizer's Model of the World*. New York: Guilford Press, 1993.

Bonillo, Marion. *"Zigeunerpolitik" im Deutschen Kaiserreich 1871–1918*. Frankfurt: Peter Lang, 2001.

Boorstin, Daniel. *The Americans: The National Experience*. New York: Knopf, 2010.

Bose, Sugata. *A Hundred Horizons: The Indian Ocean in the Age of Global Empire*. Cambridge: Harvard University Press, 2006.

Bowersox, Jeff. *Raising Germans in the Age of Empire: Youth and Colonial Culture, 1871–1914*. Oxford: Oxford University Press, 2013.

– "Classroom Colonialism: Race, Pedagogy, and Patriotism in Imperial Germany." In *German Colonialism in a Global Age*, edited by Bradley Naranch and Geoff Eley, 170–86. Durham: Duke University Press, 2014.

Bracken, Gregory. "Treaty Ports in China: Their Genesis, Development, and Influence." *Journal of Urban History* 45, no. 1 (2019): 168–76.

Brown, Walter Thaddeus. "A Pre-colonial History of Bagamoyo: Aspects of the Growth of an East African Coastal Town." PhD diss., Boston University, 1971.

Brubaker, Rogers. *Citizenship and Nationhood in France and Germany*. Cambridge: Harvard University Press, 1992.

Bückendorf, Jutta. *"Schwarz-weiß-rot über Ostafrika!" Deutsche Kolonialpläne und afrikanische Realität*. Münster: Lit, 1997.

Burbank, Jane, and Frederick Cooper. *Empires in World History: Power and the Politics of Difference*. Princeton: Princeton University Press, 2010.

Burton, Antoinette. *Burdens of History: British Feminists, Indian Women, and Imperial Culture, 1865–1915*. Chapel Hill: The University of North Carolina Press, 1994.

Byrnes, Giselle. *Boundary Markers: Land Surveying and the Colonisation of New Zealand*. Wellington: Bridget Williams, 2001.

Caillou, Alan. *South from Khartoum: The Story of Emin Pasha*. New York: Hawthorn, 1974.

Calloway, Colin G., Gerd Gemünden, and Susanne Zantop, eds. *Germans & Indians: Fantasies, Encounters, Projections*. Lincoln: University of Nebraska Press, 2002.

Carter, Paul. *The Road to Botany Bay: An Exploration of Landscape and History*. New York: Knopf, 1988.

Carter, Sarah. *Lost Harvests: Prairie Indian Reserve Farmers and Government Policy*. Montreal: McGill-Queen's University Press, 1990.

Certeau, Michel de. *The Practice of Everyday Life*. Translated by Steven F. Rendall. Berkeley: University of California Press, 1984.

Chall, Inka, and Sonja Mezger. "Die Perspektive der Sieger: Der Maji-Maji-Krieg in der Kolonialen Presse." In *Der Maji-Maji-Krieg in Deutsch-Ostafrika, 1905–1907*, edited by Felicitas Becker and Jigal Beez, 143–53. Berlin: Ch. Links, 2005.

Chamberlain, M.E. *The Scramble for Africa*. 3rd ed. London: Routledge, 2010.

Chandra, Satish, ed. *The Indian Ocean: Explorations in History, Commerce and Politics*. New Delhi: SAGE, 1987.

Chatterjee, Piya. "'Secure this Excellent Class of Labour': Gender and Race in Labor Recruitment for British Indian Tea Plantations." *Bulletin of Concerned Asian Scholars* 27, no. 3 (1995): 43–56.

Chaudhuri, K.N. *Trade and Civilization in the Indian Ocean: An Economic History from the Rise of Islam to 1750*. Cambridge: Cambridge University Press, 1985.

– *Asia before Europe: Economy and Civilization of the Indian Ocean from the Rise of Islam to 1750*. Cambridge: Cambridge University Press, 1990.

Chickering, Roger. *We Men Who Feel Most German: A Cultural Study of the Pan-German League*. Boston: Allen & Unwin, 1984.

Chittick, Neville. *Kilwa: An Islamic Trading City on the East African Coast*. 2 vols. Nairobi: The British Institute in Eastern Africa, 1974.

Ciarlo, David. *Advertising Empire: Race and Visual Culture in Imperial Germany*. Cambridge: Harvard University Press, 2011.

Clancy-Smith, Julia, and Frances Gouda, eds. *Domesticating the Empire: Race, Gender, and Family Life in French and Dutch Colonialism*. Charlottesville: University Press of Virginia, 1998.

Clarence-Smith, William Gervase. "The British 'Official Mind' and Nineteenth-Century Islamic Debates over the Abolition of Slavery." In *Slavery, Diplomacy and Empire: Britain and the Suppression of the Slave Trade, 1807–1975*, edited by Keith Hamilton and Patrick Salmon, 125–42. Brighton: Sussex Academic Press, 2013.

Coates, Colin M. *The Metamorphoses of Landscape and Community in Early Quebec*. Montreal: McGill-Queen's University Press, 2000.

Collins, Robert O., and James M. Burns. *A History of Sub-Saharan Africa*. Cambridge: Cambridge University Press, 2007.

Comaroff, Jean, and John L. Comaroff. *Of Revelation and Revolution*. Vol. 1, *Christianity, Colonialism, and Consciousness in South Africa*. Chicago: University of Chicago Press, 1991.

Confino, Alon. *The Nation as a Local Metaphor: Württemberg, Imperial Germany, and National Memory, 1871–1918*. Chapel Hill: University of North Carolina Press, 1997.

Conklin, Alice. *A Mission to Civilize: The Republican Idea of Empire in France and West Africa, 1895–1930*. Stanford, CA: Stanford University Press, 1997.

Conrad, Sebastian. *Globalisation and the Nation in Imperial Germany*. Translated by Sorcha O'Hagan. Cambridge: Cambridge University Press, 2010.
– *German Colonialism: A Short History*. Translated by Sorcha O'Hagan. Cambridge: Cambridge University Press, 2012.
Cooper, Frederick. *Plantation Slavery on the East Coast of Africa*. New Haven: Yale University Press, 1977.
– *Decolonization and African Society: The Labor Question in French and British Africa*. Cambridge: Cambridge University Press, 1996.
– "Conditions Analogous to Slavery: Imperialism and Free Labor Ideology in Africa." In *Beyond Slavery: Explorations of Race, Labor and Citizenship in Postemancipation Societies*, edited by Cooper, Thomas C. Holt, and Rebecca J. Scott, 107–49. Chapel Hill: University of North Carolina Press, 2000.
Craib, Raymond. *Cartographic Mexico: A History of State Fixations and Fugitive Landscapes*. Durham: Duke University Press, 2004.
Das Gupta, Ashin. *India and the Indian Ocean World: Trade and Politics*. Oxford: Oxford University Press, 2004.
Das Gupta, Ashin, and M.N. Pearson. *India and the Indian Ocean, 1500–1800*. Calcutta: Oxford University Press, 1987.
Davies, Norman. *Europe: A History*. Oxford: Oxford University Press, 1996.
Davis, Diana K. *Resurrecting the Granary of Rome: Environmental History and French Colonial Expansion in North Africa*. Athens: Ohio University Press, 2009.
Davis, Kathleen. *Periodization & Sovereignty: How Ideas of Feudalism & Secularization Govern the Politics of Time*. Philadelphia: University of Pennsylvania Press, 2008.
Davis, Mike. *Late Victorian Holocausts: El Niño Famines and the Making of the Third World*. London: Verso, 2000.
Demhardt, Imre Josef. *Deutsche Kolonialgrenzen in Afrika. Historisch-geographische Untersuchungen ausgewählter Grenzräume von Deutsch-Südwestafrika und Deutsch-Ostafrika*. Hildesheim: Georg Olms, 1997.
Deutsch, Jan-Georg. "Inventing an East African Empire: The Zanzibar Delimitation Commission of 1885/1886." In *Studien zur Geschichte des deutschen Kolonialismus in Afrika: Festschrift zum 60. Geburtstag von Peter Sebald*, edited by Peter Heine and Ulrich van der Heyden, 210–19. Pfaffenweiler: Centaurus, 1995.
– *Emancipation without Abolition in German East Africa c. 1884–1914*. Oxford: James Currey, 2006.
Dreier, Horst. "Wirtschaftsraum – Großraum – Lebensraum. Facetten eines belasteten Begriffs." In *Raum und Recht. Festschrift 600 Jahre Würzburger Juristenfakultät*, edited by Horst Dreier, Hans Forkel, and Klaus Laubenthal, 47–84. Berlin: Duncker & Humblot, 2002.
Driver, Felix. *Geography Militant: Cultures of Exploration and Empire*. London: Blackwell, 2001.

Edney, Matthew. *Mapping an Empire: The Geographical Construction of British India, 1765–1843*. Chicago: University of Chicago Press, 1997.

Eley, Geoff. *Reshaping the German Right: Radical Nationalism and Political Change after Bismarck*. New Haven: Yale University Press, 1980.

Escobar, Arturo. *Encountering Development: The Making and Unmaking of the Third World*. Princeton: Princeton University Press, 1995.

Etheringon, Norman, ed. *Mapping Colonial Conquest: Australia and Southern Africa*. Crawley: University of Western Australia Press, 2007.

– *Theories of Imperialism: War, Conquest and Capital*. Routledge Revivals ed. London: Routledge, 2014.

Evans, Richard. *The Coming of the Third Reich*. London: Penguin, 2004.

Fabian, Johannes. *Out of Our Minds: Reason and Madness in the Exploration of Central Africa*. Berkeley: University of California Press, 2000.

– *Time and the Other: How Anthropology Makes Its Object*. New York: Columbia University Press, 2014.

Feierman, Steven. *The Shambaa Kingdom: A History*. Madison: University of Wisconsin Press, 1974.

– "A Century of Ironies in East Africa (c. 1780–1890)." In *African History: From Earliest Times to Independence*, 2nd ed., edited by Philip Curtin, Feierman, Leonard Thompson, and Jan Vansina, 352–76. London: Longman, 1995.

– "Political Culture and Political Economy in Early East Africa." In *African History: From Earliest Times to Independence*, 2nd ed., edited by Philip Curtin, Feierman, Leonard Thompson, and Jan Vansina, 129–51. London: Longman, 1995.

Ferguson, James. *The Anti-Politics Machine: "Development," Depoliticization, and Bureaucratic Power in Lesotho*. Minneapolis: University of Minnesota Press, 1994.

Firth, Stewart. "Governors versus Settlers: The Dispute over Chinese Labour in German Samoa." *New Zealand Journal of History* 11, no. 2 (1977): 155–79.

Fitzpatrick, Matthew P. *Liberal Imperialism in Germany: Expansionism and Nationalism, 1848–1884*. New York: Berghahn, 2008.

Förster, Stig, Wolfgang Mommsen, and Ronald Robinson, eds. *Bismarck, Europe and Africa: The Berlin Africa Conference, 1884–1885, and the Onset of Partition*. London: Oxford University Press, for the German Historical Institute, 1988.

Friedland, Roger, and Deirdre Boden, eds. *NowHere: Space, Time, and Modernity*. Berkeley: University of California Press, 1995.

Friedrichsmeyer, Sara, Sara Lennox, and Susanne Zantop, eds. *The Imperialist Imagination: German Colonialism and Its Legacy*. Ann Arbor: University of Michigan Press, 1998.

Fritsch, Kathrin. " 'You Have Everything Confused and Mixed Up …!' Georg Schweinfurth, Knowledge and Cartography of Africa in the 19th Century." *History in Africa* 36 (2009): 87–101.

Germana, Nicholas. *The Orient of Europe: The Mythical Image of India and Competing Images of German National Identity.* Newcastle upon Tyne: Cambridge Scholars Publishing, 2009.

Giblin, James, and Jamie Monson. "Introduction." In *Maji Maji: Lifting the Fog of War,* edited by Giblin and Monson, 1–30. Leiden: Brill, 2010.

Gilbert, Erik. *Dhows and the Colonial Economy of Zanzibar, 1860–1870.* Athens: Ohio University Press, 2004.

Giloi, Eva. *Monarchy, Myth, and Material Culture in Germany 1750–1950.* Cambridge: Cambridge University Press, 2011.

Gissibl, Bernhard. *The Nature of German Imperialism: Conservation and the Politics of Wildlife in Colonial East Africa.* New York: Berghahn, 2016.

Glassman, Jonathon. *Feasts and Riot: Revelry, Rebellion, and Popular Consciousness on the Swahili Coast, 1856–1888.* Portsmouth: Heinemann, 1995.

Gollwitzer, Heinz. *Die Gelbe Gefahr. Geschichte eines Schlagworts. Studien zum imperialistischen Denken.* Göttingen: Vandenhoeck & Ruprecht, 1962.

Gong, Gerrit W. *The Standard of "Civilization" in International Society.* Oxford: Clarendon Press, 1984.

Graebel, Carsten. *Die Erforschung der Kolonien. Expeditionen und koloniale Wissenskultur deutscher Geographen, 1884–1919.* Bielefeld: Transfer, 2015.

Grant, Kevin. *A Civilised Savagery: Britain and the New Slaveries in Africa.* New York: Routledge, 2004.

Greenberg, Amy. *Manifest Manhood and the Antebellum American Empire.* Cambridge: Cambridge University Press, 2005.

Gregory, Derek. "Edward Said's Imaginative Geographies." In *Thinking Space,* edited by Mike Crang and Nigel Thrift, 302–48. London: Routledge, 2000.

Gründer, Horst. *Geschichte der deutschen Kolonien.* Paderborn: Ferdinand Schöningh, 1985.

Guettel, Jens-Uwe. "From the Frontier to German South-West Africa: German Colonialism, Indians, and American Westward Expansion." *Modern Intellectual History* 7, no. 3 (2010): 523–52.

Guyer, Jane I., and Samuel M. Eno Belinga. "Wealth in People as Wealth in Knowledge: Accumulation and Composition in Equatorial Africa." *Journal of African History* 36, no. 1 (1995): 91–120.

Gwassa, G.C.K. "The German Intervention and African Resistance in Tanzania." In *A History of Tanzania,* edited by I.N. Kimambo and A.J. Temu, 85–122. Nairobi: East African Publishing House, 1969.

Hagenlücke, Heinz. *Deutsche Vaterlandspartei: Die nationale Rechte am Ende des Kaiserreiches.* Düsseldorf: Droste, 1997.

Hämäläinen, Pekka. *The Comanche Empire.* New Haven: Yale University Press, 2008.

Hannoum, Abdelmajid. "Translation and the Colonial Imaginary: Ibn Khaldûn Orientalist." *History and Theory* 42 (February 2003): 61–81.

Hansen, Jason D. *Mapping the Germans: Statistical Science, Cartography, and the Visualization of the German Nation, 1848–1914*. Oxford: Oxford University Press, 2015.

Hargreaves, J.D. "The Making of the Boundaries: Focus on West Africa." In *Partitioned Africans*, edited by A.I. Asiwaju, 19–27. London: C. Hurst & Company, 1985.

Harley, J.B. *The New Nature of Maps: Essays in the History of Cartography*. Edited by Paul Laxton. Baltimore: The Johns Hopkins University Press, 2001.

Harries, Patrick. "Roots of Ethnicity." *African Affairs* 87 (1988): 25–52.

Harris, Cole. *The Resettlement of British Columbia: Essays on Colonialism and Geographical Change*. Vancouver: UBC Press, 1997.

Harvey, David. *The Condition of Postmodernity: An Enquiry into the Origins of Cultural Change*. Malden: Wiley-Blackwell, 1991.

Headrick, Daniel. *The Tools of Empire: Technology and European Imperialism in the Nineteenth Century*. New York: Oxford University Press, 1981.

Heller, Henry. "Bodin on Slavery and Primitive Accumulation." *The Sixteenth Century Journal* 25, no. 1 (1994): 53–65.

Hering, Rainer. *Konstruierte Nation: Der Alldeutsche Verband 1890 bis 1939*. Hamburg: Hans Christians, 2003.

Herwig, Holger. "Geopolitik: Haushofer, Hitler and Lebensraum." *Journal of Strategic Studies* 22, no. 2–3 (1999): 218–41.

Hobsbawm, Eric, and Terence Ranger, eds. *The Invention of Tradition*. Cambridge: Cambridge University Press, 1983.

Horton, Mark, and John Middleton. *The Swahili: The Social Landscape of a Mercantile Society*. Oxford: Blackwell, 2000.

Howard, Allen. "Nodes, Networks, Landscapes, and Regions: Reading the Social History of Tropical Africa, 1700s–1920." In *The Spatial Factor in African History: The Relationship of the Social, Material, and Perceptual*, edited by Howard and Richard M. Shain, 21–140. Leiden: Brill, 2005.

– "Re-Marking the Past: Spatial Structures and Dynamics in the Sierra Leone-Guinea Plain, 1860–1920s." In *The Spatial Factor in African History: The Relationship of the Social, Material, and Perceptual*, edited by Howard and Richard M. Shain, 291–348. Leiden: Brill, 2005.

Howard, Douglas, and Luise White. "Introduction: Sovereignty and the Study of States." In *The State of Sovereignty: Territories, Laws, Populations*, edited by Howard and White, 1–18. Bloomington: Indiana University Press, 2009.

Hull, Isabel. *Absolute Destruction: Military Culture and the Practices of War in Imperial Germany*. Ithaca: Cornell University Press, 2005.

Iggers, Georg G. *The German Conception of History: The National Tradition of Historical Thought from Herder to the Present*. Middletown: Wesleyan University Press, 1968.

Iliffe, John. *A Modern History of Tanganyika*. Cambridge: Cambridge University Press, 1979.

– *Africans: History of a Continent*. 2nd ed. Cambridge: Cambridge University Press, 2007.

Jäckel, Eberhard. *Hitler's World View: A Blueprint for Power*. Cambridge: Harvard University Press, 1981.

Jackson, Robert D. "Resistance to the German Invasion of the Tanganyikan Coast." In *Protest and Power in Black Africa*, edited by A.A. Mazuri and R.I. Rotberg, 37–79. New York: Oxford University Press, 1970.

Jameson, Frederic. *Postmodernism, or, the Cultural Logic of Late Capitalism*. Durham: Duke University Press, 1991.

Jeal, Tim. *Stanley: The Impossible Life of Africa's Greatest Explorer*. London: Faber and Faber, 2007.

– *Livingstone*. Revised edition. New Haven: Yale University Press, 2013.

Jones, Adam, and Isabel Voigt. "'Just a First Sketchy Makeshift': German Travellers and Their Cartographic Encounters in Africa, 1850–1914." *History in Africa* 39 (2012): 9–39.

Jones, Roger. *The Rescue of Emin Pasha: The Story of Henry M. Stanley and the Emin Pasha Relief Expedition, 1887–1889*. London: Allison and Busby, 1972.

Katzenellenbogen, Simon. "It Didn't Happen at Berlin: Politics, Economics and Ignorance in the Setting of Africa's Colonial Boundaries." In *African Boundaries: Barriers, Conduits and Opportunities*, edited by Paul Nugent and A.I. Asiwaju, 21–34. London: Pinter, 1996.

Kearney, Milo. *The Indian Ocean in World History*. New York: Routledge, 2004.

Keese, Alexander. *Ethnicity and the Colonial State*. Leiden: Brill, 2016.

Kennedy, Dane. *The Last Blank Spaces: Exploring Africa and Australia*. Cambridge: Harvard University Press, 2013.

Kirk-Greene, Anthony. "Heinrich Barth: An Exercise in Empathy." In *Africa and Its Explorers: Motives, Methods, and Impact*, edited by Robert Rotberg, 13–38. Cambridge: Harvard University Press, 1970.

Klein, Martin. *Slavery and Colonial Rule in French West Africa*. Cambridge: Cambridge University Press, 1998.

Kontje, Todd. *German Orientalisms*. Ann Arbor: University of Michigan Press, 2004.

Koponen, Juhani. *Development for Exploitation: German Colonial Policies in Mainland Tanzania, 1884–1914*. Helsinki: Tiedekirja, 1995.

Kopp, Kristin. *Germany's Wild East: Constructing Poland as Colonial Space*. Ann Arbor: University of Michigan Press, 2012.

Kuss, Susanne. *German Colonial Wars and the Context of Military Violence*. Translated by Andrew Smith. Cambridge: Harvard University Press, 2017.

Lacoste, Yves. *Ibn Khaldun: The Birth of History and the Past of the Third World*. Translated by David Macey. New York: Verso, 1984.

Lahme, Rainer. *Deutsche Aussenpolitik 1890–1894: Von der Gleichgewichtspolitik Bismarcks zur Allianzstrategie Caprivis*. Göttingen: Vandenhoek & Ruprecht, 1990.

Lambert, David. *Mastering the Niger: James MacQueen's African Geography and the Struggle over Atlantic Slavery*. Chicago: University of Chicago Press, 2013.

Landau, Paul. *Popular Politics in the History of South Africa, 1400–1948*. Cambridge: Cambridge University Press, 2010.

Langbehn, Volker M. *German Colonialism, Visual Culture, and Modern Memory*. New York: Routledge, 2010.

Lange, Karl. "Der Terminus 'Lebensraum' in Hitlers 'Mein Kampf.' " *Vierteljahrsheft für Zeitgeschichte* 13, no. 4 (1965): 426–37.

Laqua, Daniel. "The Tensions of Internationalism: Transnational Anti-Slavery in the 1880s and 1890s." *The International History Review* 33, no. 4 (December 2011): 705–26.

– *The Age of Internationalism and Belgium, 1880–1930: Peace, Progress and Prestige*. Manchester: University of Manchester Press, 2013.

LaViollette, Adria, and Stephane Wynne-Jones. "The Swahili World." In *The Swahili World*, edited by Wynne-Jones and LaViollette, 1–14. London: Routledge, 2018.

Lefebvre, Camille. *Frontières de sable, frontières de papier. Histoire de territoires et de frontières, du jihad de Sokoto à la colonisation française du Niger, XIXe–XXe siècles*. Paris: Sorbonne, 2015.

Lefebvre, Henri. *The Production of Space*. Translated by Donald Nicholson-Smith. Oxford: Basil-Blackwell, 2001.

Lewis, Martin W., and Kären E. Wigen. *The Myth of Continents: A Critique of Metageography*. Berkeley: University of California Press, 1997.

Leys, Colin. *The Rise and Fall of Development Theory*. London: James Currey, 1996.

Liebhafsky Des Forges, Alison. *Defeat Is the Only Bad News: Rwanda under Musinga, 1896–1931*. Edited by David Newbury. Madison: University of Wisconsin Press, 2011.

Liebowitz, Daniel, and Charles Pearson. *The Last Expedition: Stanley's Mad Journey through the Congo*. New York: Norton, 2005.

Liulevicius, Vejas Gabriel. *War Land on the Eastern Front: Culture, National Identity and German Occupation in World War I*. Cambridge: Cambridge University Press, 2004.

– *The German Myth of the East: 1800 to Present*. Oxford: Oxford University Press, 2009.

Lombard, Denys, and Jean Aubin, eds. *Marchands et hommes d'affaires asiatiques dans l'Océan Indien et la Mer de Chine 13e–20e siècles*. Paris: EHESS, 1988.

Lovejoy, Paul E. *Transformations in Slavery: A History of Slavery in Africa*. 3rd ed. New York: Cambridge University Press, 2012.

Lynch, Gabrielle. *I Say to You: Ethnic Politics and the Kalenjin in Kenya*. Chicago: University of Chicago Press, 2011.

MacArthur, Julie. *Cartography and the Political Imagination: Mapping Community in Colonial Kenya*. Athens: Ohio University Press, 2016.

Mamdani, Mahmood. *Citizen and Subject: Contemporary Africa and the Legacy of Late Colonialism*. Princeton: Princeton University Press, 1996.

– *When Victims Become Killers: Colonialism, Nativism, and the Genocide in Rwanda*. Princeton: Princeton University Press, 2001.

– *Saviors and Survivors: Darfur, Politics, and the War on Terror*. New York: Doubleday, 2009.

– *Define and Rule: Native as Political Identity*. Cambridge: Harvard University Press, 2012.

Mann, Gregory. *From Empires to NGOs in the West African Sahel: The Road to Nongovernmentality*. New York: Cambridge University Press, 2014.

Manning, Olivia. *The Reluctant Rescue: The Story of Stanley's Rescue of Emin Pasha from Equatorial Africa*. Garden City: Doubleday, 1947.

Marchand, Suzanne. "Leo Frobenius and the Revolt against the West." *Journal of Contemporary History* 32, no. 2 (1997): 153–70.

– *Down from Olympus: Archaeology and Philhellenism in Germany, 1750–1970*. Princeton: Princeton University Press, 2003.

– *German Orientalism in the Age of Empire: Religion, Race, and Scholarship*. Washington: German Historical Institute, 2009.

McClendon, Thomas V. *White Chief, Black Lords: Shepstone and the Colonial State in Natal, South Africa, 1845–1878*. Rochester: University of Rochester Press, 2010.

McClintock, Anne. *Imperial Leather: Race, Gender, and Sexuality in the Colonial Contest*. London: Routledge, 1995.

McMeekin, Sean. *The Berlin–Baghdad Express: The Ottoman Empire and Germany's Bid for World Power*. Cambridge: Belknap, 2010.

McPherson, Kenneth. *The Indian Ocean: A History of People and the Sea*. Delhi: Oxford University Press, 1993.

Mehnert, Ute. *Deutschland, Amerika und die "Gelbe Gefahr."* Stuttgart: Steiner, 1998.

Middleton, John. *African Merchants of the Indian Ocean: Swahili of the East African Coast*. Long Grove: Waveland Press, 2004.

Miers, Suzanne. *Britain and the Ending of the Slave Trade*. New York: Africana, 1975.

Mignolo, Walter. *Local Histories/Global Designs: Coloniality, Subaltern Knowledges, and Border Thinking*. Princeton: Princeton University Press, 2000.

Miller, Joseph C. *Way of Death: Merchant Capitalism and the Angolan Slave Trade 1730–1830*. Madison: University of Wisconsin Press, 1988.

– "History and Africa/Africa and History" *American Historical Review* 104, no. 1 (February 1999): 1–32.

Mitchell, Timothy. *Colonising Egypt*. Berkeley: University of California Press, 1988.

– "Introduction." In *Questions of Modernity*, edited by Mitchell, xi–xiii. Minneapolis: University of Minnesota Press, 2000.

– "The Stage of Modernity." In *Questions of Modernity*, edited by Mitchell, 1–34. Minneapolis: University of Minnesota Press, 2000.

Moeller, Robert G. *German Peasants and Agrarian Politics, 1914–1924: The Rhineland and Westphalia*. Chapel Hill: University of North Carolina Press, 2010.

Mommsen, Wolfgang J., ed. *Leopold von Ranke und die moderne Geschichtswissenschaft*. Stuttgart: Klett-Cotta, 1988.

Mongia, Radhika V. "Historicizing State Sovereignty: Inequality and the Form of Equivalence." *Comparative Studies in Society and History* 49, no. 2 (2007): 384–411.

Moore-Harell, Alice. *Egypt's African Empire: Samuel Baker, Charles Gordon & the Creation of Equatoria*. Brighton: Sussex Academic Press, 2012.

Moretti, Franco. *Atlas of the European Novel, 1800–1900*. London: Verso, 1998.

Mosse, George L. *The Crisis of German Ideology: Intellectual Origins of the Third Reich*. New York: Grosset & Dunlap, 1964.

Moyd, Michelle R. *Violent Intermediaries: African Soldiers, Conquest, and Everyday Colonialism in German East Africa*. Athens: Ohio University Press, 2014.

Müller, Fritz Ferdinand. *Deutschland-Zanzibar-Ostafrika: Geschichte einer deutschen Kolonialeroberung 1884–1890*. Berlin: Rütten & Loening, 1959.

Murphy, David Thomas. " 'Retroactive Effects': Ratzel's Spatial Dynamics and the Expansionist Imperative in Interwar Germany." *Journal of Historical Geography* 61 (2018): 86–90.

Naranch, Bradley. "Introduction: German Colonialism Made Simple." In *German Colonialism in a Global Age*, edited by Naranch and Geoff Eley, 1–18. Durham: Duke University Press, 2014.

Newbury, Catherine. *The Cohesion of Oppression: Clientship and Ethnicity in Rwanda, 1860–1960*. New York: Columbia University Press, 1988.

Newbury, David. *The Land beyond the Mists: Essays on Identity and Authority in Precolonial Congo and Rwanda*. Athens: Ohio University Press, 2009.

Nipperdey, Thomas. *Deutscche Geschichte, 1866–1918*. Vol. 1, *Arbeitswelt und Bürgergeist*. Munich: Beck, 1990.

– *Deutsche Geschichte 1866–1918*. Vol. 2, *Machtstaat vor der Demokratie*. Munich: Beck, 1995.

– *Germany from Napoleon to Bismarck 1800–1866*. Translated by Daniel Nolan. Princeton: Princeton University Press, 1996.

Noyes, John. *Colonial Space: Spatiality in the Discourse of German South West Africa 1884–1915*. Chur: Harwood Academic Publishers, 1992.

Nugent, Paul. "Putting the History Back into Ethnicity: Enslavement, Religion and Cultural Brokerage in the Construction of Mandinka/Jola and Ewe/Agotime Identities in West Africa c. 1650–1930." *Comparative Studies in Society and History* 50, no. 4 (2008): 920–48.

Nugent, Paul, and A.I. Asiwaju. "Introduction: The Paradox of African Boundaries." In *African Boundaries: Barriers, Conduits and Opportunities*, edited by Nugent and Asiwaju, 1–17. London: Pinter, 1996.

Nuhn, Walter. *Flammen über Deutschost: Der Maji-Maji-Aufstand in Deutsch-Ostafrika, 1905–1906: die erste gemeinsame Erhebung schwarzafrikanischer Völker gegen weisse Kolonialherrschaft: ein Beitrag zur deutschen Kolonialgeschichte.* Bonn: Bernard & Graefe, 1998.

Nurse, Derek, and Thomas Spear. *The Swahili: Reconstructing the History and Language of an African Society, 800–1500.* Philadelphia: University of Pennsylvania Press, 1985.

Olusoga, David, and Casper Erichsen. *The Kaiser's Holocaust: Germany's Forgotten Genocide and the Colonial Roots of Nazism.* London: Faber and Faber, 2011.

Osterhammel, Jürgen. "'The Great Work of Uplifting Mankind.' Zivilisierungsmission und Moderne." In *Zivilisierungsmissionen. Imperiale Weltverbesserung seit dem 18. Jahrhundert*, edited by Osterhammel and Boris Barth, 363–425. Konstanz: UVK-Verlagsgesellschaft, 2005.

O'Tuathail, Gearoid. *Critical Geopolitics: The Politics of Writing Global Space.* Minneapolis: University of Minnesota Press, 1996.

Parsons, Timothy. "Local Responses to the Ethnic Geography of Colonialism in the Gusii Highlands of British-Ruled Kenya." *Ethnohistory* 58 (2011): 491–523.

– "Being Kikuyu in Meru: Challenging the Tribal Geography of Colonial Kenya." *Journal of African History* 53 (2012): 65–86.

Pearson, Michael. *Port Cities and Intruders: The Swahili Coast, India, and Portugal in the Early Modern Era.* Baltimore: Johns Hopkins University Press, 1998.

– *The Indian Ocean.* London: Routledge, 2003.

– *The World of the Indian Ocean, 1500–1800.* Aldershot: Ashgate, 2005.

Perry, Adele. *On the Edge of Empire: Gender, Race, and the Making of British Columbia, 1849–1871.* Toronto: University of Toronto Press, 2001.

Pesek, Michael. "Eine Gründungsszene des deutschen Kolonialismus – Peters' Expedition nach Usagara, 1884." In *Die (koloniale) Begegnung: AfrikanerInnen in Deutschland 1880–1945, Deutsche in Afrika 1880–1918*, edited by Marianne Bechhaus-Gerst and Reinhard Klein-Arendt, 255–68. Frankfurt am Main: Peter Lang, 2003.

– *Koloniale Herrschaft in Deutsch-Ostafrika: Expeditionen, Militär und Verwaltung seit 1880.* Frankfurt am Main: Campus, 2005.

Peterson, Derek. *Ethnic Patriotism and East African Revival: A History of Dissent, c. 1935–1972.* Cambridge: Cambridge University Press, 2012.

Peterson, Derek, and Giacomo Macola, eds. *Recasting the Past: History Writing and Political Work in Modern Africa*. Athens: Ohio University Press, 2009.

Pettit, Clare. *Dr. Livingstone, I Presume? Missionaries, Journalists, Explorers, and Empire*. Cambridge: Harvard University Press, 2007.

Pomeranz, Kenneth. "Empire & 'Civilizing' Missions, Past and Present." *Daedalus* 134, no. 2 (2005): 34–45.

Pouwels, Randall L. *Horn and Crescent: Cultural Change and Traditional Islam on the East African Coast, 800–1900*. Cambridge: Cambridge University Press, 1987.

– "Eastern Africa and the Indian Ocean to 1800: Reviewing Relations in Historical Perspective." *The International Journal of African Historical Studies* 35, no. 2/3 (2002): 385–425.

Pratt, Mary Louise. *Imperial Eyes: Travel Writing and Transculturation*. London: Routledge, 1992.

Press, Steven. *Rogue Empires: Contracts and Conmen in Europe's Scramble for Africa*. Cambridge: Harvard University Press, 2017.

Prestholdt, Jeremy. "On the Global Repercussions of East Africa Consumerism." *American Historical Review* 109, no. 3 (June 2004): 755–81.

– *Domesticating the World: African Consumerism and the Genealogies of Globalization*. Berkeley: University of California Press, 2008.

Prunier, Gérard. *The Rwanda Crisis: History of a Genocide*. New York: Columbia University Press, 1995.

Ptak, Roderich, and Dietmar Rothermund, eds. *Emporia, Commodities and Entrepreneurs in Asian Maritime Trade, c. 1400–1750*. Stuttgart: Franz Steiner, 1992.

Puhle, Hans-Jürgen. *Agrarische Interessenpolitik und preussischer Konservatismus im wilhelminischen Reich (1893–1914): Ein Beitrag zur Analyse des Nationalismus in Deutschland am Beispiel des Bundes der Landwirte und der deutsch-konservativen Partei*. Hanover: Verlag für Literatur und Zeitgeschehen, 1966.

Puschner, Uwe. *Die völkische Bewegung im wilhelminischen Kaiserreich. Sprache – Rasse – Religion*. Darmstadt: Wissenschaftliche Buchgesellschaft, 2001.

Ranger, Terence. "Connexions between 'Primary Resistance' Movements and Modern Mass Nationalism in East and Central Africa." *Journal of African History* 9 (1968): 437–53, 631–41.

Reagin, Nancy. *Sweeping the German Nation: Domesticity and National Identity in Germany, 1870–1945*. Cambridge: Cambridge University Press, 2007.

Redmayne, Alison. "Mkwawa and the Hehe Wars." *Journal of African History* 9, no. 3 (1968): 409–36.

Reid, Anthony. *Southeast Asia in the Age of Commerce, 1450–1680*. Vol. 1, *The Lands below the Winds*. New Haven: Yale University Press, 1988.

Retallack, James. *The German Right, 1860–1920: Political Limits of the Authoritarian Imagination*. Toronto: University of Toronto Press, 2006.

– *Red Saxony: Election Battles and the Spectre of Democracy in Germany, 1860–1918*. Oxford: Oxford University Press, 2017.

Ribi Forclaz, Amalia. *Humanitarian Imperialism: The Politics of Anti-Slavery Activism, 1880–1940*. Oxford: Oxford University Press, 2015.

Richter, Daniel K. *Facing East from Indian Country: A Native History of Early America*. Cambridge: Harvard University Press, 2001.

Ritzheimer, Kara. *"Trash," Censorship, and National Identity in Early Twentieth-Century Germany*. Cambridge: Cambridge University Press, 2016.

Roberts, Allen F. *A Dance of Assassins: Performing Early Colonial Hegemony in the Congo*. Bloomington: Indiana University Press, 2013.

Rockel, Stephen J. *Carriers of Culture: Labor on the Road in Nineteenth-Century East Africa*. Portsmouth: Heinemann, 2006.

Rodney, Walter. *How Europe Underdeveloped Africa*. London: Bogle-L'Ouverture, 1972.

Rohkrämer, Thomas. *A Single Communal Faith? The German Right from Conservatism to National Socialism*. New York: Berghahn, 2007.

Röhl, John C.G. *Wilhelm II: The Kaiser's Personal Monarchy, 1888–1900*. Translated by Sheila de Bellaigue. Cambridge: Cambridge University Press, 2004.

Ryan, Simon. *The Cartographic Eye: How Explorers Saw Australia*. Cambridge: Cambridge University Press, 1996.

Saavedra Casco, José Arturo. *Utenzi, War Poems, and the German Conquest of East Africa: Swahili Poetry as a Historical Source*. Trenton: Africa World Press, 2007.

Said, Edward. *Orientalism*. New York: Penguin, 1978.

Sanders, Ethan R. "Missionaries and Muslims in East Africa before the Great War." Paper presented at the Henry Martyn Seminar, Westminster College, Cambridge, March 9, 2011.

Sarkin-Hughes, Jeremy. *Germany's Genocide of the Herero: Why Kaiser Wilhelm II Gave the Order*. Oxford: James Currey, 2011.

Schäbler, Birgit. *Moderne Muslime. Ernest Renan und die Geschichte der ersten Islamdebatte 1883*. Paderborn: Ferdinand Schöningh, 2016.

Schillings, Pascal. *Der letzte weiße Flecken. Europäische Antarktisreisen um 1900*. Göttingen: Wallstein, 2016.

Schmidt, Georg. "Friedrich Meineckes Kulturnation. Zum historischen Kontext nationaler Ideen in Weimar-Jena um 1800." *Historische Zeitschrift* 284 (2007): 597–622.

Schmitt, Carl. *The Nomos of the Earth*. Translated by G.L. Ulmen. New York: Telos, 2003.

Schubert, Michael. *Der schwarze Fremde: Das Bild des Schwarzafrikaners in der parlamentarischen und publizistischen Kolonialdiskussion in Deutschland von der 1870er bis in die 1930er Jahre*. Stuttgart: Steiner, 2003.

Schulte-Varendorff, Uwe. *Kolonialheld für Kaiser und Führer. General Lettow-Vorbeck – Mythos und Wirklichkeit*. Berlin: Links, 2011.

Scott, James C. *Seeing Like a State: How Certain Schemes to Improve the Human Condition Have Failed*. New Haven: Yale University Press, 1998.

Shapiro, Michael J. *Violent Cartographies: Mapping Cultures of War*. Minneapolis: University of Minnesota Press, 1997.

Sharpe, Barrie. "Ethnography and a Regional System: Mental Maps and the Myth of States and Tribes in North-Central Nigeria." *Critique of Anthropology* 6 (1986): 33–65.

Sheehan, James. "The Problem of Sovereignty in European History." *American Historical Review* 111 (2006): 1–15.

Sheriff, Abdul. *Slaves, Spices and Ivory in Zanzibar*. London: James Currey, 1987.

– *Dhow Cultures of the Indian Ocean: Cosmopolitanism, Commerce and Islam*. New York: Columbia University Press, 2010.

Short, John Phillip. *Magic Lantern Empire: Colonialism and Society in Germany*. Ithaca: Cornell University Press, 2012.

Shorter, Aylward. "Nyungu-Ya-Mawe and the 'Empire of the Ruga-Ruga.' " *Journal of African History* 9, no. 2 (1968): 235–59.

Sibeud, Emmanuelle. *Une Science impériale pour l'Afrique? La Construction des Savoirs africanistes en France, 1878–1930*. Paris: Éditions de l'École des Hautes Études en Sciences Sociales, 2002.

Sippel, Harald. "Recht und Herrschaft in kolonialer Frühzeit: Die Rechtsverhältnisse in den Schutzgebieten der Deutsh-Ostafrikanishen Gesellschaft (1885–1890)." In *Studien zur Geschichte des deutschen Kolonialismus*, edited by Peter Heine and Ulrich van der Heyden, 466–94. Pfaffenweiler: Centaurus, 1995.

Smith, Bonnie G. *The Gender of History: Men, Women, and Historical Practice*. Cambridge: Harvard University Press, 1998.

Smith, Helmut Walser. *German Nationalism and Religious Conflict: Culture, Ideology, Politics, 1870–1914*. Princeton: Princeton University Press, 1995.

– "Monuments, Kitsch, and the Sense of Nation in Imperial Germany." *Central European History* 49, no. 3–4 (2016): 322–40.

Smith, Iain. *The Emin Pasha Relief Expedition, 1886–1890*. Oxford: Clarendon, 1972.

Smith, Woodruff. *The German Colonial Empire*. Chapel Hill: University of North Carolina Press, 1978.

– "Friedrich Ratzel and the Origins of Lebensraum." *German Studies Review* 3, no. 1 (February 1980): 51–68.

– *The Ideological Origins of Nazi Imperialism*. Oxford: Oxford University Press, 1989.

Soja, Edward W. *Seeking Spatial Justice*. Minneapolis: University of Minnesota Press, 2010.

Spear, Thomas. "Neo-Traditionalism and the Limits of Invention in British Colonial Africa," *Journal of African History* 44 (2003): 3–27.

Spengler, Oswald. *The Decline of the West: Form and Actuality.* Translated by Charles Francis Atkinson. New York: Alfred A. Knopf, 1927.

Steinmetz, George. *The Devil's Handwriting: The German Colonial State in Qingdao, Samoa, and Southwest Africa.* Chicago: University of Chicago Press, 2007.

Stoecker, Helmuth, ed. *German Imperialism in Africa: From the Beginnings until the Second World War.* London: Hurst, 1986.

Stoler, Ann Laura. "Making Empire Respectable: The Politics of Race and Sexual Morality in 20th-Century Colonial Cultures." *American Ethnologist* 16, no. 4 (November 1989): 634–60.

– "Rethinking Colonial Categories: European Communities and the Boundaries of Rule." *Comparative Studies in Society and History* 31, no. 1 (January 1989): 134–61.

Strachey, Lytton. *Eminent Victorians: Cardinal Manning, Florence Nightingale, Dr. Arnold, General Gordon.* London: Continuum, 2002.

Stuchtey, Benedikt, and Peter Wende, eds. *British and German Historiography 1750–1950: Traditions, Perceptions and Transfers.* Oxford: Oxford University Press, 2000.

Sweeney, Dennis. "Pan-German Conceptions of Colonial Empire." In *German Colonialism in a Global Age,* edited by Bradley Naranch and Geoff Eley, 265–82. Durham: Duke University Press, 2014.

Tilley, Helen. *Africa as a Living Laboratory: Empire, Development, and the Problem of Scientific Knowledge, 1870–1950.* Chicago: University of Chicago Press, 2011.

Townsend, Mary Evelyn. *The Rise and Fall of Germany's Colonial Empire, 1884–1918.* New York: Howard Fertig, 1966.

Tuan, Yi-Fu. *Space and Place: The Perspective of Experience.* Minneapolis: University of Minnesota Press, 2001.

Turner, Frederick Jackson. *The Frontier in American History.* New York: Henry Holt and Company, 1921.

Unangst, Matthew. "Men of Science and of Action: The Celebrity of Explorers and German National Identity." *Central European History* 50, no. 3 (September 2017): 305–27.

Uzoigwe, G.N. "Spheres of Influence and the Doctrine of the Hinterland in the Partition of Africa." *Journal of African Studies* 3, no. 2 (1976): 183–203.

Vail, Leroy, ed. *The Creation of Tribalism in Southern Africa.* London: James Currey, 1989.

Vansina, Jan. *Antecedents of Modern Rwanda: The Nyiginya Kingdom.* Madison: University of Wisconsin Press, 2004.

Vascik, George. "Agrarian Conservatism in Wilhelmine Germany: Diederich Hahn and the Agrarian League" In *Between Reform, Reaction and Resistance: Studies in the History of German Conservatism from 1789 to 1945*, edited by Larry Eugene Jones and James N. Retallack, 229–60. Providence: Berg, 1993.

Velkley, Richard L. *Being after Rousseau: Philosophy and Culture in Question.* Chicago: University of Chicago Press, 2002.

Vink, Markus P.M. "Indian Ocean Studies and the 'New Thalassology.' " *Journal of Global History* 2, no. 1 (2007): 41–62.

Vogt, Emily A. "*Civilisation* and *Kultur*: Keywords in the History of French and German Citizenship." *Ecumene* 3, no. 2 (April 1996): 125–45.

Voigt, Isabel. "Die 'Schneckenkarte' – Mission, Kartographie und transkulturelle Wissensaushandlung in Ostafrika um 1850." *Cartographica Helvetica* 45 (2012): 27–38.

Walkenhorst, Peter. *Nation – Volk – Rasse. Radikaler Nationalismus im Deutschen Kaiserreich 1890–1914.* Göttingen: Vandenhoeck & Ruprecht, 2007.

Walther, Daniel Joseph. *Creating Germans Abroad: Cultural Policies and National Identity in Namibia.* Athens: Ohio University Press, 2002.

Wanklyn, Harriet. *Friedrich Ratzel: A Biographical Memoir and Bibliography.* London: Cambridge University Press, 1961.

Wcislo, Francis W. *Tales of Imperial Russia: The Life and Times of Sergei Witte, 1849–1915.* New York: Oxford University Press, 2011.

Weaver, John C. *The Great Land Rush and the Making of the Modern World, 1650–1900.* Montreal: McGill-Queen's University Press, 2003.

Weber, Max. "Entwickelungstendenzen in der Lage der ostelbischen Landarbeiter." In *Gesammelte Aufsätze zur Sozial- und Wirtschaftgeschichte*, edited by Marianne Weber, 470–507. Tübingen: J.C.B. Mohr, 1924.

– "Die ländliche Arbeitsverfassung" In *Gesammelte Aufsätze zur Sozial- und Wirtschaftgeschichte*, edited by Marianne Weber, 444–69. Tübingen: J.C.B. Mohr, 1924.

Wehler, Hans-Ulrich. *Bismarck und der Imperialismus.* Cologne: Kiepenheuer & Witsch, 1969.

– "Bismarck's Imperialism 1862–1890." *Past & Present* 48 (August 1970): 119–55.

– *The German Empire 1871–1918.* Translated by Kim Traynor. Dover: Berg, 1985.

Welsh, David. *The Roots of Segregation; Native Policy in Colonial Natal, 1845–1910.* New York: Oxford University Press, 1971.

Wesseling, H.L. *Divide and Rule: The Partition of Africa, 1880–1914.* Translated by Arnold J. Pomerans. Westport: Praeger, 1996.

White, Richard. *The Middle Ground: Indians, Empires, and Republics in the Great Lakes Region, 1650–1815.* 20th anniversary edition. New York: Cambridge University Press, 2011.

– *Railroaded: The Transcontinentals and the Making of Modern America.* New York: W.W. Norton, 2011.

Winkler, Martin M. *Arminius the Liberator: Myth and Ideology.* New York: Oxford University Press, 2016.

Wippermann, Wolfgang. *Wie die Zigeuner: Antisemitismus und Antiziganismus im Vergleich*. Berlin: Elefanten, 1997.

Wirz, Albert, Andreas Eckert, and Katrin Bromber, eds. *Alles unter Kontrolle: Disziplinierungsprozesse im kolonialen Tansania (1850–1960)*. Cologne: Rüdiger Köppe, 2003.

Wisnicki, Adrian. "Charting the Frontier: Indigenous Geography, Arab-Nyamwezi Caravans, and the East African Expedition of 1856–59," *Victorian Studies* 51, no. 1 (2008): 103–37.

Wokoeck, Ursula. *German Orientalism: The Study of the Middle East and Islam from 1800 to 1945*. London: Routledge, 2009.

Wolfe, Patrick. "Settler Colonialism and the Elimination of the Native." *Journal of Genocide Research* 8, no. 4 (December 2006): 387–409.

Wood, Denis. *Rethinking the Power of Maps*. New York: Guilford, 2010.

Woodward, David, and Malcolm Lewis. *Cartography in the Traditional African, American, Arctic, Australian, and Pacific Societies*. History of Cartography, vol. 2, book 3. Chicago: University of Chicago Press, 1998.

Wright, Donald. "'What Do You Mean There Were No Tribes in Africa?': Thoughts on Boundaries – and Related Matters – in Precolonial Africa." *History in Africa* 26 (1999): 409–26.

Wright, Marcia. *German Missions in Tanganyika, 1891–1941: Lutherans and Moravians in the Southern Highlands*. Oxford: Clarendon, 1971.

– "Maji Maji Prophecy and Historiography." In *Revealing Prophets: Prophecy in Eastern African History*, edited by David Anderson and Douglas H. Johnson, 124–42. London: James Currey, 1995.

Wynn, Grame. "Settler Societies in Geographical Focus." *Historical Studies* 20, no. 80 (1983): 353–66.

Zimmerer, Jürgen. *Deutsche Herrschaft über Afrikaner: Staatlicher Machtanspruch und Wirklichkeit im kolonialen Namibia*. Münster, Lit, 2004.

Zimmerer, Jürgen, and Joachim Zeller, eds. *Genocide in German South-West Africa: The Colonial War (1904–1908) in Namibia and Its Aftermath*. Monmouth: Merlin, 2008.

Zimmerman, Andrew. *Anthropology and Antihumanism in Imperial Germany*. Chicago: The University of Chicago Press, 2001.

– *Alabama in Africa: Booker T. Washington, the German Empire and the Globalization of the New South*. Princeton: Princeton University Press, 2010.

Ziolkowski, Theodore. *Clio the Romantic Muse: Historicizing the Faculties in Germany*. Ithaca: Cornell University Press, 2004.

Žižek, Slavoj. *Violence: Six Sideways Reflections*. London: Macmillan, 2008.

Index

Note: Page numbers in **bold** refer to illustrations

GERMAN AND EUROPEAN STUDIES

General Editor: Jennifer L. Jenkins